Metallurgy for the Non-Metallurgist™

by

Harry Chandler

First printing, March 1998

Comments, criticisms, and suggestions are invited, and should be forwarded to ASM International.

Library of Congress Cataloging-in-Publication Data

Chandler, Harry
Metallurgy for the nonmetallurgist / Harry Chandler.
Includes bibliographical references and index.
1. Metallurgy—Popular works.
I. Title.
TN667.C43 1998 669—DC21 98-4664
ISBN 0-87170-652-0
SAN 204-7586

ASM International®
Materials Park, OH 44073-0002

Printed in the United States of America

Preface

I submit that I am specially qualified to represent the nonmetallurgist in writing this book, because I, too, am a nonmetallurgist.

As an outsider, I do not feel bound to follow in the footsteps of tradition or practice. Textbook formats are avoided in favor of a narrative style based on the notion that a book does not always have to be dull and/or difficult to be technical.

Selection of material is based largely on a personal predilection for keeping a single spotlight on a few well-chosen, related topics. In this regard there is a general leaning toward subjects dealing with the use of properties of metals and their alloys.

Depth of coverage of topics is guided by the assumption that if the subject is the tree, it is not necessary to describe an entire forest.

The goal is deceptively modest, using the carefully planned, well-guided plant tour as a model. To wit: provide the reader with the understanding needed to feel comfortable in the presence of subjects that were foreign to him or her prior to the tour. There is no intent to confer a degree in metallurgy.

By necessity Chapters 5 and 6 in places are heavy in technical detail and jargon. A suggestion is to skim through these pages on first reading to get a nodding acquaintance with the subject matter. Then proceed to Chapter 7. After finishing the book, go back and reread Chapters 5 and 6.

As a nonmetallurgist, I am gratefully indebted to a number of authoritative sources of information, as listed in the Bibliography.

—Harry Chandler

Contents

Chapter 1

The Accidental Birth of a No-Name Alloy

In 1906, three years after the first successful flight of the Wright brothers, a German research metallurgist, Dr. Alfred Wilm, was commissioned by the Prussian government to invent an alternative to the metal then used in cartridge cases. Their preference was for a type of aluminum harder and stronger than anything on the market.

The story begins on a Saturday morning in 1906 in Dr. Wilm's laboratory. At this point, he had concluded that pure aluminum was too soft for the application, and had ruled out a variety of copper-zinc alloys (bronze) because they were not heat treatable. Heat treating was one of the methods being used to upgrade the hardness and strength of materials. Wilm knew that some of the aluminum alloys were heat treatable, and he chose this route.

Beyond this, some—but not quite all—of the facts are known. For instance, it is known that for his experiment Dr. Wilm:

- Made an aluminum-copper-magnesium alloy that contained 4% copper and 0.5% magnesium
- Heat treated the metal at a temperature of 520 °C (1970 °F) in a salt bath furnace (see Fig. 1-1)
- Cooled (quenched) the metal rapidly in water down to room temperature, following heat treatment

Wilm's next step is open to conjecture. However, it is possible that while the metal was still at room temperature he rolled it to a thinner gage, called sheet. This process is known as *cold rolling,* or *cold working*.

Following these preliminaries, Wilm gave a sample of the alloy to his assistant Jablonski, with instructions to run it through a series of tests to determine its properties (Fig. 1-2). The time was shortly before noon, and Jablonski asked to put off testing until the following Monday because he had an appointment to keep. Wilm persuaded him to do a quick hardness test before leaving. Results were less than encouraging. Both hardness and strength were much lower than Wilm had hoped for.

On the following Monday, much to the surprise and pleasure of both men, the properties of the metal were better than expected or hoped for; in fact, they now were outstanding. Strength, for instance, was ten times greater on Monday than on Saturday morning. Subsequently, it was learned that maximum properties were obtained with the passage of four or five days.

The discovery was accidental—a common occurrence due to the manner in which research was conducted those days—and Wilm was unable to explain how and why he had arrived at the result he obtained, also a common occurrence. The science of metallurgy was in its early stages; and in this case, a scientific explanation for what had happened was unavailable until the latter days of the next decade. (Chapter 2 is devoted to this and related subjects.)

Further, the alloy made by the new heat treating process did not have a name, nor would it have a known use for several years. Its eventual name, Duralumin, was selected by Wilm's next employer, Dürener Metallwerke, to whom he sold his patent rights in 1909.

Dürener put Duralumin on the market as an experimental alloy, selling for $200 per pound ($440 per kilogram). (By 1964 the price of aluminum would be down to 24 cents per pound and would become even cheaper with the adoption of recycling.) Initially, consistency in quality was a problem with Duralumin. Copper in the alloy was detrimental to its corrosion resistance. The metal had a tendency to disintegrate in random spots, which ultimately transformed to a white powder.

Today, a modified version of Duralumin exists as alloy 2017, an aluminum-copper alloy in the 2000 series (Fig. 1-3). These alloys are characterized by strength higher than that of plain carbon steel and by improved machinability. On the negative side, corrosion resistance and weldability are limited. The nominal composition of alloy 2017 consists of 5.8% alloying elements, the balance aluminum. Two of the alloying elements, silicon and manganese, were not present in Duralumin. Generally, manganese increases strength and hardness, while silicon increases toughness and ductility. Usage of alloy 2017 today is limited. Rivets represent its chief application in airplane construction.

Curiously, aluminum was destined to become the number one material in aircraft construction, a distinction it still holds today. Its first significant application, however, was as a construction material for dirigibles, in such forms as strip, girders, rivets, and other parts.

Besides inventing an alloy, Wilm unknowingly had discovered a process called *solution treating and natural aging*. This process is still extensively used today in treating aluminum and other nonferrous and ferrous alloys.

Some Definitions

Ferrous is a Latin word meaning iron. Iron is the chief component of this family of metals, which includes wrought iron, carbon steels, alloy steels, tool steels, and stainless steels.

Nonferrous metals, then, are the "other metals" that do not belong to the ferrous family. These include aluminum and aluminum alloys, copper and copper alloys, beryllium copper and other beryllium-containing alloys, magnesium and magnesium alloys, tin and tin alloys, lead and lead alloys, refractory metals and alloys (niobium, tantalum, molybdenum, tungsten, and rhenium), titanium and titanium alloys, zirconium and hafnium, the precious metals (silver and silver alloys, gold and gold alloys, platinum and platinum alloys, palladium and palladium alloys), uranium and uranium alloys, beryllium, rare earth metals, germanium and germanium

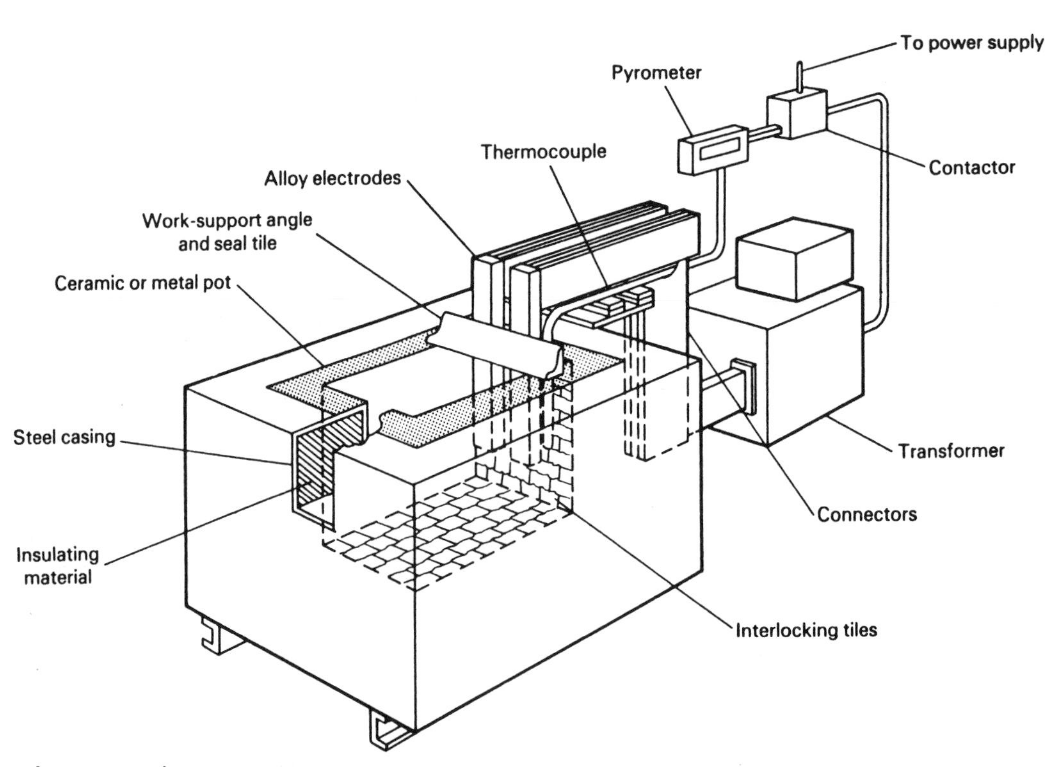

Fig. 1-1 Wilm annealed his aluminum-copper-magnesium alloy in a salt bath furnace. Shown here is a type of heat treating furnace in use today.

compounds, gallium and gallium compounds, indium, and bismuth.

Some nonferrous alloys contain ferrous metals as alloying elements.

The Status of Metallurgy at the Turn of the Century (1900)

In the first decade of the 20th century, metallurgy stood essentially at the same spot it occupied at the end of the preceding century. Metallurgists mainly were concerned about making established metals available in greater quantity, of better quality, and at lower cost.

Materials of construction were largely limited to those that had been around for centuries: wood, stone, leather, brass, bronze, copper, and cast or wrought iron. Less attention was paid to innovation than to making marginal improvements in what was available. Normally, doing research meant working in isolation on subjects chosen without regard for the marketplace.

Lacking the benefits of established metallurgical science, the metallurgist had to rely on experience, intuition, hunches, and/or luck. Innovation was a trial-and-error process. Up to the turn of the century, for example, steel was being heat treated without the support of metallurgical theory. In 1900, the science of making tool steel was virtually unknown. Researchers, mainly chemists, relied chiefly on chemical analysis, which did not address most of the properties of interest. In those days, the saying was, "Science follows technology," the opposite of the situation today. Theory now precedes innovation. Metallurgy has become a knowledge-based profession.

In the decade from 1900 to 1910, upgrades in the hardness and strength of metals were common goals. Usual routes were:

- *Alloying,* a process that involves coming up with different combinations of metals (i.e., aluminum and copper) or metal/nonmetal combinations (i.e., iron and carbon—which is known as steel). Making adjustments in the amounts of individual metals in an alloy is another approach.
- *Cold working*, or deforming metals by such means as rolling, bending, or stretching. This is an alternative method of strengthening.
- *Heat treating,* which is often used in conjunction with alloying or cold working. In this instance, desired results are obtained by heating and cooling solid metals.

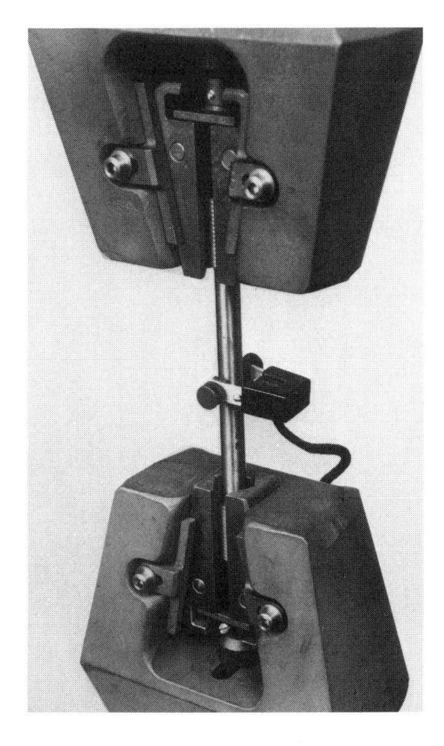

Fig. 1-2 Wilm's assistant tested the new aluminum alloy for both hardness and strength. Shown here is a tensile tester for steel.

Four Turning Points in Technology

Today's awesome metals technology—the results of which are so much around us daily that we tend to take them for granted—can trace its beginnings to the following events in the first decade of the 20th century:

Fig. 1-3 Microstructure of an aluminum-copper-magnesium alloy (alloy 2024) similar in composition to Wilm's alloy (now 2017). In this instance, the alloy was cold rolled, solution annealed, and aged.

- *1900*: Frederick Taylor and Maunsel White of Bethlehem Steel Company invented a high-impact tungsten carbide tool steel that outperformed the competition worldwide.
- *1903*: Orville and Wilbur Wright succeeded in flying their airplane, a wood, cloth, and metal structure.
- *1906*: Dr. Alfred Wilm invented a new process for heat treating aluminum alloys.
- *1908*: Henry Ford introduced his Model T, which featured a strong, weight-saving vanadium alloy.

The Story of a New Tool Steel

Taylor and White demonstrated their tungsten carbide alloy at the Paris Exposition of 1900. To show it, plain carbon steel forgings were machined in a lathe under normal working conditions. The heat generated in machining turned the tool steel red, but it did not lose its edge. Tungsten carbide is one of the hardest substances known.

A German firm, Ludwig Loewe Company, was impressed and took several of the new tools back to Berlin for testing. The tools were installed in a lathe and a drill press and operated under conditions intended to determine maximum performance. It is reported that in less than a month the lathe and drill press were reduced to scrap. The tools, however, were still in good shape. In one stroke, every machine tool in the world became obsolete. The machines did not have the capability needed to fully utilize the new alloy. Existing machine tools had to be redesigned to live with the stresses of high-speed metal removal.

The Taylor-White steel was a modified version of a tungsten alloy invented by Mushet in 1868. Tungsten is an excellent metal for the application because tool steels are subjected to high temperatures in service, and tungsten has the highest melting point of any metal. It melts at 3410 °C (6170 °F), is extremely hard (to resist wear), and is two and a half times heavier than iron (a property taken advantage of in balancing the wings of jet fighter planes—being superheavy, a small amount of the metal does the job).

Mushet had used 9% tungsten in his composition. Taylor and White doubled the tungsten, increased the chromium content, and used a higher hardening temperature in heat treating. Their composition included 18% tungsten, 4.25% chromium, 1.10% vanadium, and 0.75% carbon (Fig. 1-4). A similar high-speed steel available today is in the T-series of alloys. Its composition includes 18% tungsten, 4% chromium, 1% vanadium, and 0.75% carbon. A

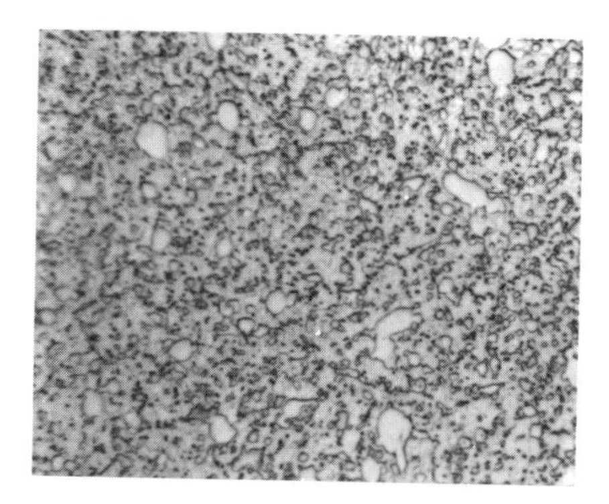

Fig. 1-4 Microstructure of an 18% tungsten high-speed tool steel of the type developed by Taylor and White

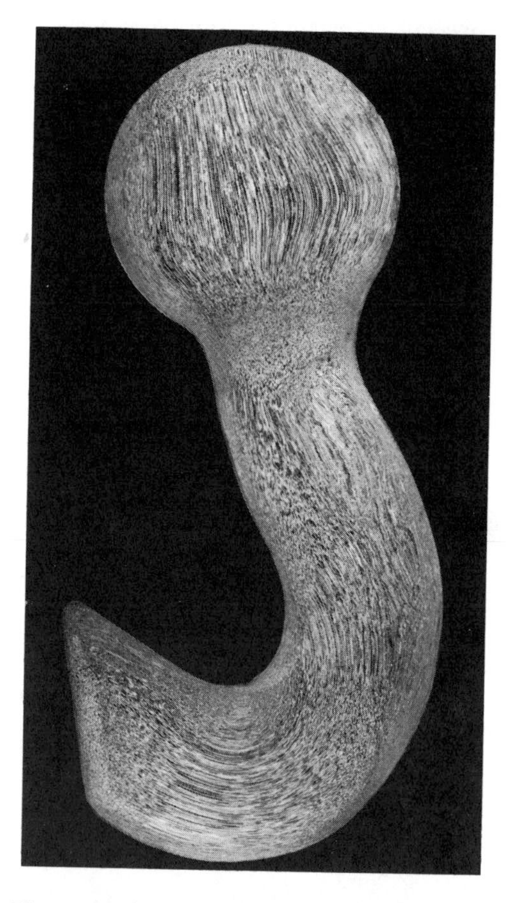

Fig. 1-5 The Wright brothers used a forged nickel-chromium alloy steel crankshaft in their first plane. Forging improves strength by orienting the structure of a metal in the direction that calls for strength in service. Note the longitudinal flow lines in this hook forging.

standard high-speed steel known as 18-4-1 is rated very high in wear resistance and resistance to *softening* at elevated temperatures.

Another Materials Story, Three Years Later

The "Wright Flyer" was essentially a wood and cloth structure with metals serving essential supporting roles: Bracing was hard drawn steel wire (cold working increases strength); engine materials included cast aluminum components (to save weight and as an alternative to machining); and the crankshaft was made of forged nickel-chromium alloy steel (forging increases strength) (Fig. 1-5).

Orville was the materials man. For wood he chose West Virginia white spruce because it is light and has a straight grain (for strength). The fabric was "Pride of the West," a quality muslin. Wooden runners—conventional buggy components—served as wheels..

After the first flight of 59 seconds on 17 December 1903, no one managed to stay aloft for more than one minute the next few years. By 1908, the Wrights were flying publicly, circling and turning for more than an hour. No one else knew how to make proper turns or efficient propellers.

In the 1920s the airplane advanced from wood and fabric construction to metal construction. The all-metal airplane appeared in the early 1930s. Incidentally, a spacecraft landed on the moon 66 years after the maiden flight of the "Wright Flyer"—a dramatic advance in technology in a mere six decades.

Dr. Wilm's Amazing Discovery

In 1910 Duralumin production amounted to only 10 tons (9.1 metric tons). By 1918 the annual total had jumped to 1000 tons (910 metric tons).

One of the first uses for Alfred Wilm's new alloy arose in 1908, when Count Ferdinand von Zeppelin built an airship, the LZ4, in Germany. It had aluminum structural members and an aluminum skin. After completing a 12-hour endurance flight covering 240 miles (384 kilometers), the craft was destroyed on the ground during a thunderstorm when it was torn loose from its mooring. Despite the incident, Count Zeppelin picked up great public support for his pioneering work and subsequently formed a private company, Lüftschiffbau Zeppelin.

By 1913 zeppelins, as they came to be known generically, had logged 400 flights. One craft, the *Victoria Louise*, had carried 8551 passengers and traveled 29,430 miles (47,000 kilometers). The *Graf Zeppelin*, the 127th to be built, had capacity for 20 passengers and 5 tons (4.6 metric tons) of freight, a cruising range of 6000 miles (9600 kilometers), and a cruising speed of 70 miles per hour (112 kilometers per hour). She made her first Atlantic crossing to New York in October 1928. Goodyear Tire Company acquired North American rights to the zeppelin in 1923.

The first applications of Duralumin in airplanes (for gas tanks and as a substitute for lumber and canvas) appeared around 1910. Volume usage of aluminum alloys started in 1916, when the French builder L. Brequet chose Duralumin for his Brequet 14:—one of the first planes with a metallic structure. Wing construction included torpedo tube struts and hollow rectangular spars, reinforced with ash linings in heavily stressed areas. The fuselage was a prismatic boom made of round Duralumin tubes, pinned in welded steel couplings.

In Spain, Hispano Suiza adopted Duralumin for auto engine crankcases in 1914 and for airplane engine blocks the following year. Horace W. Clarke, a British metallurgist, pioneered the application of Duralumin in England and also introduced it in France and the United States. He was awarded a knighthood and an honorary degree for his achievements.

In England, some of the copper in Duralumin was replaced by nickel in Y-series alloys, improving their high-temperature properties and making them suitable for pistons in airplane engines. The alloy, which contained 4% copper, 2% nickel, and 1.5% magnesium, came to be known as L35. With age hardening (Wilm's process), strength shot up by 50%. This alloy became the basis for succeeding British alloys. One, RR 58, contained 2% copper, 1.5% magnesium, 1.5% iron, and 1% nickel; it was used throughout the structure and in surface cladding for the Concorde supersonic jet.

Alloying reduces the natural corrosion resistance of aluminum, a problem solved in the United States with a composite called Alclad, which is composed of a skin of high-purity aluminum (a nonalloy) over an aluminum alloy core. Duralumin and Alclad sheet were major materials of construction for the Ford Tri-motor airplane. In the tri-motor, the sheet was corrugated to make it stiffer and thus stronger.

The all-metal airplane made its appearance in the 1920s. By the 1930s aluminum alloys were firmly established as the principal material of the aircraft industry. In the mid-1980s, for example, 80% of all airframes (excluding equipment) were made of aluminum alloys.

Henry and His Famous Flivver

The Ford Model T, also known as the Flivver or Tin Lizzie, has been called the most important car in the history of motordom. The 20-horsepower Model T, introduced in 1908, boasted several automotive breakthroughs. In addition to pioneering the use of vanadium alloys in the United States, the car had left-hand drive (versus right-hand drive in Europe), an improved planetary transmission, three-point motor suspension, a detachable cylinder head, a simple but workable magneto, a double system of braking (pedal and emergency brake), and a host of similar fine engineering touches.

Two other firsts are associated with the Model T. The production line was introduced, which in combination with high-speed machining marked the beginning of mass production in the United States and ultimately in other parts of the world. In addition, Ford set a new standard for bluecollar worker pay in auto plants: $5 per day, replacing $2.41 per day. The intention of the latter move was to reduce manufacturing costs via increased worker production. Net result of both innovations is that the Model T became more affordable to more people through lower prices. In time, some models sold for as little as $260.

Henry Ford built his first gasoline engine-powered car in 1902. He was encouraged to become an automaker by Thomas Edison, his boss at Detroit Electric Company, where Ford worked as a machinist and engineer. Ford Motor Company was founded in 1903.

The Foundation Was in Place

The history-making events of 1900 to 1910 attributed to Taylor and White, the Wright brothers, Wilm, and Ford should not be regarded as bolts out of the blue. By and large, these singular achievements represent the continuation of work with metals in the preceding century. The marvel of it all is the timing: the concurrence of events of such magnitude in the same time frame.

During the early part of the 20th century, articles and books written about metallurgy reveal great interest in alloying as a way of upgrading the properties of alloys, including:

- *Mechanical properties,* such as strength, hardness, and elongation
- *Physical properties,* such as electrical conductivity, thermal conductivity, density, and melting point
- *Fabrication properties,* such as weldability, formability, and machinability

Early Work on Tool Steels

In 1868 Robert Mushet invented a tool steel alloy containing 9% tungsten. Tools containing tungsten wore better in machining than did their carbon steel counterparts. Any gains in performance, however, tended to be minor. A breakthrough in this field did not arrive until 1900, when Taylor and White introduced their new alloy. Like Mushet's, their composition was based on tungsten—an extremely hard material with the highest melting point of any metal (3410 °C, or 6170 °F). Resistance to the high heat generated in machining is an important property. A proprietary heat treating process described only as "cooling from high temperature" supposedly was part of the secret of success.

In 1899 Elwood Haynes and Apperson introduced an alloy steel to the auto industry that was based on nickel for the purpose of improving high-temperature toughness and ductility and resistance to corrosion. The metallurgists equipped a car with nickel alloy steel axles, with steel made by Bethlehem Steel Company. The axles survived a 1000-mile (1600-kilometer) trip from Kokomo, Indiana, to New York City over roads generally unsuited for travel by car—"without serious breakage of anything."

Meanwhile, interest in vanadium as an alloying element in steel began to grow. At the close of the 19th century, the French started to use vanadium in alloys either alone or in combination with nickel and chromium. Armor plate was their first application. In England, John O. Arnold and Kent-Smith were also experimenting with vanadium in alloys.

In 1900 Kent-Smith, a metallurgist, found that adding small amounts of vanadium to plain carbon steels increased their strength, making

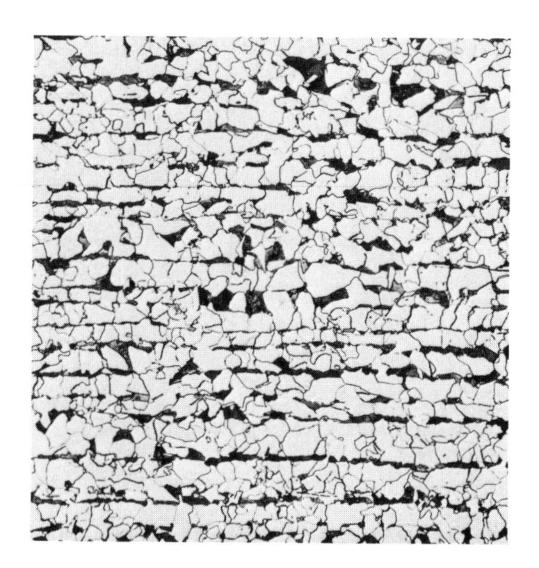

Fig. 1-6 Microstructure of a high-strength low-alloy steel of the type used by Ford. Today's versions include compositions containing vanadium, aluminum, and nitrogen, in addition to iron.

them suitable for service in high-speed steam engines. In a paper presented in Paris in 1903, Leon Guillet reported that a metallurgist named Choubly had used vanadium additions to boost the resistance of carbon steels to shock, and to increase their strength.

Henry Ford, who is credited with the development of vanadium alloy steels in the United States, first learned about the alloy around 1900. While examining a wrecked French racecar at Palm Beach, Florida, he picked up a damaged part and noticed that it was both light in weight and tough (bending before breaking). He sent a specimen back to his laboratory and learned that the steel contained vanadium. Following an unsuccessful search in the United States for a metallurgist with experience with vanadium, Ford invited Kent-Smith to cross the ocean and join him in his quest for chromium-vanadium alloys to replace the carbon-manganese alloys then favored for such parts as crankshafts and springs. Manganese increases strength and hardness.

The Ford vanadium alloy was introduced in the first Model T, the wonder car said to be "made possible" by the new alloy. The vanadium alloy provided twice the strength at half the weight of carbon steels (Fig. 1-6). Weight saving was imperative. With the alloys in use up to that time, cars would have been so massive and heavy that it would have been impossible to build engines with enough muscle to move them. High strength was also a must because of the abuse cars had to take from primitive roadways.

Early applications of alloy steels in cars included transmission gears and pinions (which require load-carrying capability); reciprocating parts, shafts, cams, and steering mechanisms (all need to be light and strong); valves (which must resist engine heat); and water pumps (which must resist leakage and corrosion).

The first record (circa 1905) of alloy steel usage in the auto industry was published in bulletins prepared by Henry Souther, a pioneer in automotive engineering. The list included carbon-chromium-nickel alloys with carbon contents ranging from 0.25 to 0.50%; one containing 0.50% carbon, a low-carbon steel; and a spring steel. Low-carbon steel generally contains less than 0.30% carbon; carbon content of medium-carbon steels ranges from about 0.30 to 0.60%; high-carbon steels contain more than 0.77%, and may be called tool steels. The upper limit of carbon in steel is about 2%. Spring steel falls into an in-between category, with carbon content ranging from 0.56 to 0.64%, for example. Note: Carbon content has a pronounced effect on the properties of steel and its response to heat treatment.

A Cross Section of Developments: 1900 to 1910

- Oldsmobile was the first automaker to have a full-time testing laboratory. It was headed by a graduate of Michigan State University, Horace T. Thomas, who was credited with, among other things, the development of a more powerful engine for the Olds.
- The first industrial lab in the United States was established in 1900 by General Electric Company. The facility was devoted exclusively to basic science.
- Carnegie Technical Schools, later to be known as Carnegie Institute of Technology, was founded in 1905. Initial enrollment totaled 126 students. The first professional degrees, including degrees in metallurgical engineering, were conferred ten years later.
- In 1905 Crucible Steel Company added the metallurgist and the microscope to its laboratory resources. With the aid of the microscope, the metallurgist discovered many unexpected characteristics of the microstructure of metals, which opened new fields of inquiry for Crucible.
- One of the first books on metallography (the science based on use of the microscope as a way of studying those features of metals invisible to the naked eye) was written by an Englishman, A.H. Hiorns.
- Ford started the practice of heat treating alloy steel car parts to make them stronger and more durable.
- In 1902 George J. Fuller of Wyman-Gordon Company, the acknowledged leader in forging crankshafts for cars, developed a heat treating schedule for this vital component. After forging, parts were brought to a bright red heat before being cooled to room temperature—a process call *drawing*, which would become standard in the metalworking industry. *Metalworking industry* is the generic term used for industries that not only make parts but also convert parts into consumer goods, as opposed to the generic term *steelmaker,* applied to producers of raw materials and semifinished materials, such as sheets and plates of steel, which are subsequently converted to parts by metalworking companies.
- The closed auto body was introduced, sprucing up the appeal of cars in the marketplace.
- The Audiffren, a sulfur dioxide compressor refrigerator developed in France, was manufactured by General Electric Company.
- An electric cooking range for the home was produced by Hotpoint.

- Tungsten ignition contacts for cars replaced platinum and silver types that cost more and had sticking problems. General Electric developed the devices.
- Resistance of steel to oxidation at elevated temperatures was enhanced by coating it with an aluminum alloy. This process, known as *calorizing,* was patented by Tycho Van Allen.
- Igor Sikorsky built the first helicopter in his native Russia. "It was a very good machine," he recalled, "but it wouldn't fly." The flying phase came 40 years later. Sikorsky is credited with an important first: the multiengine plane (circa 1913).
- In France, Philipp Monnarts discovered the "stainlessness" of stainless steel. Léon B. Guillet, a fellow countryman, reputedly made the alloy a few years earlier.
- Most steelmakers were advertising alloy steels—chromium, nickel, and vanadium types, as well as tungsten and silicon-manganese alloys.
- White, the truckmaker, introduced the four-speed transmission.
- General Bakelite Company started the production of Bakelite, a nonmetal (plastic).
- In Sweden, Kjellerberg developed the coated welding electrode—a major advance in the evolution of welding technology.
- In the United States, Robert H. Goddard started his pioneering work on rockets.
- A German, W. von Bolton, found a way to add ductility to the metal tungsten being used in filaments for electric light bulbs, making them more resistant to breakage.
- The first heat of silicon iron electrical steel was produced by General Electric in the United States. The silicon-iron alloy upgraded the efficiency of electric motors and transformers.
- A nickel-iron storage battery was invented by Thomas A. Edison.
- The first motorized machine tool (presumably electric) was introduced.
- The oxyacetylene torch for cutting steel was invented.
- James J. Wood was awarded patents for stationary and revolving fans.
- Packard introduced the steering wheel, replacing the tiller as the standard way of steering cars.
- A Frenchman, Paul Herout, invented electric steelmaking, which continues to play a major role in the production of quality steels and alloys.
- The first flight in a dirigible by Count von Zeppelin was recorded.

The Age of Innovation

The accomplishments of Taylor and White, the Wrights, Wilm, and others in the first decade of the 20th century lit the fuse for unprecedented innovation in science and technology that continues unabated to this day. In tracking the course of innovation, it is interesting to note that advances typically come in bunches, or clusters, following the time-tested principle that "one thing leads to another."

For instance, the National Academy of Sciences pointed out in its COSMAT Report (1975) that "alloy steel permitted development of the automobile . . . titanium was fundamental in launching the Space Program . . . the discovery of gallium [a metal] was essential to the growth of the laser . . ."

A Wyman-Gordon publication stated: "The manufacture and heat treatment of crankshafts and other auto parts beginning around 1902 probably started the real development of heat treatable steel. . . . In 1903 the first powered flight of the Wright brothers was a catalyst that launched the closed die forging industry with previously undreamed of activities." The publication also cited a symbiotic relationship between the forging process and heat treating, noting that "dramatic changes in both came hand in hand."

Alloy steels did not become a major material in airplane construction until aircraft became established as a vehicle of commerce and airline metallurgists started to think in terms of reducing empty weight to increase useful payload. Alloy steels offered the opportunity to push up strength and push down weight at the same time. A 1948 article in *Metal Progress* magazine pointed out, "Today, alloy steel engine mounts, landing gear structures, and attachment fitting . . . provide a gain of as much as 4 to 1 in strength to weight ratio over plain carbon steels."

The Age of Abundance

Our Age of Abundance has been made possible by the coexisting Age of Innovation. Mass production, one of the master keys to abundance, leads to mass consumption and was made possible largely by the assembly line and high-speed production machines. The values of the concept are summed up in the advertising cliché, "better, faster, cheaper":

- Better = better product quality
- Faster = high-volume production, which tends to lower both manufacturing costs and selling prices

- Cheaper = making the end product more affordable to more people

The Age of Abundance is a generic term for the many and varied ripple effects of the Age of Innovation. The following incident, though possibly apocryphal, condenses the meaning and magnitude of ripple effect into ten words.

The story goes: Early in the 20th century, a representative of the U.S. Department of Agriculture was touring the country to find out to what extent farm families were taking advantage of modern conveniences. In this instance, the man was interviewing a farmer's wife. The official observed, "I see you have a Model T parked out by the barn, but you still have outdoor plumbing. I don't understand." The farmer's wife replied, "You can't go to town on Saturday in a bathtub."

The point is rounded out by a comment from another source: "Although the automobile was at first the toy of the rich and the mechanically minded, it soon became a symbol of cheap, reliable transportation for everyone."

The Metallurgist-Innovator

Today, as in Dr. Wilm's day, metallurgy is a knowledge-based profession. Back then, though, knowledge was limited to that acquired through practical experience. Now, knowledge is based on the science of metallurgy.

The modern metallurgist-innovator occasionally works alone like Wilm, but usually will also function as a member of a team responsible for innovations that often fall outside the scope of his or her more usual and accepted responsibilities. Teams may be made up of technical and nontechnical people, or technical people who represent a variety of disciplines. A mixed team may include metallurgists, design engineers, production engineers, marketing specialists, financial experts, and purchasing staff. Depending on the assignment, an interdisciplinary team may include metallurgists, metallurgical scientists, research engineers, materials engineers, and physicists.

In addition, the metallurgist's knowledge base has been expanded beyond ferrous and nonferrous metals and alloys to include nonmetals, such as plastics and ceramics. The metallurgist is expected to have a sufficient grasp of these new subjects to contribute input at meetings of interdisciplinary teams.

In other words, the metallurgist is presumed to know something about all the engineering materials used in manufacturing today. Add to that knowledge of extractive metallurgy (extraction of metals from ores); mill metallurgy (production of metals and semifinished products, such as sheet and plate converted to end products at the manufacturing plant); plus supporting technology, such as that for the testing and inspection of materials and the know-how needed to convert metals and other engineering materials into articles of commerce at the manufacturing plant. This book, incidentally, is oriented toward the metallurgist in the manufacturing industry, with the focus on ferrous and nonferrous metals and alloys.

One more major development needs to be accounted for: Science and engineering have joined forces, resulting in a new discipline called materials science and engineering (MSE for short). It should be noted that "metallurgist" is still a working title. Metallurgy and MSE coexist. Both disciplines are implicit in the team approach to innovation.

The National Academy of Sciences report quoted earlier also had this to say: "The old idea of separate and individual inventors working in an isolated fashion on problems of their own selection has been largely replaced by R&D groups working toward a defined objective in an industrial research laboratory, a university laboratory, a government laboratory, or a laboratory located at a manufacturing plant."

MSE has been defined in this manner: The science part of MSE seeks to discover, analyze, and understand the nature of materials, to provide coherent explanations of the origin of properties used; while the engineering aspect takes this basic knowledge and whatever else is necessary (not the least of which is experience) to develop, prepare, and apply materials for specified needs. In other words, technology can be made to order; humans now have the capability of imitating nature. Such was the intent of researchers at General Electric when they set out to produce the synthetic diamond. Such was the intent of the interdisciplinary team at Bell Labs that invented the transistor.

The case history of an innovation (involving razor blades) provides further insight into the MSE concept and into how an interdisciplinary team operates. The assignment was to come up with a substantial improvement in razor blade performance. The resulting R&D program required considerable science and sophisticated techniques. Experts on the team were specialists in biophysics, physical chemistry, physical metallurgy, and life sciences. A then new scientific diagnostic instrument, the scanning electron microscope, provided a way of explaining differences in the performance of various razor blade materials. A key finding (that a thin layer of plastic dramatically improved performance) was partially accidental—not explained in the account—but the investigators recognized the importance of the discovery and were able to exploit it fully. The incident, it was explained, illustrates that "materials R&D seldom runs a neatly plotted course; it requires perception, training, and freedom to capitalize on unexpected events or observations."

Innovation is a way of life for the metallurgist. As the president of a metalworking company has observed, "Many of our present customers came to us before they were customers because they were having problems dealing with metallurgy. They are customers now because our metallurgists were able to solve their problems."

The metallurgist is the general practitioner of metallurgical science. Like his counterpart in medicine, his many good deeds are lost in anonymity, as the following quotes concerning the lack of public status of scientists, engineers, metallurgists, and others of their ilk bear out.

"Nearly all engineers, scientists, and technicians plus what they do are unknown by most of the public. A Nobel Prize winner in physics, chemistry, or biology may get a one minute mention on the evening news, while Oscar winners get two hour specials."

"Without question, the accomplishment of these people is enormous. If the engineers lay low, their works, in contrast, loom large. From skyscrapers, bridges, subways, dams, and expressways to supertankers, atomic powerplants, and supersonic jets . . . the products of the engineer's mind dominate our technitronic civilization and affect our lives. . . ."

Looking Ahead to Chapters 2 and 3

In both instances, these chapters are mostly about theory: theory that reveals and explains unseen and mysterious happenings taking place inside all solid metals and alloys—ferrous and nonferrous alike—while they are at rest (equilibrium), at room temperature, in the process of changing from solid to liquid at rising temperatures, or from liquid to solid at falling temperatures, or when they are being cooled slowly or rapidly back to room temperature.

The intent of Chapter 2 is to equip the nonmetallurgist with the metallurgical knowledge needed to follow and understand the technical explanation of what happened to Dr. Wilm's alloy while it was "just sitting around" in the lab from Saturday morning to the following Monday. For this reason, chapter 2 is slanted toward nonferrous metals, primarily aluminum.

In Chapter 3, the same general treatment (heavy on theory) will be applied to ferrous metals, primarily steel. The intent is the same as that of Chapter 2: building up a required background for the nonmetallurgist.

Chapter 2

Dr. Wilm's Mystery: What Happened?

On a Saturday morning in 1906 somewhere in Germany, Dr. Alfred Wilm designed an experimental aluminum-copper-magnesium alloy and then heat treated it. Before lunch, Dr. Wilm asked his assistant, Jablonski, to run a quick hardness test on the alloy to determine its properties (both hardness and strength). Results were less than encouraging. Nothing further was done from Saturday to Monday, when another test was run. This time the results were hard to believe: a major gain in hardness and a threefold gain in strength (hardness roughly equates to strength).

The discovery was purely accidental, and the unnamed alloy did not have a known use. Wilm was at a loss to explain what had happened over the weekend. In fact, at the time no supporting science existed.

The alloy was to become known as Duralumin, named for the German company that bought Wilm's patent. For the next 13 years and without the benefit of theory, metallurgists put the new alloy to use.

The mystery was solved in 1919 in the United States by a team of metallurgists headed by Paul D. Merica and including R.G. Walthenberg, Howard Scott, and J.B. Freeman, all employees of what was then the Washington-based U.S. National Bureau of Standards, which is now known as the National Institute of Standards and Technology. These metallurgists provided the theory that explains what happened from Saturday to Monday, revealing that the process used to heat treat the alloy, not the alloy itself, was responsible for the quantum improvements in properties.

Today, the Wilm process is a widely used standard heat treatment known by a variety of names. Perhaps the most technically accurate is *solution heat treating and natural aging*. The alternative treatment is called *solution treating and artificial aging*. In natural aging, the desired end result takes place over a matter of days; by comparison, in artificial aging the process is speeded up considerably. Examples of both processes are presented at the end of this chapter.

Any further discussion as to what happened in Dr. Wilm's lab from Saturday to Monday calls for a cram course in Lite Metallurgy 101. The quest for knowledge begins with a profile of the atom.

Profile of the Atom

The atom is the basic component, or building block, of all matter, including metals. Although several models have been proposed, no one is quite sure what an atom actually looks like. In 1911, Niels Bohr developed a solar system model that looked something like the model shown in Fig. 2-1. Another model that is consistent with theory today shows the subatomic components of the atom (Fig. 2-2), which include negatively charged electrons and a positively charged nucleus.

The atom is the basic component of a given metal that has all the properties of that metal. For instance, an aluminum atom has all the properties (such as hardness and strength) of the metal aluminum.

An atom is unbelievably small: It would take 10 million atoms spread out to cover the period at the end of this sentence, or a stack of 10 million atoms to equal the thickness of this piece of paper. Atoms exist in solid metals and in metals in transition from solid to liquid or liquid to solid. They are in a constant vibratory motion. The degree of movement of an atom is temperature dependent: the higher the temperature, the greater the movement, and vice versa. Movement of atoms is confirmed by the fact that metals expand when hot, and shrink when cool.

Movement of an atom is at its lowest level when a metal is at room temperature. At such time the atom is almost, but not quite, at rest—a condition called *equilibrium*. Even at equilibrium the atom continues to function in the manner of nature's Energizer battery: It just keeps going and going.

Like Atoms in Groups

In addition to acting alone, atoms function in groups. In the cubic structure shown in Fig. 2-3, an atom is located at each of the eight corners of a cube. Each location is called a *site*. The three-dimensional structure is called a *space lattice*. Most atoms are housed in three different types of space lattices: Two are cubic and one is hexagonal. No metal has the atomic structure shown in Fig. 2-3. The purpose of this illustration is to show the basic structure of the space lattice together with examples of atom sites.

The *body-centered cubic (bcc) lattice,* (Fig. 2-4) houses nine atoms, one more than shown in Fig. 2-3. The ninth atom is located in the center of the cube. Metals with this structure include niobium, chromium, lithium, molybdenum, tungsten, and vanadium.

The second cubic lattice is called the *face-centered cubic (fcc) lattice* (Fig. 2-5). Unlike the bcc type shown in Fig. 2-4, this lattice does not have an atom in the center of the cube. Instead, an additional atom is placed in the center of each of the six faces of the cube, making a total of 14 atoms in the lattice. Of the three types of lattice structures, the atoms in the fcc lattice are packed the closest. Both aluminum and copper (two of the three elements in Wilm's alloy) have this structure. Other metals with this structure include gold, lead, nickel, platinum, and silver.

The third common lattice structure is the *hexagonal close-packed (hcp)* type (Fig. 2-6). Six atoms are located at the top of this structure and six at the bottom. A straight line drawn between either set of atoms will form a hexagon. Addi-

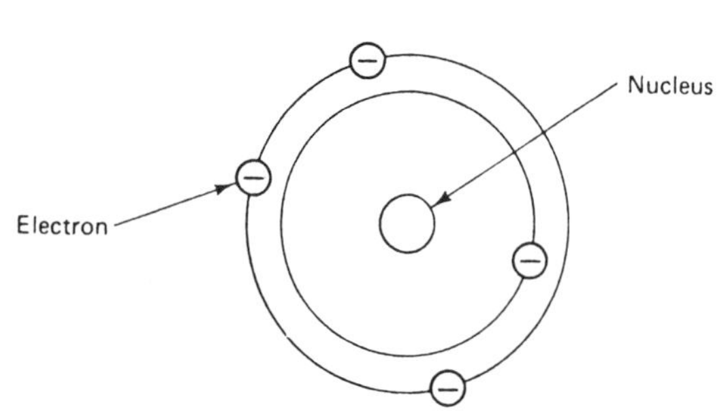

Fig. 2-1 Planetary model of an atom

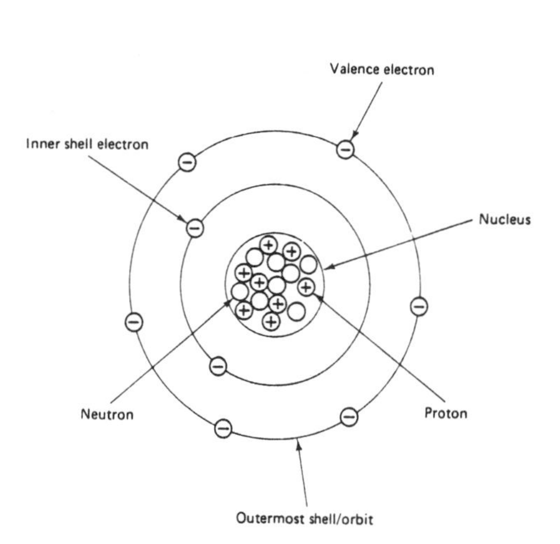

Fig. 2-2 Subatomic parts of an atom. This atom is shown in an equilibrium, balanced, or electrically neutral state.

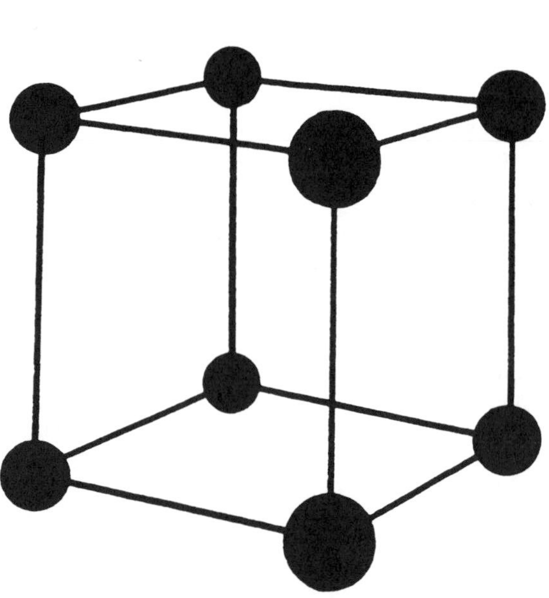

Fig. 2-3 Simple cubic lattice. No metal has this atomic structure.

tionally, three atoms in the form of an equilateral triangle are located between the top and bottom hexagons, for a total of 15 atoms. Metals with the hcp structure include magnesium, cadmium, cobalt, zinc, beryllium, titanium, and zirconium. Wilm's alloy also contained magnesium.

Next Size Up: Grains and Grain Boundaries

Grains are buildups of individual space lattices; *grain boundaries* fill the open spaces between grains (Fig.2-7 and 2-8). Grains and grain boundaries provide opportunities for the metallurgist to see what cannot be seen with the naked eye—namely, to see "inside" solid metals and to photographically record what is seen. This photograph is called a *photomicrograph*, or *micro* for short (Fig. 2-9), and this branch of metallurgy is known as *metallography*.

Behavior of Atoms

As stated previously, atoms are in constant movement regardless of the condition in which the metal housing them happens to be. In the heat treating of metals and alloys, the primary concerns are:

- The behavior of atoms in solid metals at room temperature
- The behavior of atoms in solid metals that are heated to elevated temperatures, just short of melting
- The behavior of atoms in metals cooled rapidly from an elevated temperature back to room temperature

However, for the purpose of this general discussion, we will start with metals at a high temperature in the liquid state, followed by a 180° change in direction—a progressively decreasing temperature all the way back down to room temperature.

Atoms making the transition from solid to liquid ultimately gain enough energy from heat to escape from both their lattice sites and the attractive forces generated by electrons within the atom. Atoms in this condition are able to move individually and freely in all directions within the extremely hot liquid metal, as depicted in Fig. 2-10.

Liquid-to-Solid Metal Transition Zone

As the temperature of the liquid metal starts to drop, a point is reached where some of the liquid metal enters the beginning stage of becoming solid. The result is a mixture of liquid metal and metal in the liquid-to-solid transition

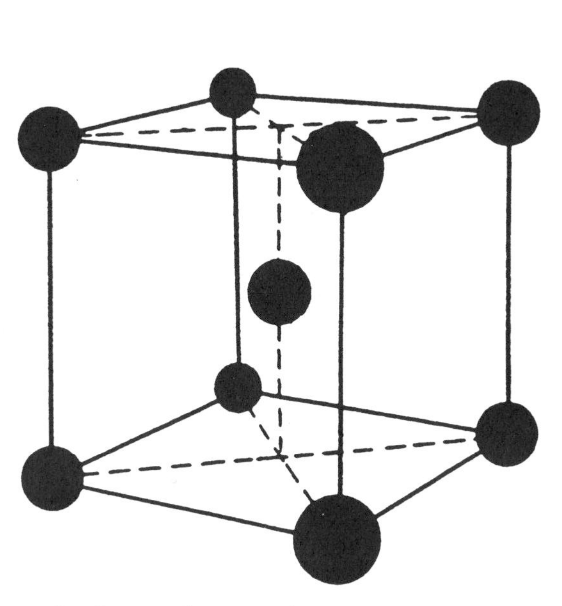

Fig. 2-4 Body-centered cubic lattice

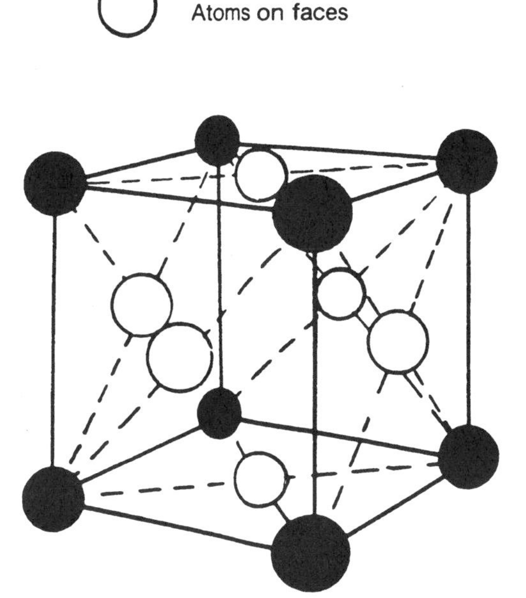

Fig. 2-5 Face-centered cubic lattice

zone. The technical term for this event is *freezing*.

A Closer Look: Liquid-to-Solid Phase

This section is prefaced with a phase diagram (Fig. 2-11) for the purpose of visualizing the mixed liquid-solid and related concepts. The nickel-copper phase diagram identifies temperatures or phases at which the alloy is liquid, is both liquid and solid, and is a solid.

- Starting in the upper left-hand corner and moving downward, phases and other areas are identified: liquid, liquid and solid, and solidus line (marks completion of transition). The solid zone, which is not labeled, is the entire area below the solidus line.
- Alloy composition is plotted on the horizontal axis at the bottom.
- Temperature is plotted on the vertical axis (extreme left).
- Broken vertical line A-B indicates a composition of 75% nickel and 25% copper.
- Broken horizontal lines C-T_1, D-T_2, and E-T_3 represent temperatures at which different phases exist:

  a. At the C-T_1 temperature, the alloy is in the solid phase.
  b. At the D-T_2 temperature, the alloy is in the liquid-solid phase.
  c. At the E-T_3 temperature, the alloy is in the liquid phase.

Differences: Pure Metals and Alloys

Metals in their pure (unalloyed) state, such as aluminum or iron, do not freeze and start nucleation in the same manner as their alloys do. Cooling curves for pure metals are shown in Fig. 2-12(a) and (b). When the temperature of the pure metal drops to its freezing point, freezing starts but cannot be maintained because the heat of fusion generated in the transition from liquid to solid is not sufficient to support the

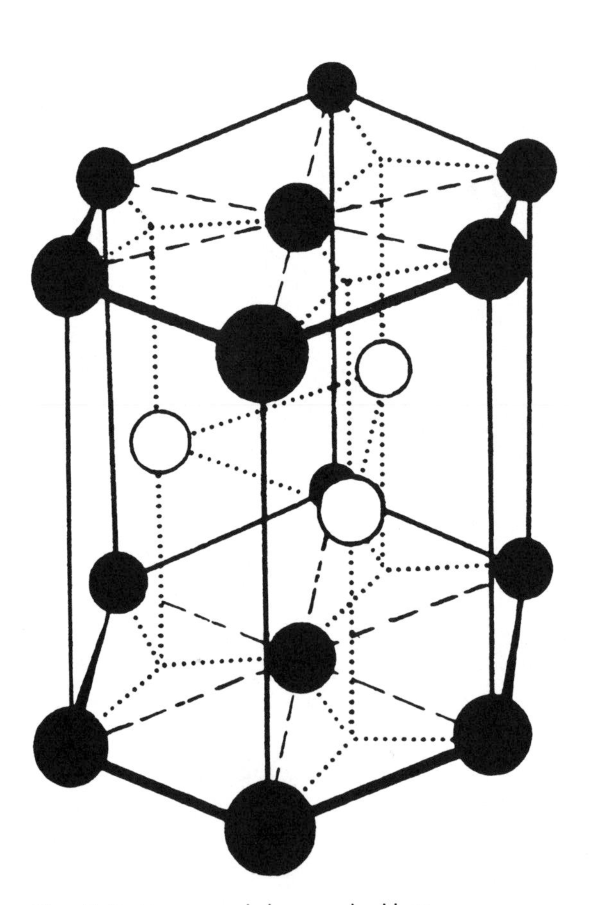

Fig. 2-6 Hexagonal close-packed lattice

reaction. As a result, the freezing process is stopped, or *arrested* (Fig. 12a).

One solution to the problem is to supercool pure metals at cryogenic temperatures. As the temperature rises back up to the freezing point, enough heat of diffusion is generated to initiate and support nucleation (Fig. 12a). The catch here is that supercooling is not feasible in commercial practice.

As shown in Fig. 12(b), a second, more economical approach is available. Minute solid impurities, such as dust particles or fragments of refractory brick, are added to the liquid metal. Free-moving atoms anchor themselves on these impurities, which serve as nucleation sites. Then only some supercooling is required (Fig. 12b).

Pure metals have only one freezing temperature. By comparison, as they freeze over a range of temperatures, alloys generate enough heat of fusion to support freezing. A typical cooling curve for a solid-solution alloy is shown in Fig. 2-13.

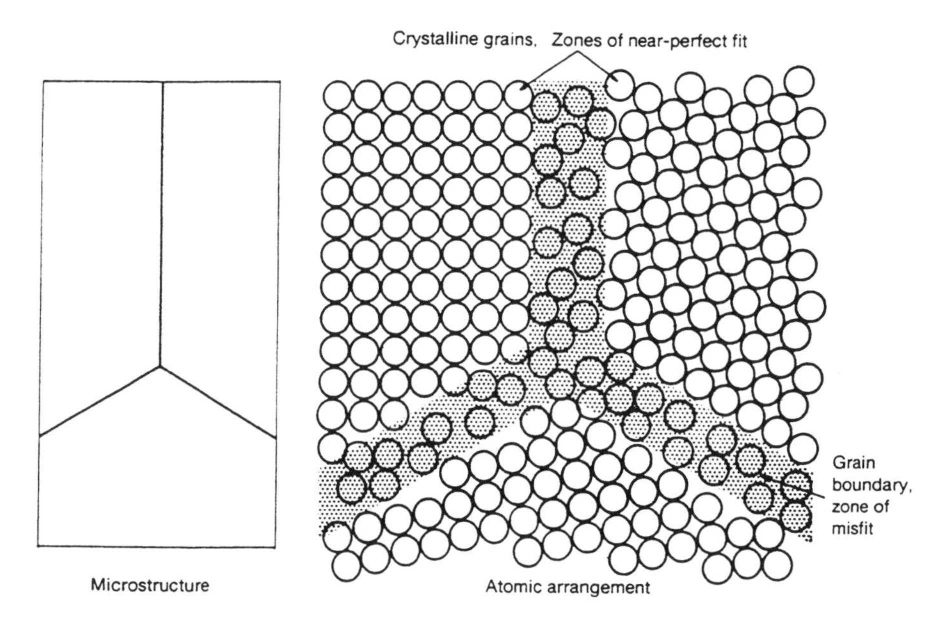

Fig. 2-7 Appearance and nature of grain boundaries

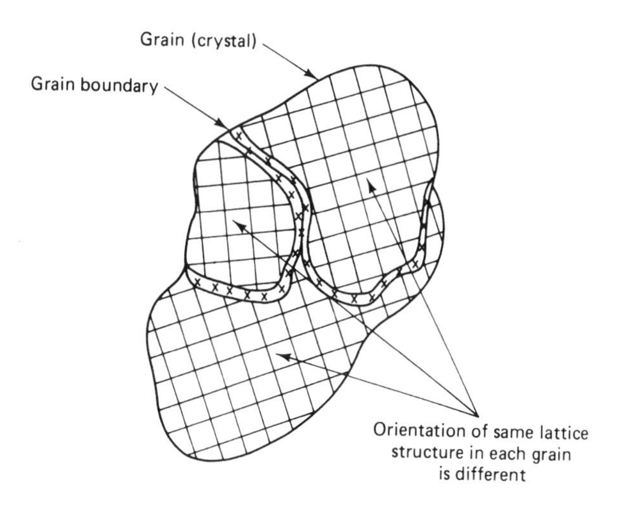

Fig. 2-8 Grain boundaries

From Freezing to Grain Growth

Once nucleation is initiated in both pure metals and alloys, grain growth proceeds in a straightforward manner. One scenario is depicted in Fig. 2-8 and 2-14.

In Fig. 2-14, the term *dendrite* is introduced. Dendrites are small crystalline needles shaped like pine trees—a pattern that usually disappears during mill processing of wrought metals, as in rolling sheet and plate.

Figure 2-8 provides more information on grains. Note the cluster of grains and grain boundaries. Grain boundaries fill in open spaces between grains. They are the last areas of the liquid metal to freeze. They are not arranged in the same orderly manner as atoms in crystal lattices in the equilibrium condition. Grain boundaries are high-energy areas that are under high stress, causing distortion (Fig. 2-7).

Fig. 2-9 Photomicrograph of brass (70% zinc + 30% copper). Courtesy of Buehler Ltd.

Because grain boundaries are distorted, they are harder and stronger than lattice structures in the equilibrium condition. Distortion of crystal lattices is also a factor in cold working of metal. The two micros in Fig. 2-15 show grain structures in a copper alloy before cold working and after cold working.

Metals and alloys in the *equilibrium* condition are not as hard or as strong as those with distorted lattice or grain structures, but they are softer (more *ductile*) and better able to withstand impacts (called *toughness*) because they bend before they break. Ductility is a preferred property in fabricating end products from metals and alloys.

Several Solidification Processes

Liquid metals and alloys poured into molds start to cool and form nuclei at the colder mold walls. The resulting columnar structure grows inward toward the center of the mold. This nucleation/grain growth process is illustrated in Fig. 2-16 and 2-17. Note the progressive growth of dendrites in Fig. 2-16(a) to (d). Figures 2-17(a) to (e) present a different perspective. A columnar structure is shown in Fig. 2-18.

A second process is for alloys that freeze in molds over a range of temperatures (Fig. 2-19). In this process, enriched liquid alloy surrounding the first crystals formed stops the growth of these crystals, creating conditions favorable for the nucleation of a second set of crystals just beyond the first set (Fig. 2-19a and b). The process is repeated as many times as necessary (Fig. 2-19c and d). Note that these crystals are not dendritic in shape. The name for the shape

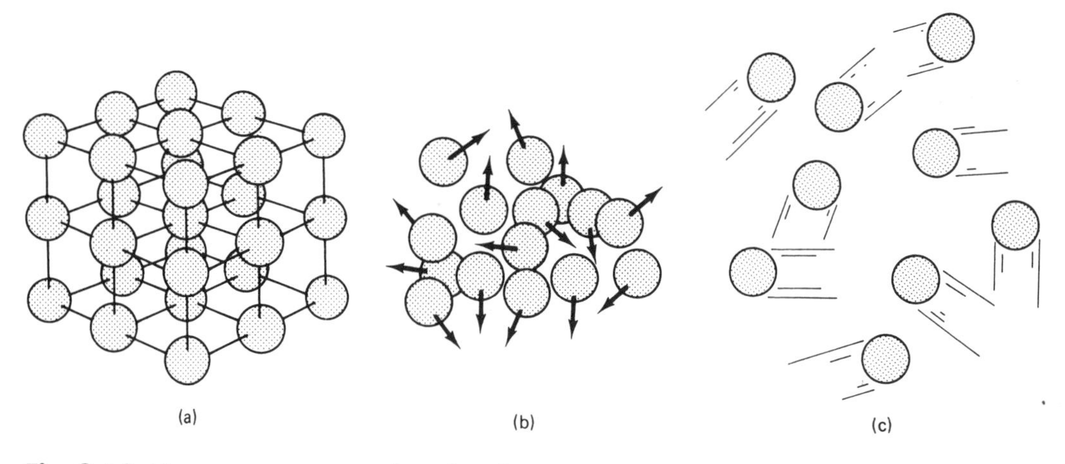

Fig. 2-10 How atoms are arranged in a liquid

pictured is *equiaxed*—meaning crystals are similar in length, breadth, and width. This stage in cooling is sometimes called *mushy freezing*. During most of the freezing period the alloy is in a semisolid condition.

Atom Movement in Solid Solutions

Convincing the nonmetallurgist that atoms move about in solid metals and alloys may be a hard sell, but it is 100% gospel metallurgy. Movement (*diffusion*) takes place in both solid metals and solid alloys in a variety of ways (Fig. 2-20).

The vacancy process is one example. Atoms move to unoccupied lattice sites. Open sites are a common occurrence.

In the second process, atoms of one element escape from their lattice structures and enter the lattice of a second element, forming a kind of partnership. The guest is called the *solute*; the host is called the *solvent*. In practical terms, the solute is dissolved into the solvent. This process takes place in either of two ways:

- The guest and the host are about the same size. The guest occupies an open atom site

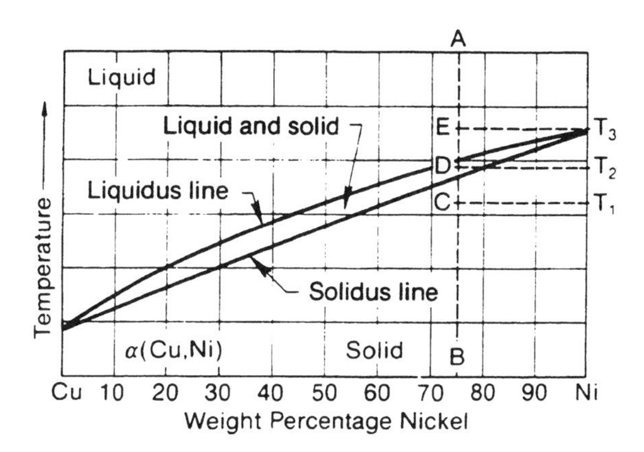

Fig. 2-11 Nickel-copper phase diagram, showing liquidus and solidus lines. Between these two lines is the freezing range. Liquid and metal coexist in the process of solidifying.

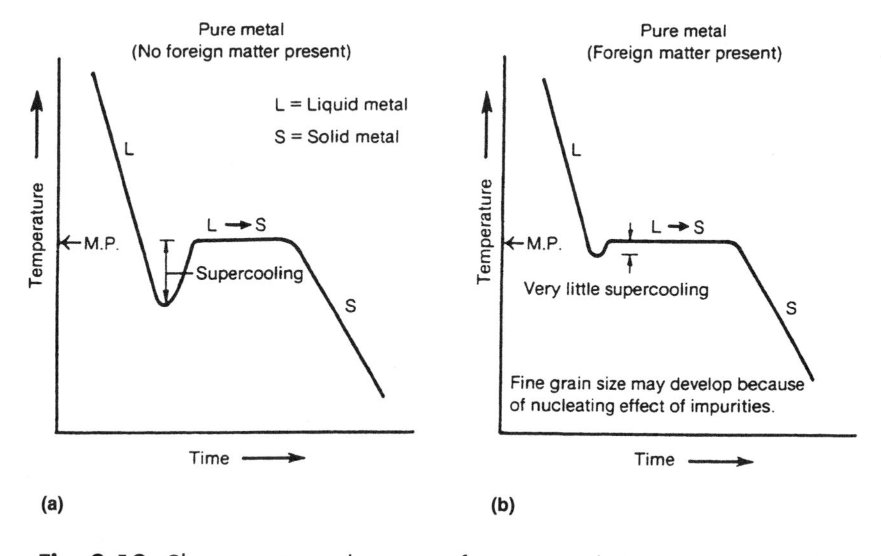

Fig. 2-12 Characteristic cooling curves for pure metals. Less supercooling develops when foreign matter is present (b), because the impurities have a strong nucleating effect. The true melting point, however, is not affected.

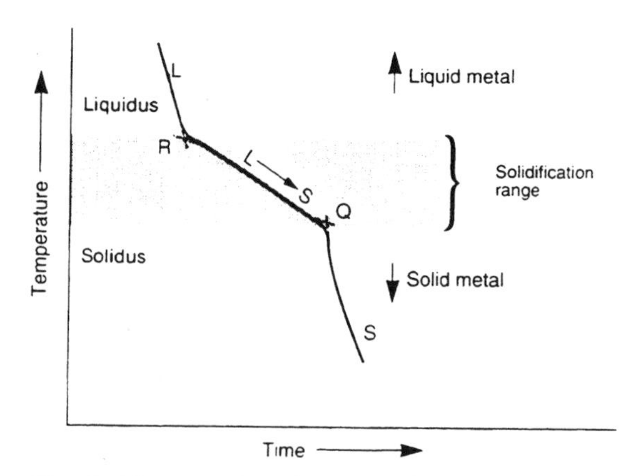

Fig. 2-13 Typical cooling curve for a solid-solution alloy. Note that freezing (L → S) occurs over a range of temperatures.

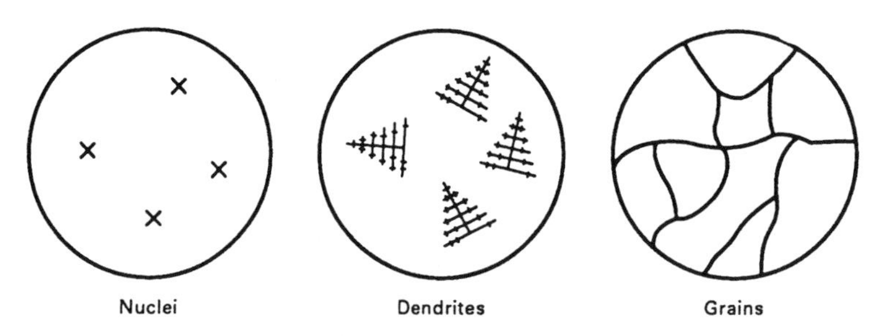

Fig. 2-14 Stages of a metal cooling through equilibrium. It first nucleates, then forms dendrites, and finally cools into a crystalline solid of many metal grains.

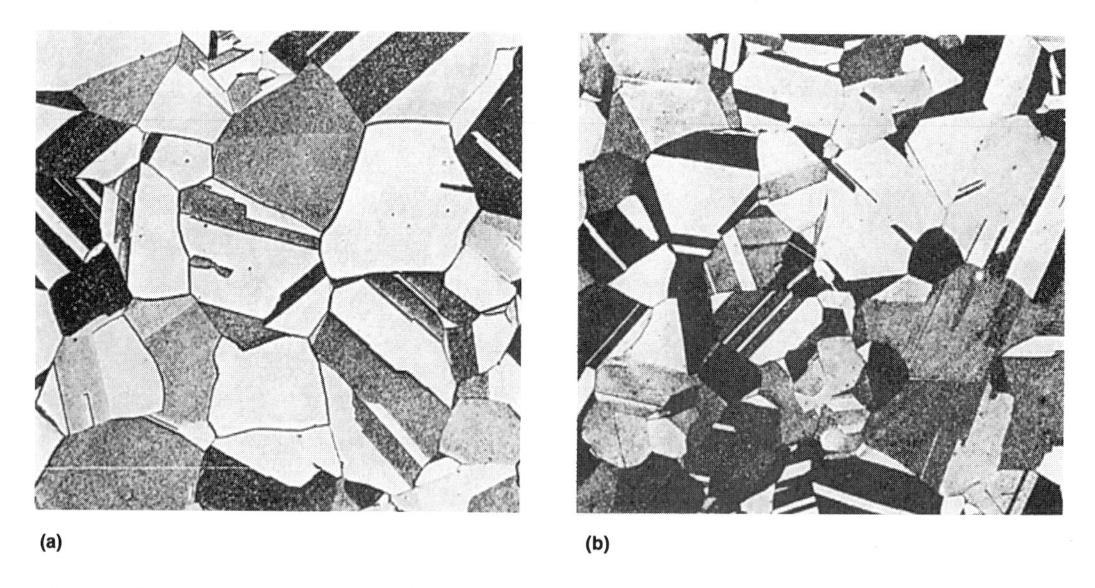

Fig. 2-15 Micrographs showing 70/30 brass at 300×. (a) Reduction of hot-worked metal is 0%. (b) Reduction cold-worked metal is 5%. Courtesy of Buehler Ltd.

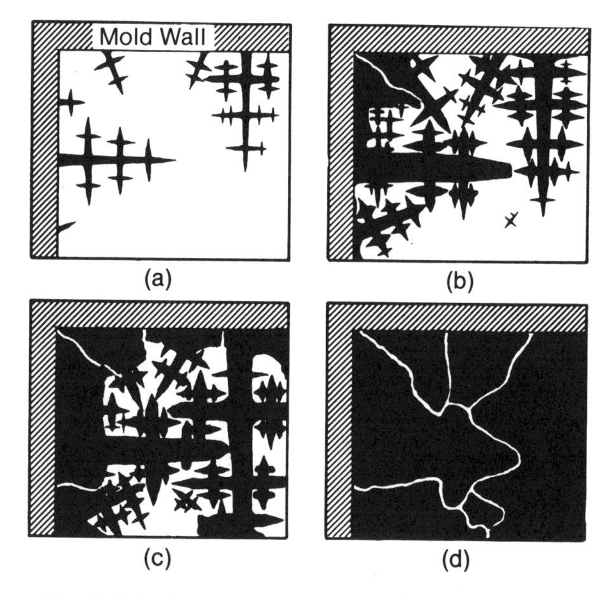

Fig. 2-16 Progressive growth of dendrites into grains. Time gradually increases from (a) to (d).

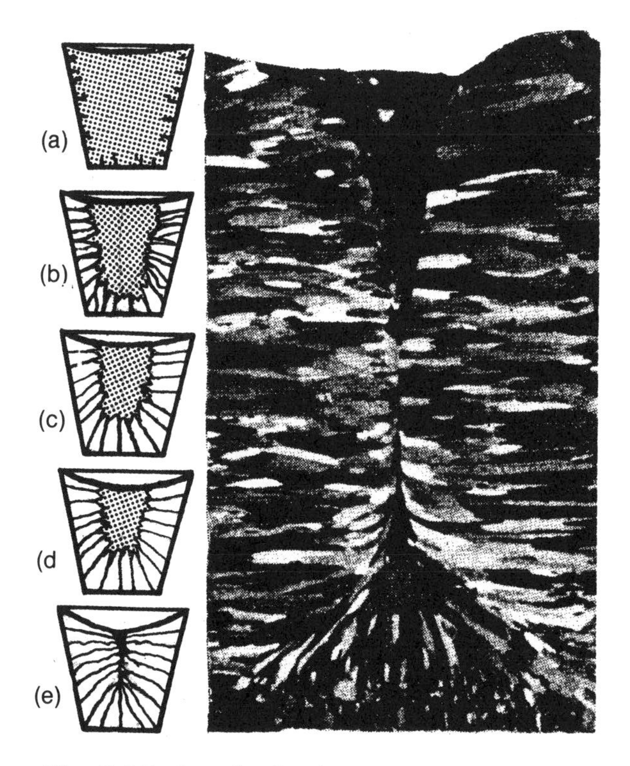

Fig. 2-17 Growth of columnar grains. As metal cools, nucleation begins at the mold walls (a), and columnar crystals grow toward the center of the metal as heat continues to be extracted through the walls (b, c, d). When freezing is complete (e), the planes of weakness are visible; this is particularly clear in the photograph at right.

in the host's lattice structure. This is called the *substitution process*.
- The guest is much smaller than the host, and the tiny guest occupies open spaces between atoms in the host's lattice structure. This is called the *interstitial process,* and the open spaces are known as *interstices*.

More About Solid Solutions

A solid solution has two or more components. There must be at least two components: the solute and the solvent. Although there is only one solvent, more than one solute may be involved. The key topic is *solubility*. The solvent is always the largest component, and the solute the smallest. Solutes are dissolved in solvents. The solubility of solutes varies with the metal. For example, copper is the solute in aluminum-copper alloys. At lower temperatures, the solubility of copper in aluminum is nominal. Solubility increases at temperatures of around 500 °C (930 °F), but is limited to the range of 4 to 5%. The degree of solubility is particularly important in such processes as heat treating, and, as indicated, temperature is always a factor.

In Fig. 2-20, note that the solute atoms in substitutional solid solutions are about the same size as the solvent atoms, and they occupy spaces in the lattice structure of the solvent. By comparison, solute atoms in interstitial solid solutions are considerably smaller than the solvent atoms.

Substitutional Solid Solutions. The comparative size of solute and solvent atoms is important here, and the difference in size should not exceed 15%. In an alloy containing 90% nickel and10% copper, 10% of the atom sites in the nickel lattice—also known as a *matrix*—must be occupied by copper atoms. The resulting alloy is stronger than plain nickel alone.

Various requirements for the formation of substitutional solid solutions must be met. For example:

- Both metals must solidify in the same type of lattice. (Note: Nickel and copper both have an fcc lattice).

Fig. 2-18 Columnar structure in solidified metal. The crystals growing from two walls intersect in planes: a clearly defined line of intersection is visible on fracturing of the metal and etching of the polished surface with acid.

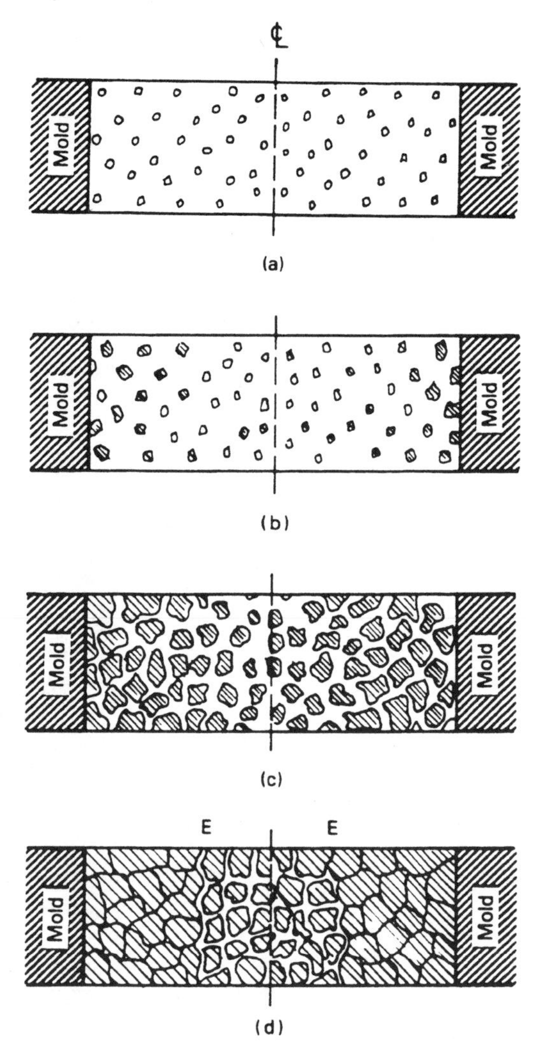

Fig. 2-19 Solidification in a long-freezing-range alloy. (a) and (b) Early stages. (c) Intermediate stage. (d) Late stage

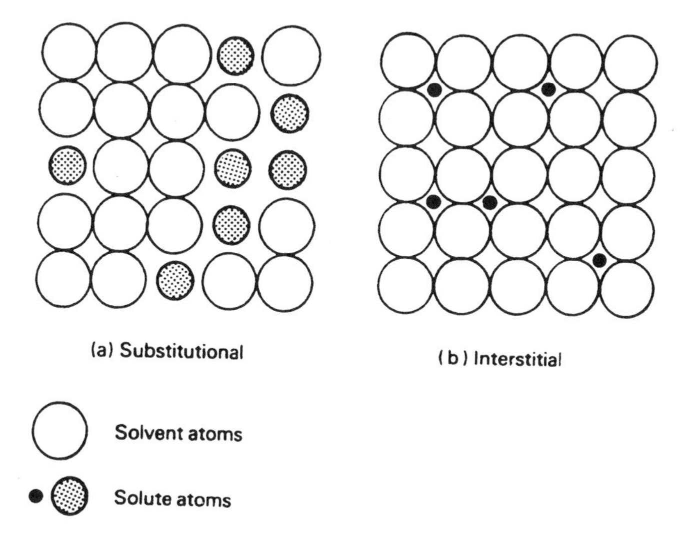

Fig. 2-20 Types of solid solutions. Two-dimensional model shows the substitutional type (a), in which the solute atoms substitute at positions of the solvent atoms, and the interstitial type (b), in which the solute atoms are much smaller than the solvent atoms and fit in spaces between the larger atoms.

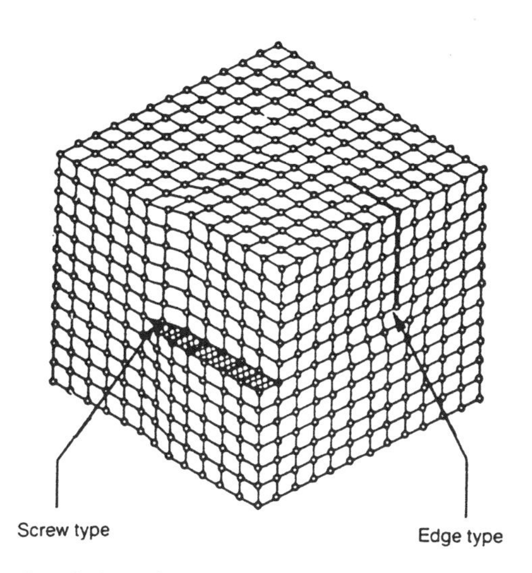

Fig. 2-21 Three-dimensional sketch of a crystal lattice showing an edge dislocation and a screw dislocation

- Both atoms must be about the same size (as mentioned previously).
- Both atoms must be similar chemically.

Interstitial Solid Solutions. In these solutions, small atoms, such as carbon, nitrogen, boron, and hydrogen, are the usual solutes. Comparative size is important because forcing oversized solutes into undersized holes will strain the lattice. In response, the solvent limits solubility to minimize lattice strain, but does not prevent it.

Solubility is limited in interstitial solid solutions. For example, in the iron-carbon phase diagram, the limit to carbon solubility is 6.67%.

A Final Word on Diffusion

Movement of atoms within a lattice structure can be impeded by lattice defects called *dislocations*. Two types of dislocations, edge and screw, are shown in Fig. 2-21.

Upgrading Pure Metals and Alloys

The focus in this section is on the cold working and annealing of pure aluminum and the heat treatment of aluminum alloys. The principles involved apply to other nonferrous metals and alloys as well as ferrous alloys, particularly when alloys are treated by the solution treating and aging processes.

First, Some Definitions

heat treating. A process of heating and then cooling a solid metal or alloy below its melting point to obtain desired results, such as improved resistance to corrosion or increased hardness, strength, and electrical conductivity.

quenching. Generally used in reference to the rapid cooling of a metal or alloy after it has been heated to a specific temperature. Slow cooling a metal or alloy in air back to room temperature can also be regarded as a form of quenching.

properties of metals. Include hardness, strength, ductility, toughness, malleability, conductivity

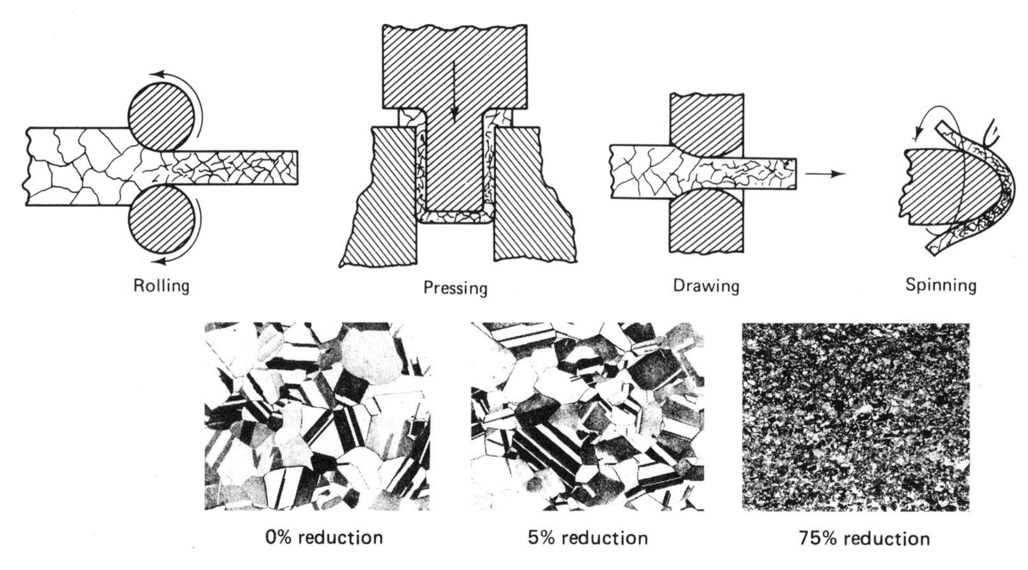

Fig. 2-22 Some work hardening (cold working) processes

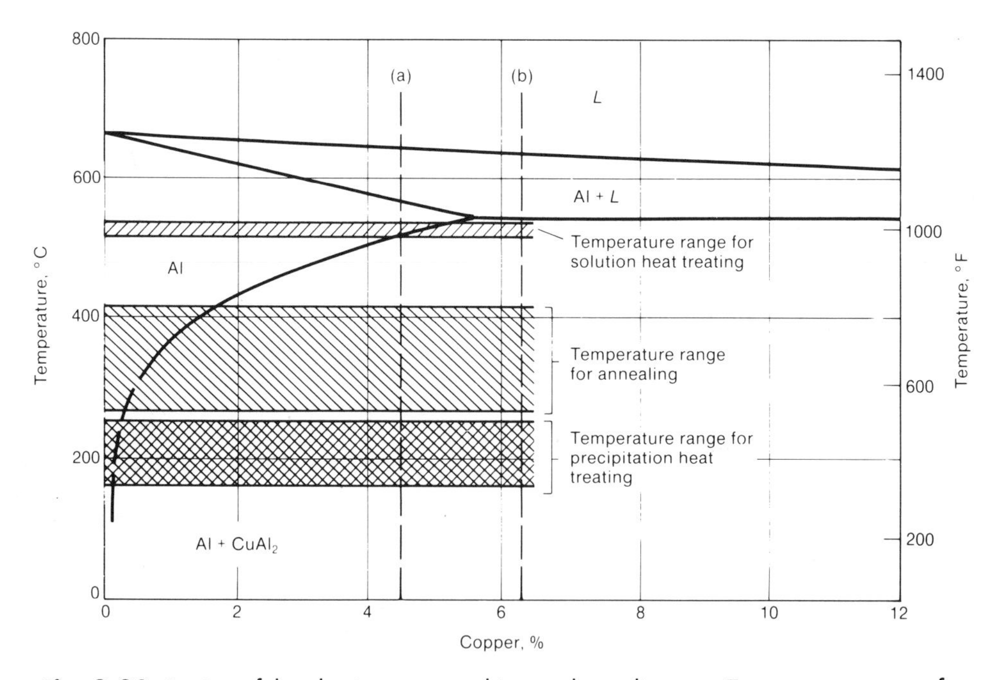

Fig. 2-23 Portion of the aluminum-copper binary phase diagram. Temperature ranges for annealing, precipitation heat treating, and solution heat treating are indicated. The range for solution treating is below the eutectic melting point of 548 °C (1018 °F) at 5.65 wt% copper. L, liquid; Al-$CuAl_2$, a mixture of aluminum and an aluminum-copper alloy

of heat and electricity, thermal expansion, and elongation.

basic properties of aluminum. Include excellent thermal conductivity, light weight (one-third the weight of steel), good strength, and resistance to corrosion.

ferrous metals. Principal component is iron—hence "ferrous." Ferrous alloys include low-carbon steel, alloy steels, tool steels, stainless steels, and high-strength steels.

nonferrous metals. An unfortunate term that implies "contains no ferrous metals. This is not true; a number of nonferrous alloys contain varying amounts of iron as an alloying agent. Nonferrous metals include aluminum, nickel, cobalt, copper, chromium, molybdenum, zinc, magnesium, zirconium, titanium, and superalloys based on heat-resisting metals such as nickel and cobalt.

pure metals. Another misleading term. Commercially, a *pure* metal is not feasible any more than it is necessarily essential technically. All metals contain some impurities, ranging from tiny fragments of dust to the so-called *tramp elements* that are found in recycled scrap metals. For example, phosphorus and sulfur are tramp elements in steel. The purest available metals are aluminum and copper, both of which can be 99.9% pure.

annealing. Heating a metal to a specific temperature and holding it there for a specific time, then cooling at a specific rate. Primary uses of annealing include softening metals prior to or during fabrication processes, such as forming. One objective in annealing nonferrous metals is to remove any undesirable effects of cold work. Grain structure is restored to its original condition.

cold working. Plastic deformation of a metal below the annealing temperature to cause permanent strain hardening.

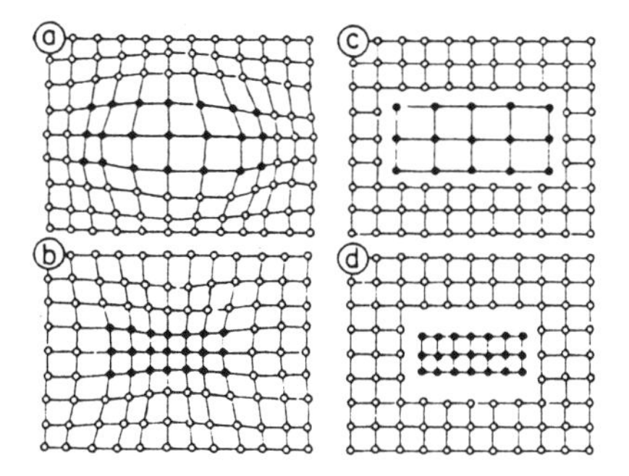

Fig. 2-24 Coherent (left) and noncoherent (right) precipitation. (a) and (b) A coherent or continuous structure forms when any precipitate is very small. (c) and (d) Coherency is lost after the particle reaches a certain size and forms its own crystal structure. Then a real grain boundary develops, and severe lattice stresses disappear.

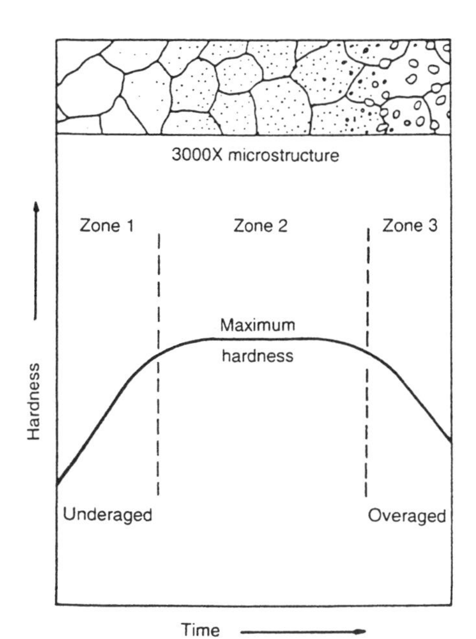

Fig. 2-25 Relation between aging curve and microstructure

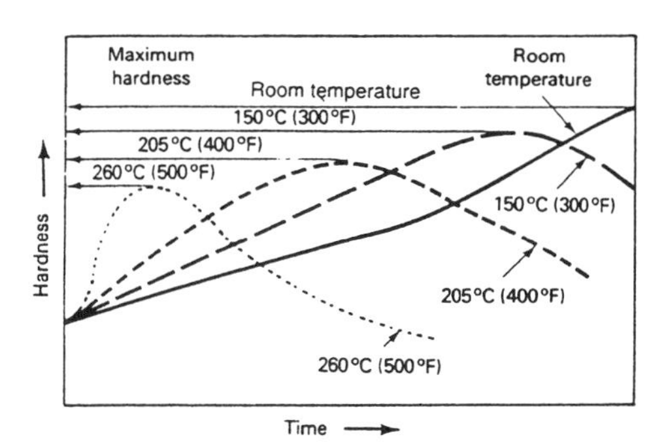

Fig. 2-26 Effect of time and temperature on the aging properties of Al-4.5Cu alloy. Maximum hardness is not achieved in reasonable time by aging at room temperature or at the lower aging temperatures. But at 205 °C (400 °F), a satisfactory high hardness is attained with aging. When higher aging temperatures are used, the maximum hardness is attained in a shorter time, but is not as high as the hardness reached with lower aging temperatures.

hot working. Plastic deformation of a metal at a rate and temperature such that strain hardening cannot occur.
plastic deformation. Permanent change in the shape or size of a solid body without fracture resulting from the application of sustained stress beyond the elastic limit.
elastic deformation. Metal returns to its original size and shape when applied force is removed.
strain hardening. Increasing the hardness and tensile strength of a metal by cold plastic deformation (*cold working*).

This chapter closes with:

- An example of improving properties by *cold working*
- A review of *precipitation hardening* (solution treating and aging)
- An example of *artificial aging*
- An example of *natural aging*, using the modern version of Dr. Wilm's Duralumin, also known as alloy 2017

Upgrading a "Pure" Metal: 1060 Aluminum

The strength of this metal is enhanced moderately by cold working. Four cold working processes are shown in Fig. 2-22. Note that cold working, also called *work hardening*, deforms grains; with increasing reductions as shown in the three micros in Fig. 2-22, grain size is reduced with increasing reductions (a desirable result in structural applications). Annealing follows work hardening. Unlike other metals in the aluminum 1000 series, 1060 may also be solution treated.

Data on 1060 Aluminum

Composition. Classified as 99.6% aluminum, 1060 also contains a number of other metals, ranging from silicon and iron to vanadium and titanium.

Products forms include sheet and plate.

Applications. This metal is typically used in products requiring good resistance to corrosion and good cold formability, and where low strength can be tolerated—as in chemical processing equipment and railway tank cars.

Thermal Properties. Liquidus temperature is 657 °C (1215 °F); solidus temperature: 646 °C (1195 °F). Note that the difference between the liquidus and solidus temperatures is only 11 °C (20 °F). In heat treating of aluminum alloys, the difference between the heat treating temperature and the solidus may be as little as 5 °C (9 °F).

The annealing temperature for 1060 is 345 °C (650 °F).

Overview of Precipitation Hardening Treatments

Some aluminum alloys can be solution annealed to a specific temperature, quenched rapidly in water, then aged naturally at room temperature for four days or more to obtain maximum properties such as hardness and strength. This process is known as *natural aging*.

The aging process is speeded to a matter of hours following solution treatment and rapid cooling by heating the alloy to a specific temperature and holding at that temperature for a specified time. This process is called *artificial aging*.

Figure 2-23, which is a portion of the aluminum-copper phase diagram, gives temperature ranges for solution heat treating, annealing, and precipitation heat treating. Solution treating takes place at a temperature below 548 °C (1018 °F). The solid solubility of copper (the solute) and aluminum (the solvent) increases with increasing temperature—from about 0.20% at 250 °C (480 °F) to 6.5% at 548 °C (1018 °F).

In solution treating, the major requirement is to develop and maintain a supersaturated solution of copper in the aluminum lattice. The result is obtained by rapid cooling (a matter of seconds) to room temperature—a temperature at which copper has very little solubility in aluminum. As a result, copper atoms (the solute) are trapped inside the aluminum solute, causing distortion of the lattice, which increases the strength of the alloy. This process is known as *precipitation hardening*, or *age hardening*.

At this point, the copper solute is locked inside the aluminum lattice (matrix), and the potential for precipitation exists. Ordinarily, the precipitate (tiny copper particles) and the parent solid solution would have totally different lattice arrangements.

During the initial stages of precipitate formation, the submicroscopic particles are forced to conform to the parent lattice (Fig. 2-24a and b). The tiny precipitate (closed dots) distorts the aluminum lattice (open dots). This condition is called *coherency*.

Ultimately, the precipitate escapes from the aluminum lattice and forms its own lattice (Fig. 2-24c and d). A grain boundary between the precipitate and aluminum lattices forms, and the severe lattice stresses disappear.

The coherency theory has three stages (Fig. 2-25):

1. Solute atoms diffuse through the solvent lattice and segregate to form a few very small particles, a stage in which solute atoms are grouped or rearranged so they can begin to form a precipitate. The result is a slight increase in hardness.
2. In this stage of aging, the solute atoms move about and form small particles attempting to form the precipitate phase. The precipitate remains coherent with the aluminum lattice. Strength increases significantly.
3. The precipitate grows in size with continued aging and tends to become noncoherent. The loss of coherency reduces lattice stresses, resulting in decreased hardness. Ductility increases.

Stages in Precipitation Hardening

In Fig. 2-25, hardness is plotted against time. The increase in hardness with time is known as precipitation hardening or age hardening. In zone 1, submicroscopic precipitate particles are being formed as the alloy attempts to reach equilibrium. Maximum hardness is reached in zone 2. If the aging temperature is held too long, overaging occurs, which reduces both hardness and strength.

In alloys that age naturally at room temperature, the matrix is relatively soft after rapid quenching. Hardness increases to its maximum value after several days.

Effects of Temperature

Precipitation is hastened by heating the alloy to a higher temperature. Figure 2-26 illustrates what happens when an aluminum alloy containing 4.5% copper is aged at temperatures above room temperature. At 150 °C (300 °F), hardness exceeds that obtained in natural aging. At 205 °C (400 °F), maximum hardness is reached in even less time. At 260 °C (500 °F), maximum hardness is reached in less time, but the level of hardness is below that obtained at 205 or 150 °C (400 or 300 °F).

Artificial Aging of Alloy 7075

Composition. This alloy contains zinc, magnesium, copper, and chromium in addition to aluminum.

Applications. Typical uses are in aircraft structural components and other highly stressed structures requiring a combination of very high strength and good corrosion resistance.

Thermal Properties. Liquidus temperature is 635 °C (1175 °F); solidus temperature is 477 °C (890 °F).

Heat Treating Practice

Solution treating temperatures range from 465 to 480 °C (870 to 900 °F), depending on the product. Sheet, for example, is solution treated at 480 °C (900 °F) to the W temper (indicates solution heat treatment).

Artificial Aging. Example: Sheet is aged at 120 °C (250 °F) to T6 temper (indicates solution treating and artificial aging). The required metal temperature must be reached as rapidly as possible and maintained within ±6 °C (±10 °F). In this application, aging time is 24 hours.

The annealing temperature for alloy 7075 is 415 °C (775 °F).

Natural Aging of Alloy 2017 (Duralumin)

Composition. The modern-day version of Wilm's alloy contains copper, magnesium, and silicon, in addition to aluminum.

Applications. The alloy has drifted into limited usage. Its major application is in rivets; other uses are in construction and transportation, screw machine products, and fittings.

General characteristics include medium strength and ductility, good machinability and formability, plus fair resistance to atmospheric corrosion. Welding is not recommended unless heat treating after welding is practical.

Product forms include forgings, extrusions, bars, rod, wire, shapes, and rivets.

Thermal Properties. Liquidus temperature is 640 °C (1185 °F); solidus temperature is 513 °C (955 °F).

Heat Treating Practice

Solution treating temperatures range from 500 to 510 °C (930 to 950 °F). During solution treating, the temperature should be maintained within ±6 °C (±10 °F) of the required treatment temperature.

Quenching is performed in water which is at room temperature and which should not be allowed to drop below 38 °C (100 °F) during the quenching cycle. Cooling from the solution treating temperature to room temperature must be completed within about 10 seconds.

Aging takes place at room temperature over a period of several days (natural aging).

To Dr. Wilm: Solute Atoms Did It

The good doctor invented the process now known as *solution heat treating and natural aging*. He also designed an alloy (Duralumin) suitable for the process.

To Explain:

When Duralumin is subjected to an elevated temperature, the solubility of solute atoms in the solvent far exceeds their solubility at room temperature, creating an unstable, supersaturated condition (Fig. 2-24a and b). Note that the solutes (called *precipitates* in the figure) are trapped inside the solvent, which distorts the aluminum (*solvent*) lattice, which in turn increases hardness and strength but decreases ductility. Solutes become trapped during the rapid (10-second) quench from 320 °C (970 °F) down to room temperature. At room temperature, the solubility of the solutes drops down to a small fraction of 1%. Eventually, as shown in Fig. 2-24 (c) and (d), the solutes escape and form their own lattice. At this time, severe lattice stresses disappear. Some hardness and strength are lost, but some ductility is gained.

Some What If's

Q: What if Dr. Wilm had solution treated and rapidly quenched the alloy, but did not follow up with cold working?

A: This treatment is defined as the W temper. It is an unstable temper that applies to any alloy that is naturally aged. In rapid quenching after solution treating, solute atoms are trapped in the solute lattice, causing distortion, which has a strengthening effect that continues for several days. In time, though, solid atoms manage to return to their equilibrium (almost at rest) condition.

Q: What if Dr. Wilm had solution heat treated, rapid quenched, cold worked, and naturally aged this alloy?

A: This treatment is defined as the T3 temper: Parts are solution heat treated, rapidly quenched, cold worked, and naturally aged to a substantially stable condition. The treatment applies to parts cold worked after solution treatment to increase hardness and strength. Cold working serves essentially the same function as precipitation hardening (Fig. 2-22). Grains are distorted by cold working. They change their size, shape, and orientation, which increases hardness and strength but reduces ductility.

Chapter 3

Steels and Cast Irons: The Why of Where They Are Used

Taylor and White elected to dramatically improve upon Mushet tool steel by doubling its tungsten content.

Why?

Orville Wright chose to use drawn steel wire for structural support of the brothers' first airplane.

Why?

Henry Ford opted for a carbon steel alloyed with a fractional amount of vanadium in preference to other structural materials available at the time.

Why?

Answer: Taylor and White. The aim here was to come up with a tool steel that held its cutting edge (did not soften) in the higher temperatures generated in high-speed machining. In metallurgical jargon, Taylor and White wanted to improve the elevated-temperature properties of the existing alloy. For this reason, the tungsten content was doubled from 9% to 18%. Tungsten was chosen because of its extremely high melting point: 3410 °C (6170 °F). In comparison, iron, although a constituent of steel, has a melting point of 1533 °C (2800 °F), and aluminum has a melting point of 660 °C (1220 °F). Tool steels also must have high hardness, resistance to abrasion, and high toughness.

The superiority of the new composition was proved in field tests with a lathe similar to that shown in Fig. 3-1 (a modern version) and with tools similar to those shown in Fig. 3-2 (modern versions).

Answer: The Wright One. This application called for the strength required to cope with repeated pulling and twisting forces experienced by the wire in flight. Drawn wire—formed by pulling steel rods through a die (Fig. 3-3)—has the highest strength of all forms of steel. Drawing or cold working is the reason. The strengthening mechanism is similar in effect to that in aging solution-treated aluminum (discussed in Chapter 2). The strength of the wire is substantially increased via distortion of the crystal structure of the metal. Drawn wire has strengths up to 600 ksi (1000 lb per square inch), or 4135 megapascals (MPa). By comparison, the strength of steel used in landing gears for today's high-performance airplanes has strength in the neighborhood of 300 ksi (2065 MPa).

Answer: Henry Ford. When the Model T was introduced in 1908, all parts of the structure of a car were subjected to the constant shake, rattle, and roll encountered in navigating deeply rutted and pothole-infested primitive dirt roadways. Therefore, the obvious property requirements were resistance to metal fatigue and failure due to repetitious abuse, as well as resistance to impacts likened to repeated blows struck with a heavy hammer. Ford chose to use a carbon steel enhanced with a fraction of 1% of vanadium. In extolling the virtues of the alloy, Ford bragged that it had "twice the strength of steel at half the weight." These virtues actually resulted in a bonus saving for the automaker. High strength means that less metal is needed to meet specific strength requirements.

Fatigue strength is defined as the maximum stress that can be sustained for a specified number of cycles without failure. Figure 3-4 shows a Charpy impact specimen and test device. Figure 3-5 shows radial marks on the fracture surface of an AISI 4140 steel load cell, and Fig. 3-6 shows the "woody" appearance of a notched impact specimen of wrought iron.

Standard Versus Tailored Properties. The preceding examples suggest an analogy between

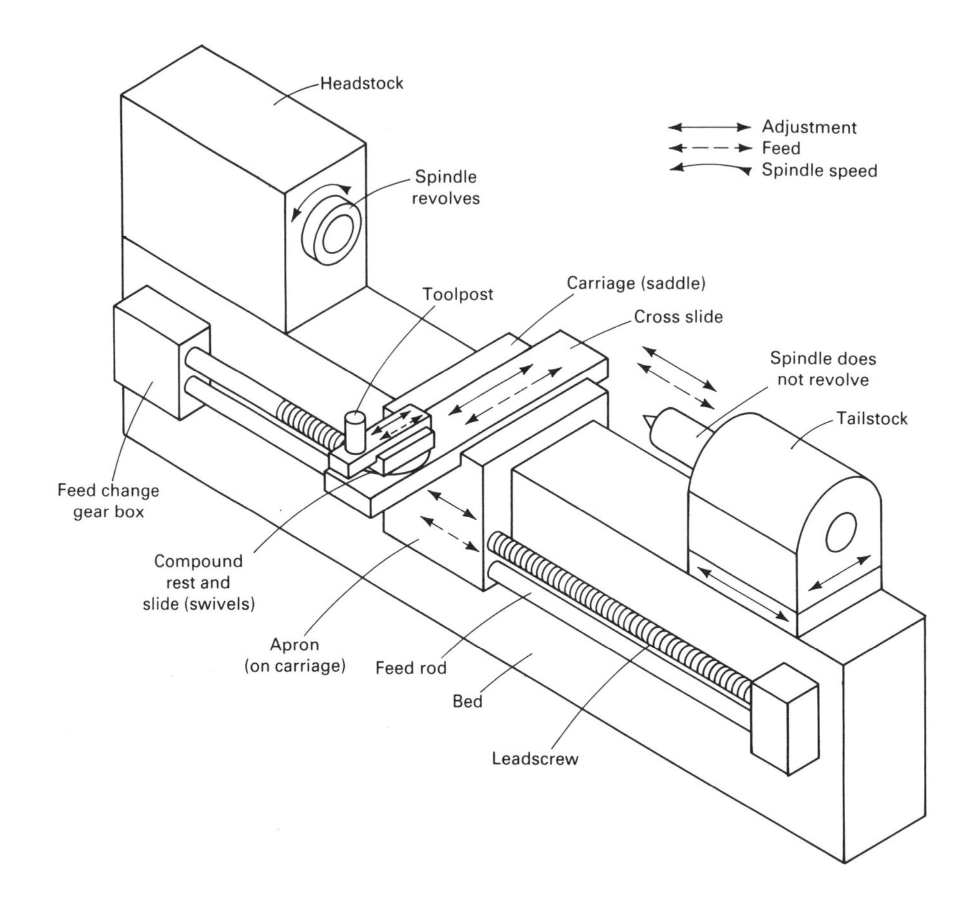

Fig. 3-1 Principal components and movements of a lathe

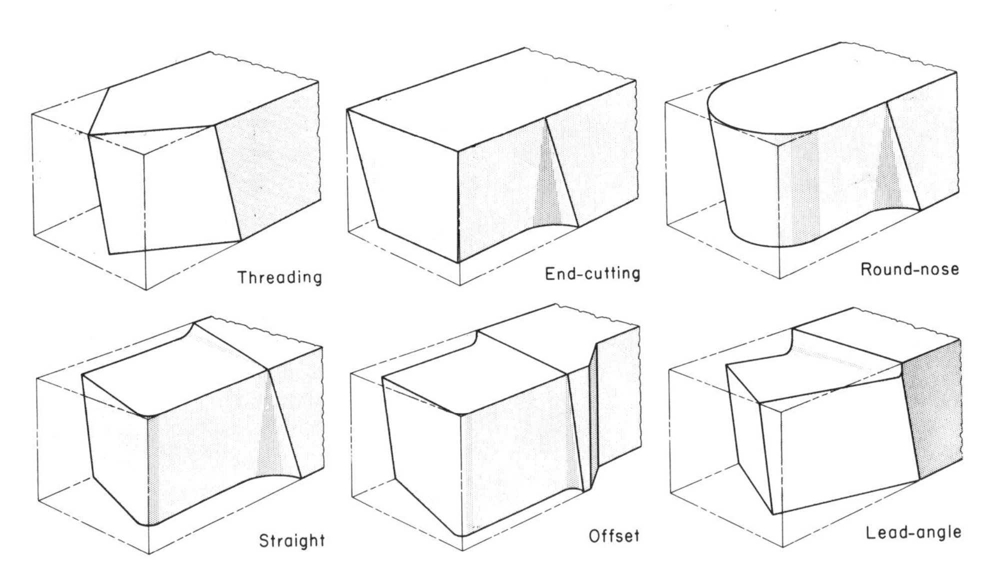

Fig. 3-2 Six common shapes of turning tools ground from solid bars

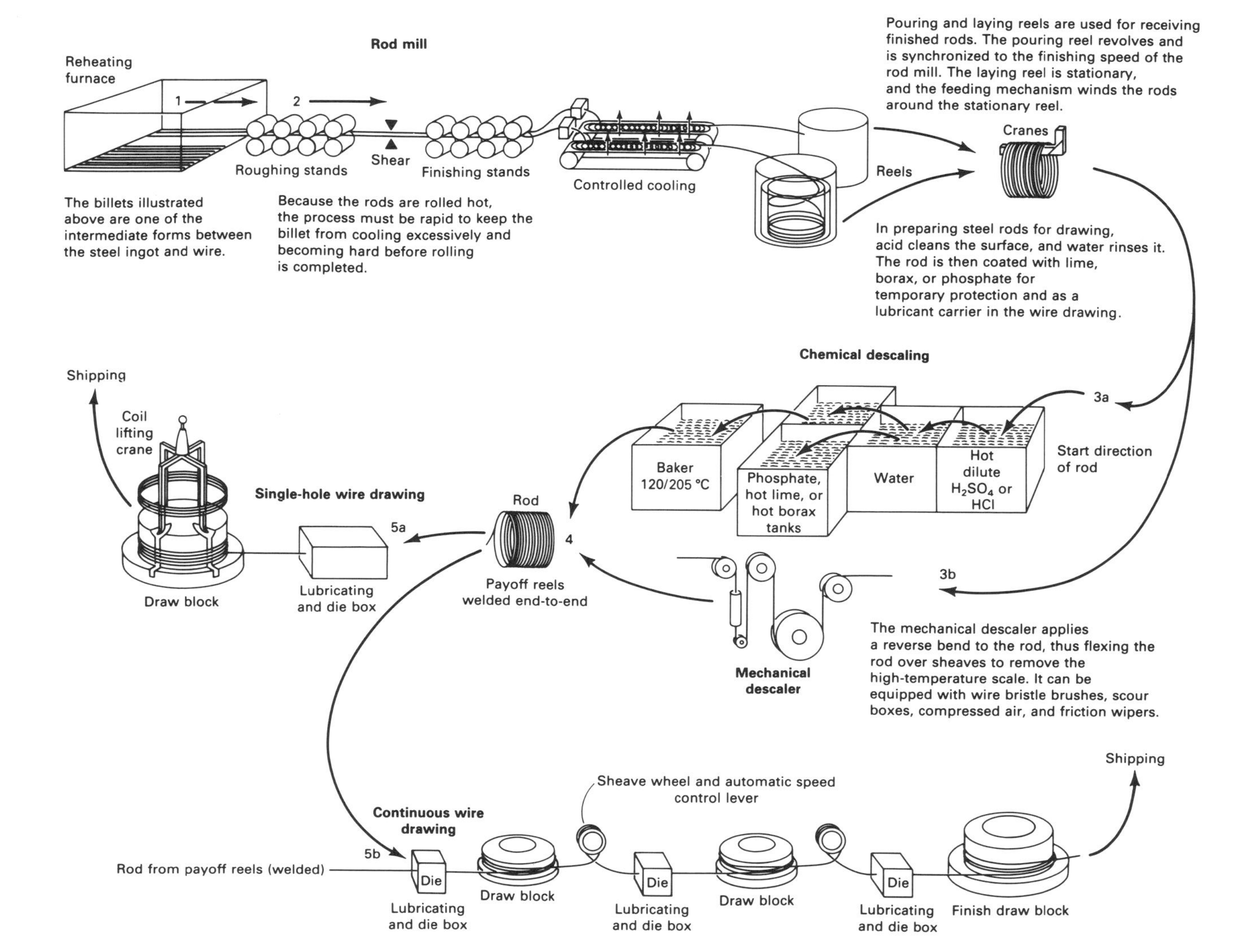

Fig. 3-3 Schematic showing how steel wire is made from rods

metal properties and men's clothing. If requirements for an application can be satisfied by standard properties, one can take advantage of an "off-the-rack" composition. If requirements are unique, on the other hand, the metallurgist can tailor properties by modifying composition (as Taylor and White did), or by alloying (as Henry Ford did), or by utilizing properties available through processing (as the Wright brothers did by cold working). Heat treating is another option for increasing strength by solution treating and aging (remember Dr. Wilm).

A Closer Look at Properties

Thus far, properties have been touched on in a general way. To gain a better understanding of the subject, some knowledge of the metallurgical underpinnings of properties is essential.

Topics that follow include mechanical and physical properties of wrought steels (alloys based on iron and carbon) and cast irons. Along the way, mention is made of processing and fabrication properties important in the manufacture of parts and end products.

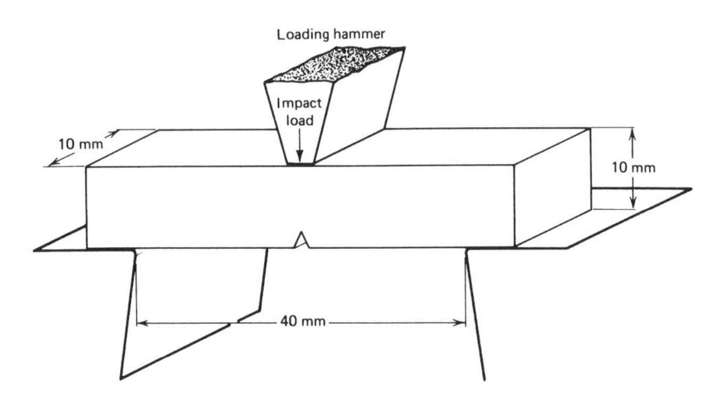

Fig. 3-4 Charpy impact specimen and test device

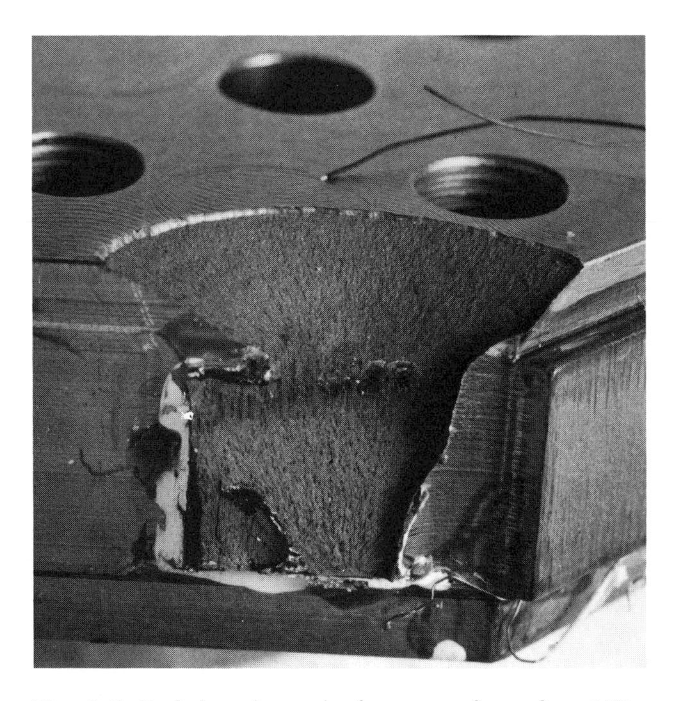

Fig. 3-5 Radial marks on the fracture surface of an AISI 4140 steel load cell

Wrought steels are purchased from steel mills and warehouses in standard forms such as sheet, plate, bar, and extrusions, which are subsequently converted into articles of commerce. Like steel, *cast iron* identifies a large family of ferrous alloys in the form of castings, which in the manufacturing cycle are close to "net shape." This means that further processing such as welding or machining may not be required to get to the desired end result.

Later in this chapter, brief reference is made to the production of steel at the mill and to the production of castings in the foundry. Both products are treated in additional detail in Chapter 7.

Fig. 3-6 "Woody" fracture appearance of a notched impact specimen of wrought iron

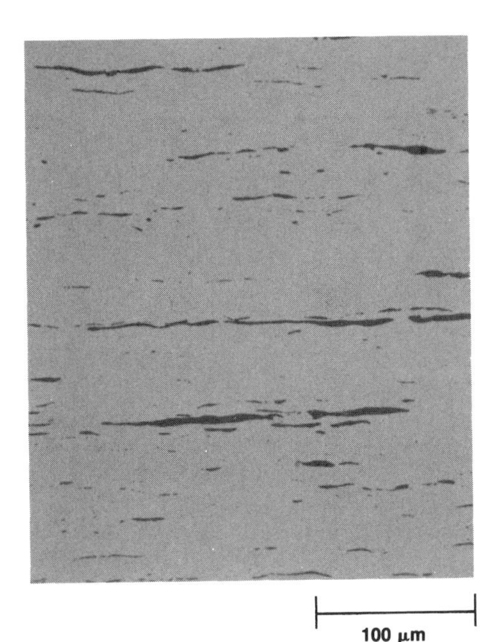

Fig. 3-7 Sulfide inclusions (dark areas in micrograph), which degrade properties of stainless steel

Profile of Steel

Iron is the basic component of steel. When carbon, a nonmetal, is added to iron in amounts up to 2%, the result is an alloy known as steel.

In addition to carbon, alloys may contain such metals as chromium and/or nickel and/or such nonmetals as silicon and nitrogen. One or more of such alloying elements may be needed to obtain special characteristics or properties required in engineering applications. In most steel, however, carbon is the principal ingredient.

So-called *plain carbon steels* contain small amounts of manganese and silicon. But the amount of carbon has the most pronounced effect on properties and in the selection of suitable heat treatments.

Almost all steels contain fractional amounts of impurities—phosphorus and sulfur, for example—which are present in steelmaking raw materials such as scrap. Generally, the impurity content is so low that its presence is not detrimental to engineering properties and, with exceptions, the economics of steelmaking rule out the removal of such impurities. An exception is shown in Fig. 3-7. In this instance, sulfide inclusions (dark areas) downgrade the properties of a stainless steel. Sulfide content must be down to thousandths of a percent—around 0.003%—in high-strength steel pipe destined for arctic service, such as the Trans-Alaska Pipeline. The pipe, which is 1220 mm (48 in.) in diameter, is joined by welding.

Steel is classified by one system on the basis of carbon content:

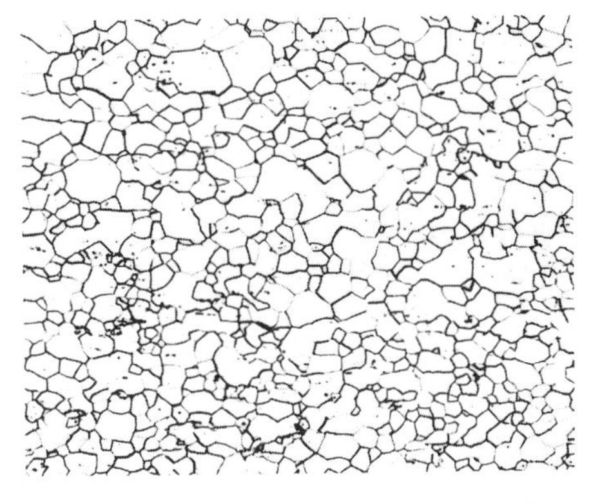

Fig. 3-8 Microstructure of low-carbon ferritic steel etched to reveal ferritic grain boundaries (dark areas)

- Steel containing less than 0.30% carbon is called *low carbon steel* or *mild steel.*
- Steel containing between 0.30 and 0.60% carbon is called *medium-carbon steel.*
- Steel containing more than 0.60% carbon is called *high-carbon steel.*
- Steel containing more than 0.77% carbon may be called *tool steel.*

The upper carbon limit for steel is about 2%. Cast irons (iron-carbon alloys) usually contain 2.30 to 4.00% carbon.

The microstructure of a low-carbon ferritic steel that has been etched to reveal ferrite grain boundaries (dark areas) is shown in Fig. 3-8. Ferrite is a solid solution—as opposed to a liquid metal—consisting of one or more elements in bcc iron; unless otherwise indicated, the other component is assumed to be carbon, a nonmetal. In this instance the solute, carbon, is dissolved in the solvent, iron. Microstructure is revealed by microscopic examination of a highly polished and etched (chemically treated) specimen—a technology called metallography. Spark patterns can be used to identify low-, medium-, and high-carbon steels (Fig. 3-9).

In Fig. 3-10, cup is being drawn from sheet steel. Formability is a fabrication property that indicates the relative ease with which a metal can be deformed.

Mechanical Properties of Steel

Mechanical properties are a measure of the ability of a material to carry or resist mechanical forces or stresses. These properties include strength, ductility, hardness, impact resistance, and fatigue resistance.

- *Stress* is the result of such forces as tension or pulling, compression or pushing, and shear or twisting or cutting.
- *Strain* is the change in size or shape of steel which results from stress.

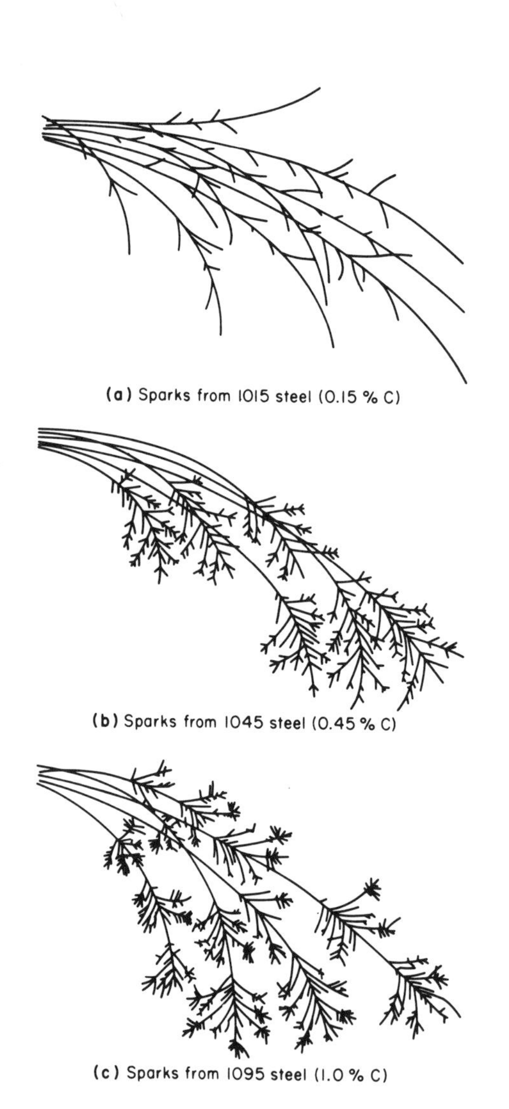

Fig. 3-9 Spark patterns used to identify low-, medium-, and high-carbon steels

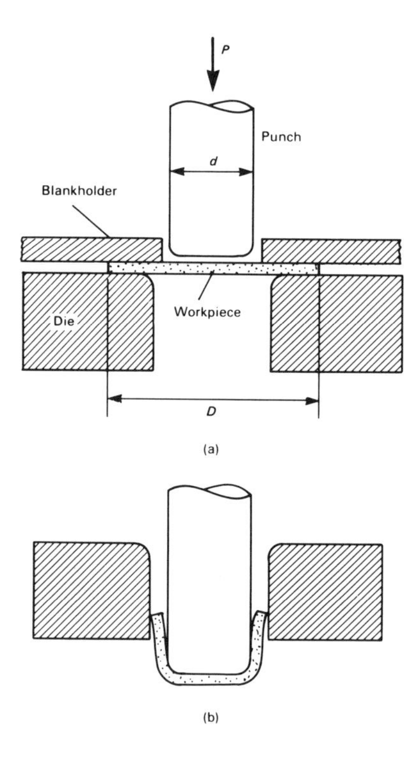

Fig. 3-10 Cup being drawn from sheet steel

Strength Properties

Elasticity, plasticity/ductility, and brittleness are the basic strength properties of steel and can be characterized as follows.

Elastic high-strength steels are used in end products ranging from bed springs to bridges—products that must maintain their size and shape in service. An elastic steel bends when a load is applied, but does not break; it returns to its original shape and size when the load is removed.

Plastic lower-strength steels typically are used in end products where the use of high-strength steels is not economical. Applications include pots, pans, and beer and pop cans formed from sheet steel. This type of steel bends when load is applied, but does not return to its original configuration when load is removed. The metal is permanently deformed, changing in shape (cross-sectional area) and size (becomes longer). It is in what is called the plastic mode, meaning it has a combination of strength and ductility that allows it to absorb a certain amount of plastic deformation prior to breaking.

Brittle steels are used in products such as wire coat hangers that break when bent back and forth briskly several times and razor blades that snap suddenly with minimum bending. Such

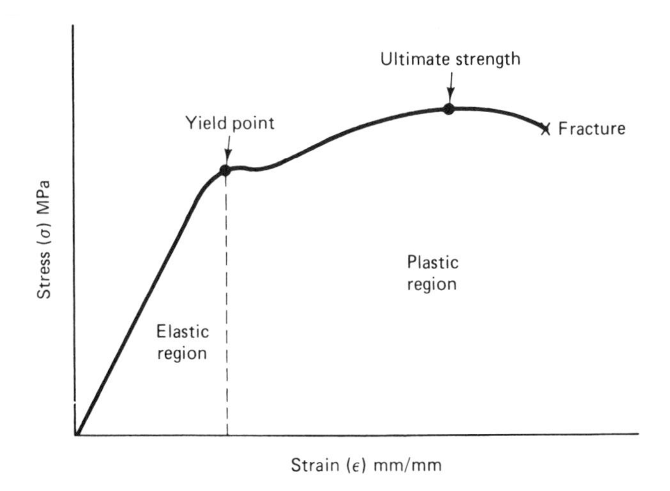

Fig. 3-11 Typical stress-strain curve for mild steel

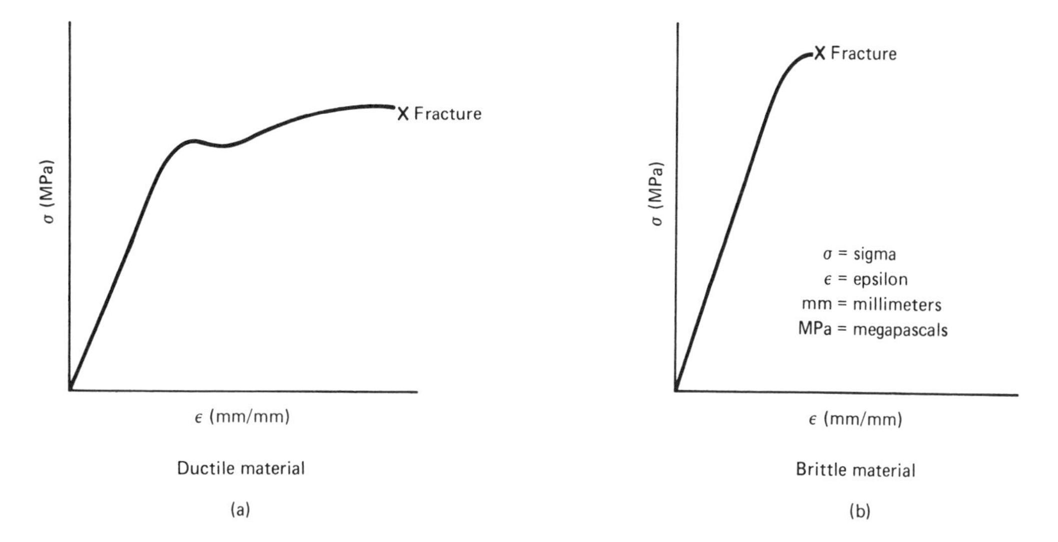

Fig. 3-12 Comparison of stress-strain curves for ductile and brittle materials

steels have little or no plasticity and are lacking in ductility.

A good overview of the elastic and plastic behavior of steel is given in Fig. 3-11, which shows a typical stress-strain curve for mild steel. Note that both the elastic and plastic regions are identified, as well as yield point (at the transition from elastic to plastic behavior). Ultimate strength is also known as *ultimate tensile strength*. Stress-strain curves for both ductile and brittle materials are given in Fig. 3-12.

The mechanical strength properties of steel include tensile strength, yield strength, percent elongation, and reduction in area. In general,

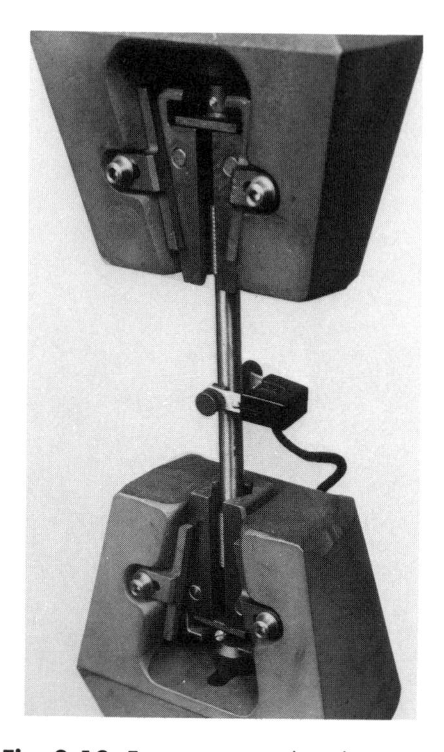

Fig. 3-13 Extensometer. This device is attached to round tensile specimens to measure changes in length (reported in percent elongation) caused by the application or removal of force.

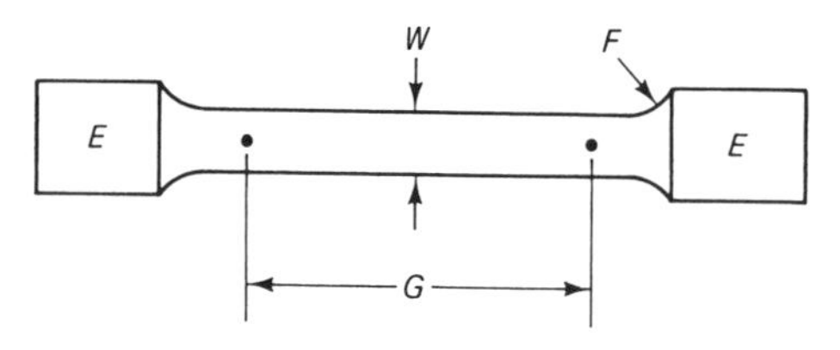

E – gripped ends, may be threaded, plain, or with hole for gripping by machine
W – reduced width to insure specimen breaks in middle – round on round specimens and flat on flat specimens
G – marked gage length to precisely measure the change in length before, during and after test
F – fillet to reduce stress concentrations

Fig. 3-14 Standard tensile test specimen

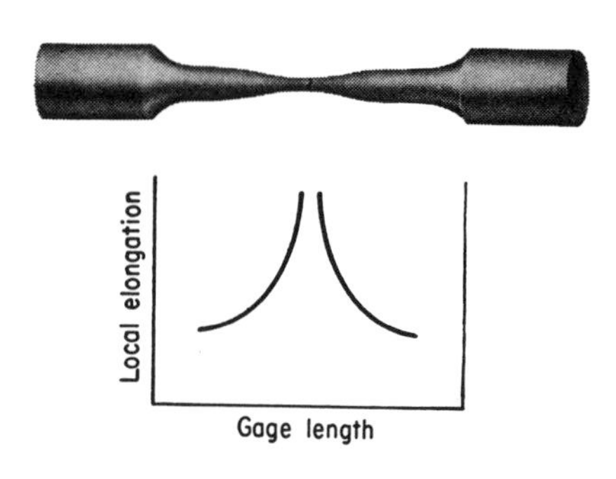

Fig. 3-15 Test specimen exhibiting local elongation (called necking)

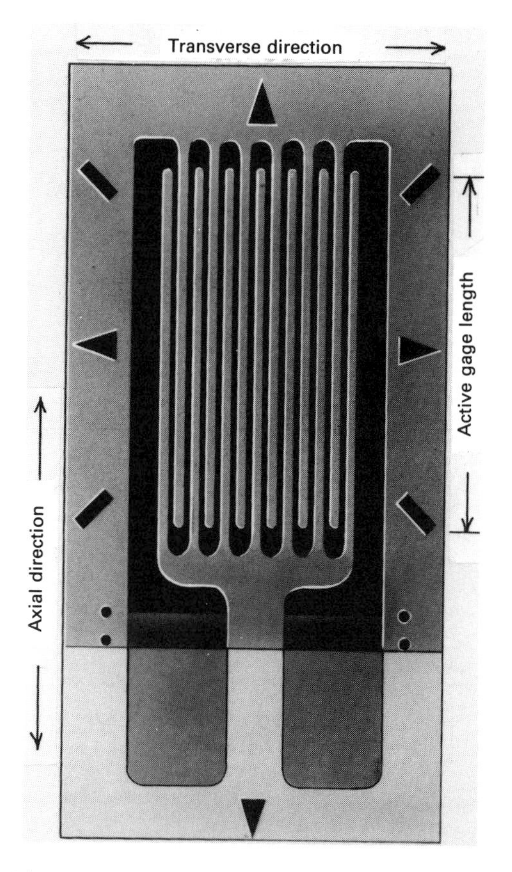

Fig. 3-16 Strain gage used in tensile and similar tests to measure small amounts of strain (deformation)

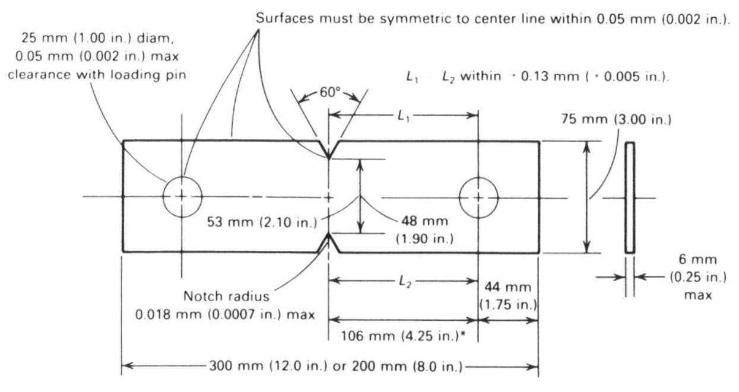

Fig. 3-17 Notched tension specimen

Fig. 3-18 Lüders lines on the surface of rimmed 1008 steel wire

the higher the tensile strength, the lower the elongation and reduction in area. Conversely, the lower the tensile strength, the higher the elongation and reduction in area. Tensile strength is always higher than yield strength.

Measuring Mechanical Strength Properties. In tensile testing, machines apply loads to specimens, and instrumentation enables any changes in the deformation of the material under load to be detected. The device shown in Fig. 3-13 measures changes in test specimen length. A standard tensile test specimen is shown in Fig. 3-14, and one that exhibits local elongation, called necking, is shown in Fig. 3-15. Figure 3-16 depicts a strain gage used in tensile and similar tests to measure small amounts of strain or deformation. A standard notched tension specimen is shown schematically in Fig. 3-17. This type of specimen is used in measuring changes of cross-sectional area.

Lüders lines (stretcher strains) on the surface of steel result when the sheet is stretched beyond its yield point during forming (Fig. 3-18). The Olsen cup test (Fig. 3-19) is used to determine the formability (ductility) of sheet steel. In this test, a piece of sheet metal, restrained except at its center, is deformed by a standard steel ball until fracture occurs. The measure of ductility is the height of the cup at the time of fracture.

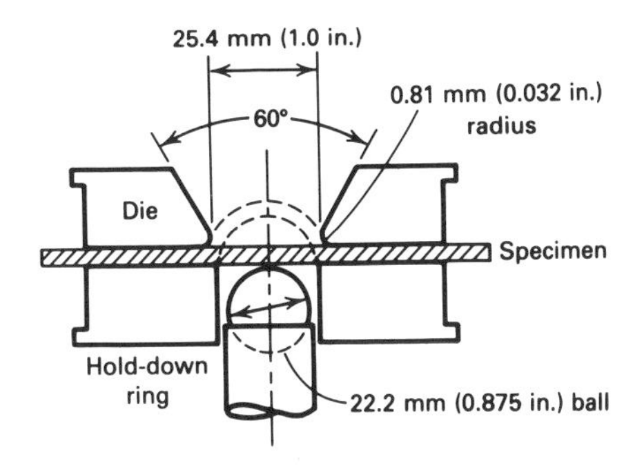

Fig. 3-19 Olsen cup test

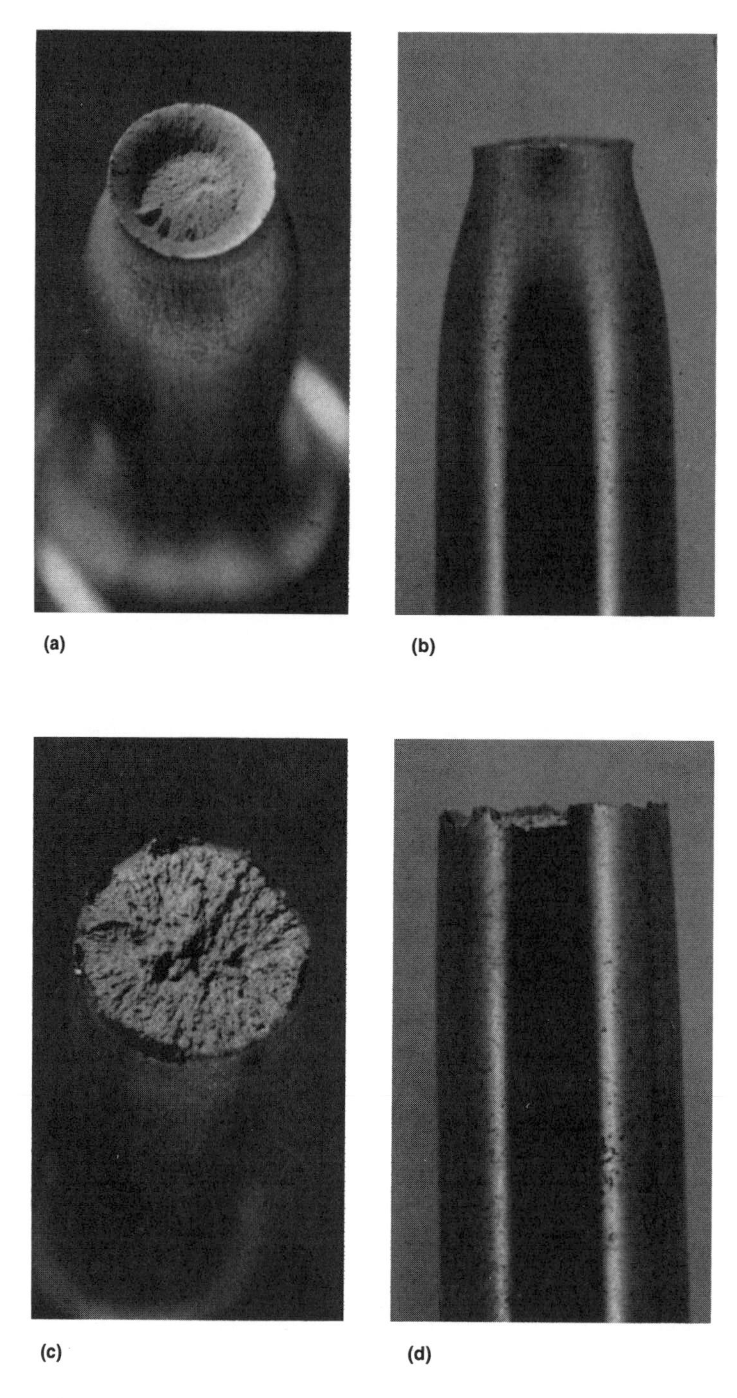

Fig. 3-20 Appearance of ductile fracture (a and b) and brittle fracture (c and d)

Ductile and brittle fractures are compared in Fig. 3-20. In Fig. 3-21 chevron patterns are the result of brittle failure of mild steel in ship plate samples. Brittleness also leads to propagation of cracks in metals.

Hardness Properties

Hardness is a measure of a steel's resistance to penetration (local plastic deformation). In the Brinell hardness test, for example, hardness is indicated by the size of the penetration made by a 10 mm (0.4 in.) steel or tungsten carbide sphere under different loads (Fig. 3-22). Rockwell, Vickers, and Knoop are similar hardness tests.

A schematic of a Rockwell hardness tester is shown in Fig. 3-23. This test uses a diamond cone Brale indenter (Fig. 3-24 and 3-25). A diamond pyramid indenter is used in Vickers hardness tests (Fig. 3-26).

The Jominy end-quench test (Fig. 3-27) is used to determine the hardenability of steel. This value is obtained by finding the relationship between the cooling rate (quenching rate) of a steel and the depth below the surface at which a given hardness value is obtained.

An empirical relationship exists between hardness and tensile strength. The relationship is proportional, as shown in Fig. 3-28. Rule of thumb: The higher the hardness, the higher the tensile strength. This relationship is useful because hardness testing equipment is not only portable, but is also far more readily available than tensile testing equipment. In addition, hardness testing is quicker and less expensive and does not require a machined specimen. In many instances, it is possible to do a hardness test on a part without destroying it. In contrast, machining a tensile specimen from the same part would in all probability render the part useless.

Hardness testing does have limitations. It relates reasonably well to tensile strength, but not to elastic limit or yield strength. Therefore, hardness testing is no substitute for tensile testing.

Hardness has a similar relationship with other properties as well. With few exceptions, resistance to abrasive wear is almost directly related to hardness. High-hardness steels resist wear. In a negative way, hardness is also a measure of machinability: Cutting speeds and tool life are almost directly related to hardness and microstructure. Briefly put, soft steels are easier to machine than hard steels.

Impact Resistance

Having a combination of strength and ductility, steel generally is considered to be a tough material. The most common measure of toughness is resistance to impact. Axle shafts and steering components for cars, for example, require toughness.

Plain carbon and low-alloy steels lose resistance to impact with decreasing temperature in service or in testing. A series of impact transi-

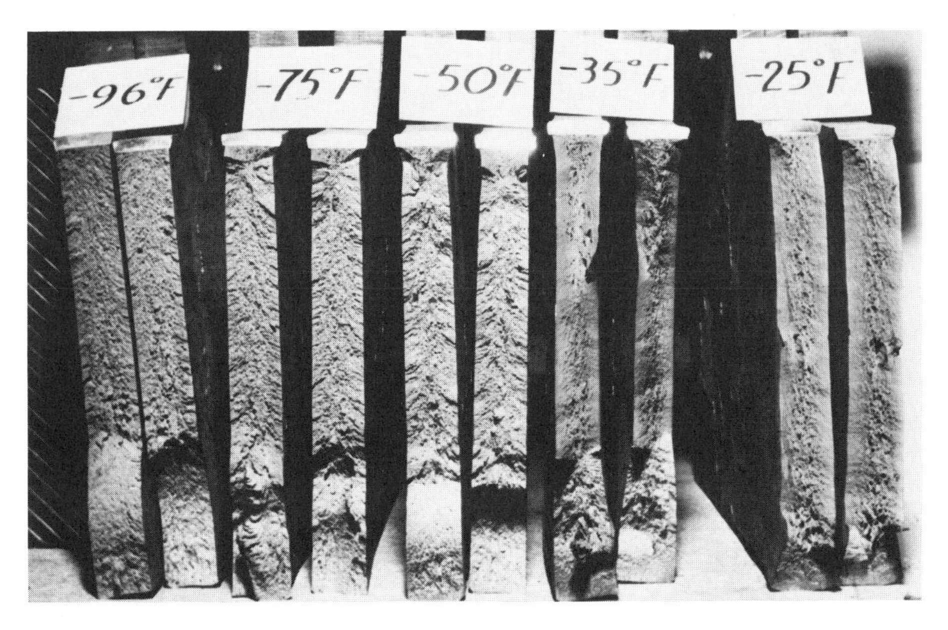

Fig. 3-21 Chevron patterns in samples of mild steel ship plate, indicating brittle fracture

tion curves, showing the transition from ductile to brittle behavior, are presented in Fig. 3-29. The figure shows (1) the effect of carbon content on the maximum amount of energy that can be absorbed in impact testing, or resistance to impact, and (2) the effect of carbon content on the transition from ductile to brittle behavior. For example, a steel containing 0.11% carbon can absorb a great amount of impact energy when tested near room temperature. This capability drops off sharply when test temperatures fall in the neighborhood of –40 °C (–40 °F)—in this instance, the ductile-to-brittle transition temperature.

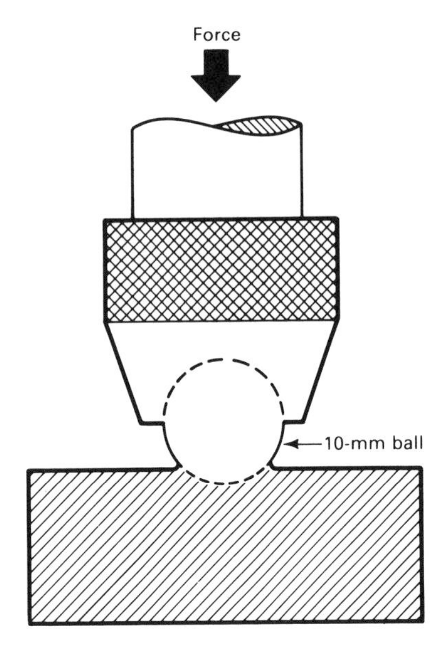

Fig. 3-22 Schematic of the Brinell hardness test

Carbon is one of the most potent elements with regard to notch toughness. Consequently, carbon content should be kept as low as possible and consistent with strength requirements.

The setup and specimen for Izod impact testing is shown in Fig. 3-30. In testing, the V-notch specimen is mounted vertically and struck by a weight at the end of a pendulum. The energy needed to break off the free end of the specimen is a measure of impact strength or *toughness,* which is a combination of strength and ductility.

Effects of Microstructure. At each carbon level, the ductile-to-brittle transition temperature is markedly influenced by the microstructure developed in steelmaking or in subsequent heat treating. Pearlite, or a mixture of ferrite and pearlite in steels containing less than 0.5% carbon, is a microstructure that develops in as-rolled bar or plate. *Pearlite* is an aggregate of ferrite and cementite. *Ferrite* is a solid solu-

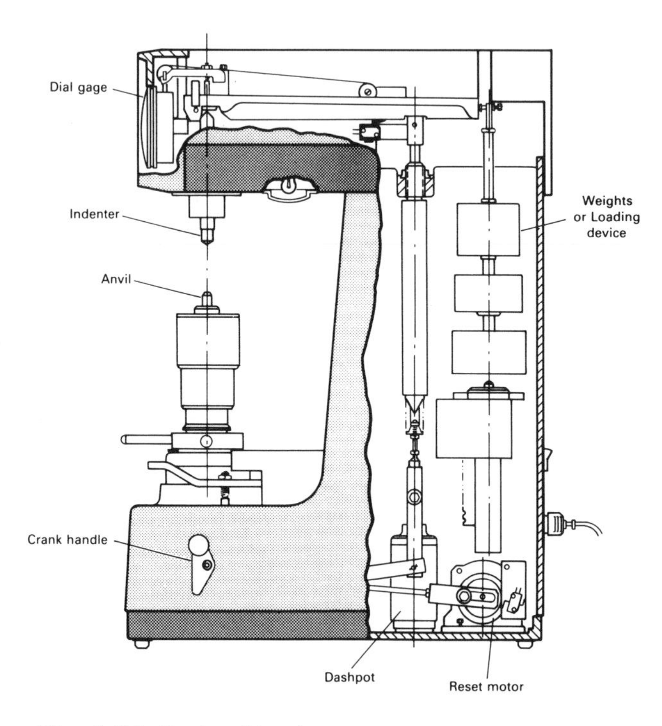

Fig. 3-23 Rockwell hardness tester

tion in bcc iron, and carbon is the typical solute. *Cementite* is composed of iron and carbon. Pearlite in rail steel is shown in Fig. 3-31. The microstructure in Fig. 3-32 is a combination of ferrite (white) and pearlite (dark) in carbon-manganese-silicon steel. These microstructures are produced at the steel mill.

By comparison, bainite and tempered martensite are produced during heat treating by the steel user or by a commercial heat treater. In some instances, bainite and martensite may be mixed with pearlite and ferrite. At all carbon levels martensite is the preferred structure, as it has the lowest ductile-to-brittle transition temperature. All alloyed steels must be heat treated in order to develop a fully martensitic structure. *Bainite* is an aggregate of ferrite and cementite. *Martensite* is a supersaturated solid solution of carbon in iron with a body-centered tetragonal (bct) lattice structure. Figure 3-33 shows lower bainite (dark plates or needles) in 4150 steel; Fig. 3-34 shows lath martensite in 4340 steel that has been quenched and tempered.

There are two caveats. A steel being considered for service such as that discussed in this section should be tested at or below its lowest anticipated service temperature, as well as at a high enough temperature to identify its ductile-to-brittle transition temperature. It is equally important to test the steel with the same microstructure as that required by the part.

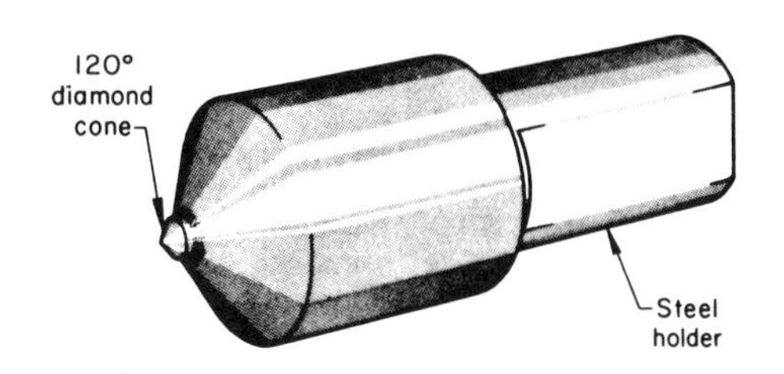

Fig. 3-24 Diamond cone Brale indenter

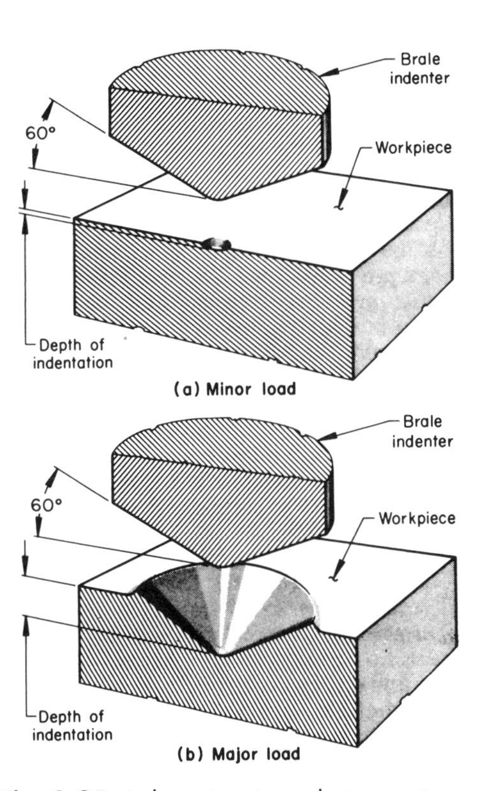

Fig. 3-25 Indentations in workpieces using a Brale indenter with minor (a) and major (b) loads

Fatigue Resistance

A steel part subjected to repeated stress can ultimately fail by fatigue even though the stress levels are well below the ultimate tensile strength of the part. For example, bending a wire coat hanger several times quickly results in fatigue failure because a large plastic strain has been placed on a part low in strength and with limited ductility. Other examples include broken car axles and broken gear teeth.

Resistance to fatigue usually is proportional to hardness and tensile strength. An exception is where tensile levels are high. The ultimate behavior of a metal subjected to cyclic stressing is influenced by a number of factors, including processing, fabrication, and heat treating techniques, surface treatment, finishing method, and service environment.

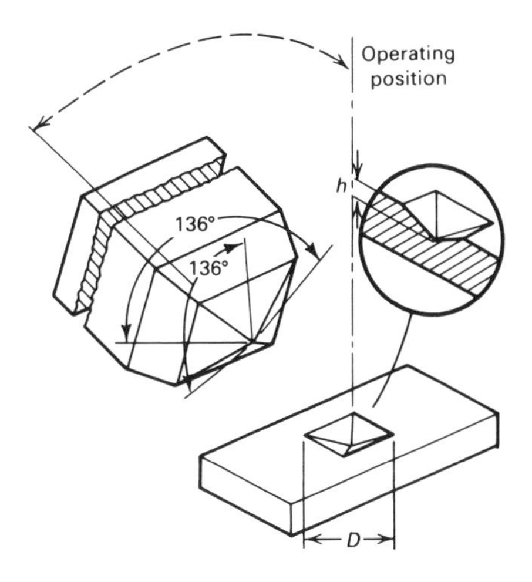

Fig. 3-26 Diamond pyramid indenter used in the Vickers hardness test. *D*, mean diagonal of the indentation in millimeters

Prediction of fatigue life is complicated because metals are sensitive to small changes in loading conditions, stress concentrations, and local characteristics. Fatigue tests on small parts do not provide a reliable indication of the fatigue life of a part, but such tests are useful in making decisions regarding which steel and which heat treatment to use to ensure resistance to fatigue failure.

Surface treatments, such as shot peening with steel shot or pellets, can extend fatigue life. Figure 3-35 compares the fatigue lives of crankshafts that were both heat treated and shot peened, or heat treated only. Similarly treated test bars also are compared. Shot peening introduces internal compressive stresses that reduce the effect of external tensile stresses.

Other variables that influence the fatigue behavior of carbon and alloy steels include strength level, ductility, and cleanliness. Any heat treatment or alloying addition can be expected to increase fatigue life.

Cleanliness means relative freedom from nonmetallic oxides (produced when oxygen combines with the steel), which are detrimental to fatigue life—particularly in long-term applications. Aggressive environments also substantially shorten fatigue life; steel with a rusted surface has a much lower fatigue life than clean, smooth steel.

Preventive measures include careful attention to design details and manufacturing processes, as well as maintenance practices. These measures are more effective in improving fatigue life than merely changing construction materials.

A design should:

- Eliminate stress raisers in a part
- Avoid sharp surface tears caused by punching, stamping, or shearing
- Prevent the development of surface discontinuities or *decarburization* (loss of carbon from the surface of steel via reaction with the furnace atmosphere)

Important maintenance measures include protection against corrosion and damage to a critical part such as a gear in normal service.

Elevated-Temperature Characteristics

Creep is defined as time-dependent strain occurring under stress. All metals undergo creep when they are stressed for extended periods. An increase in stress or temperature can accelerate the creep process.

Carbon and low-alloy construction steels are used extensively in applications involving elevated temperatures, such as electric power generating equipment, chemical processing equipment, aircraft powerplants, automotive exhaust

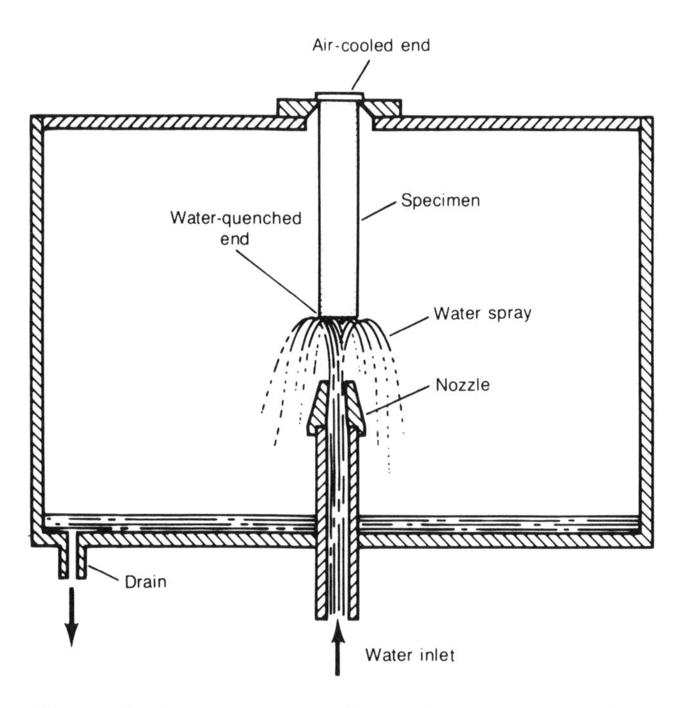

Fig. 3-27 Jominy end-quench test for determining hardenability of steel

systems, home furnaces, toasters, and ovens. Steels have more stability at high temperatures than the common nonferrous metals and polymers. For example, aluminum and copper-base alloys tend to lose strength faster than steel in environments involving increasing temperatures.

Creep is accelerated by increases in stress or temperature. A slow process of plastic deformation takes place when metal is subjected to constant loading below its normal yield strength at any temperature. At low temperatures, *slip* (movement of dislocations in grain structures) is impeded by impurity atoms and by grain boundaries. At high temperatures, the diffusion of atoms and vacancies permit dislocations to move around impurity atoms and beyond grain boundaries, resulting in higher creep rates. Figure 3-36 shows the four stages of plastic deformation caused by the formation of a slip dislocation. Some materials subject to creep can fail; this is called *creep rupture*. Boiler tubes can sag, which can cause overheating and their eventual collapse or rupture.

Steels suitable for service at elevated temperatures generally have microstructures that are stable at service temperatures. Steels heat treated for high strength at room temperature will not be as stable at elevated temperatures as steels that have been *tempered* (reheated after hardening treatment to reduce hardness and increase toughness) for long exposure above the service temperature. Rule of thumb: Temper above the intended maximum service temperature.

Alloying elements that help steel resist softening at tempering temperatures will contribute to elevated-temperature strength and resistance to creep. Such alloying elements include chromium and molybdenum, or small additions of vanadium.

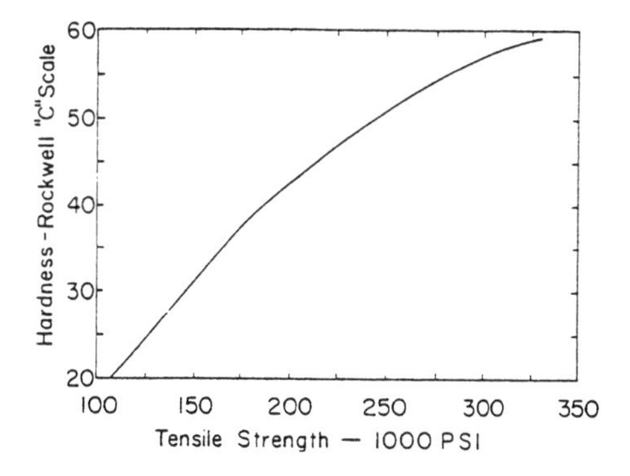

Fig. 3-28 Approximate relationship between hardness and tensile strength

Physical Properties of Steel

Physical properties are relatively insensitive to structure and can be measured without the application of force. They include density (mass per unit volume), electrical conductivity, coefficient of thermal expansion, and magnetic permeability. The list does not include chemical reactivity.

Generally, physical properties do not change significantly with changes in microstructure, as do mechanical properties. Physical properties are controlled more by interatomic forces at the submicroscopic level, such as the diffusion of atoms to fill vacancies in lattice structures.

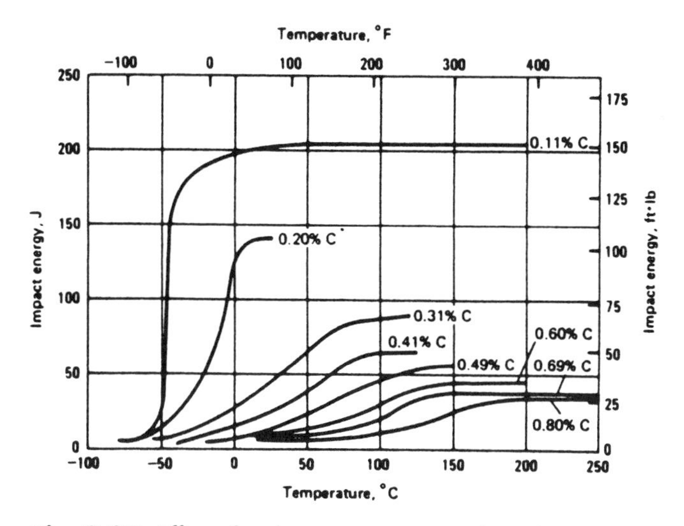

Fig. 3-29 Effect of carbon content on metal toughness

The high-temperature properties of steel are due in large part to its high melting point. Some comparisons:

Metal	Melting point °C	Melting point °F
Iron	1535	2800
Copper	1085	1980
Aluminum	660	1220
Zinc	420	785

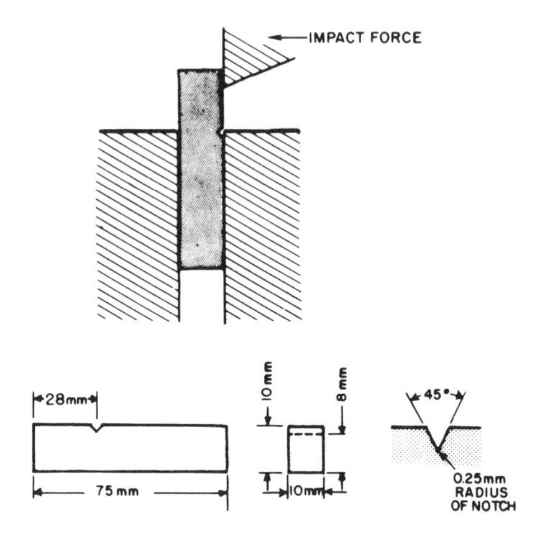

Fig. 3-30 Setup and specimen for Izod impact testing

All metals soften at some fraction of their melting point, beyond which they melt. This is called the *liquidus temperature,* the point at which atoms break loose from their lattices and transform from solid to liquid; heat stimulates atom movement. Alloying can delay softening at elevated temperatures to some extent.

Thermal Expansion

All metals expand when they are heated and shrink with exposure to low temperatures. At room temperature, movement of atoms is minimal.

Each metal expands at a different rate, called the *coefficient of thermal expansion.* Changes in unit length per unit of temperature (e.g., μin./in. · °F) are reported. The thermal expansion of iron and steel is low in comparison with the same properties of aluminum, aluminum alloys, copper, brass, and zinc.

Construction materials low in thermal expansion are favored for applications where service temperatures fluctuate—for example, a car that must function in desert heat as well as at subzero temperatures. Under such circumstances, the fit of mating parts and their clearances must be maintained.

Another example is chromium plating on zinc die castings, such as car door handles. Zinc has a very high thermal expansion coefficient, more

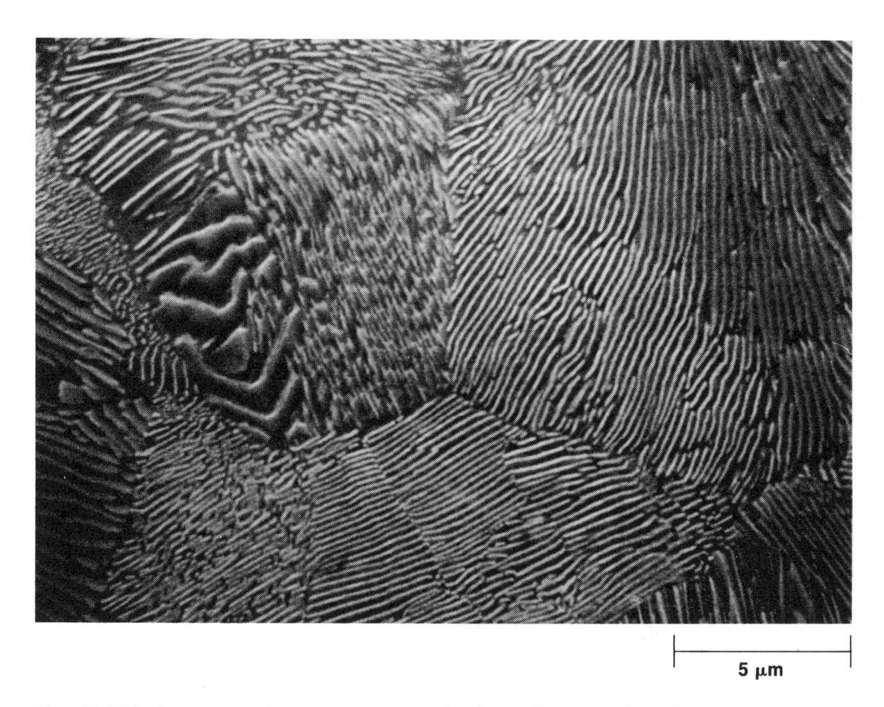

Fig. 3-31 Scanning electron micrograph of pearlite in rail steel

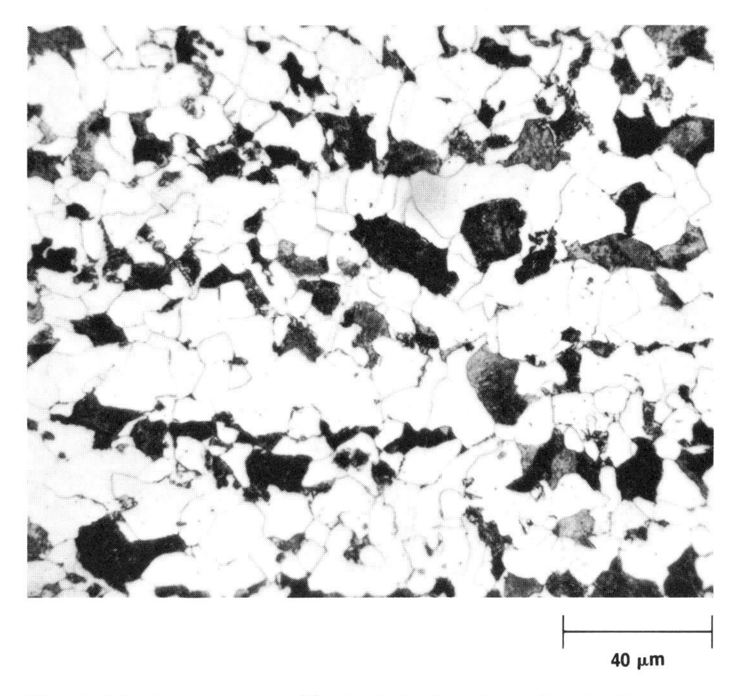

Fig. 3-32 Combination of ferrite (white) and pearlite (dark) in carbon-manganese-silicon steel

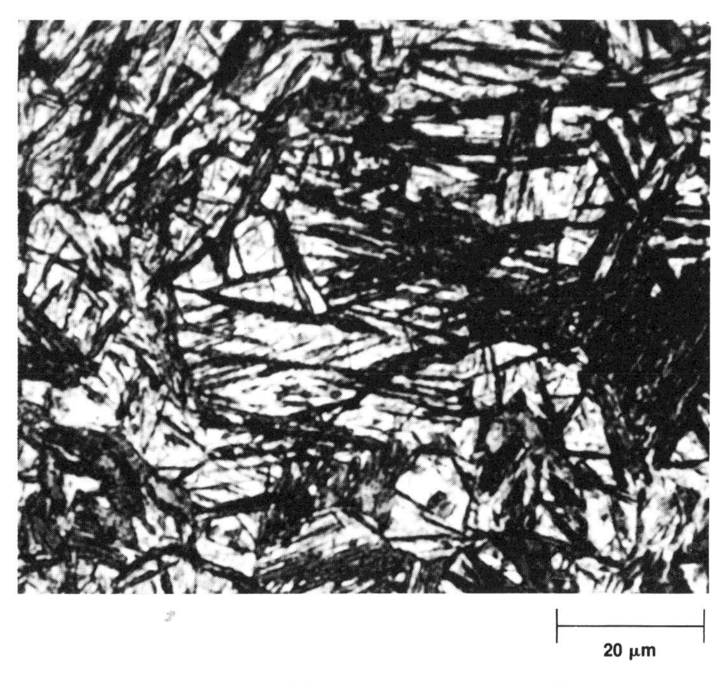

Fig. 3-33 Lower bainite (dark plate or needles) in 4150 steel

than six times greater than that of chromium. The zinc, which is brittle, cracks in normal thermal cycling and then corrodes (evidenced by an unsightly white powder). Ultimately, the chromium plating may peel off.

Electrical Conductivity

For a given application, electrical conductivity may be an important factor in materials selection. Copper, for example, is a better conductor of electricity than aluminum. Electrical steel sheet is shown in Fig. 3-37.

Resistance, the opposite of conductivity, is a factor in selecting a metal for use in heating elements for appliances ranging from toasters to sophisticated induction heating equipment. A variety of induction heating coils are illustrated in Fig. 3-38.

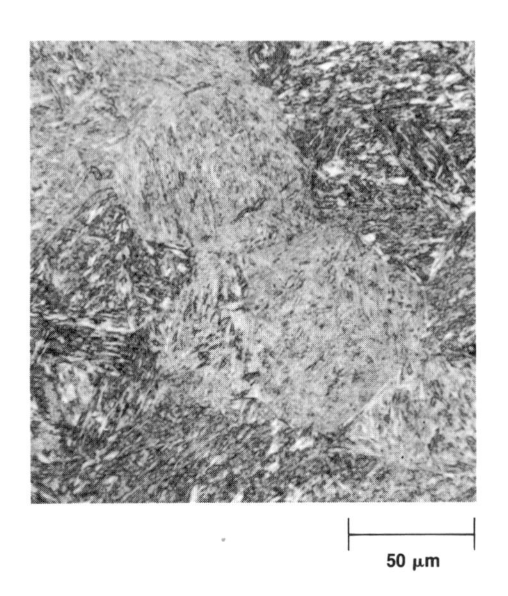

Fig. 3-34 Lath martensite in 4340 steel that has been quenched and tempered

Steel Mill Products

Most steels are made from *pig iron,* the first product obtained when iron is extracted from its ore. Figure 3-39 outlines the steps in making pig iron and steel.

Steel is primarily an alloy of iron and carbon; alloying elements aid in processing or are important in producing desired properties. Various steel product forms, called *mill products,* include:

- Sheets and coils of low-carbon, cold-reduced steel for applications ranging from cars to refrigerators
- Wire, used in such products as bed springs, wire rope, nails, and wire fabric
- Structural shapes used in buildings and bridges
- Bars for forgings and machined parts, such as gears and bearings

The flow diagram in Fig. 3-40 shows the processes used to convert raw steel into a variety of mill products.

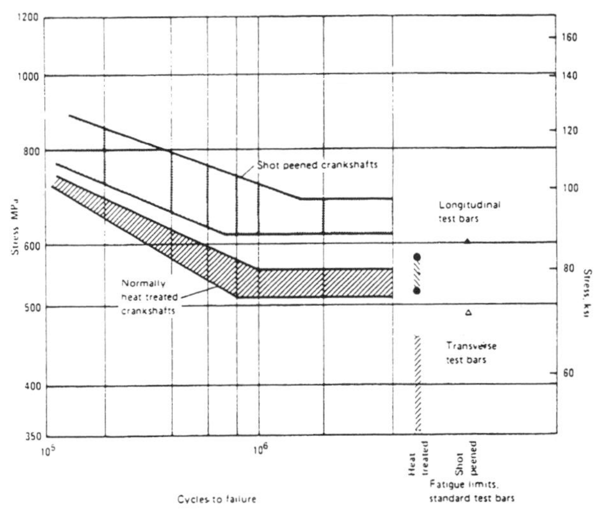

Comparison between fatigue limits of crankshafts (*S-N* bands) and fatigue limits for separate test bars, which are indicated by plotted points at right. Steel was 4340.

Fig. 3-35 Effect of shot peening on fatigue behavior

Profile of Cast Irons

The cast iron family encompasses a large number of ferrous alloys used in the form of castings. Cast irons usually contain more than 2% carbon and 1 to 3% silicon. Variations in properties are obtained by alloying with different metallic and nonmetallic elements and by varying melting, casting, and heat treating practices. As liquid iron is cooled, its capacity to dissolve carbon decreases, and carbon must precipitate out in other forms. This means cast irons have a matrix of steel (iron plus carbon), and precipitated carbon is distributed within the matrix in different forms. Four basic types are white iron, gray iron, ductile iron, and malleable iron.

White iron and *gray iron* derive their names from the appearance of their fracture surfaces: White iron has a white crystalline surface; gray iron has a gray fracture surface with exceedingly tiny facets. *Ductile iron* gets its name from the fact that in the as-cast form it has measurable ductility—a property not shared by white or gray iron. *Malleable iron* is cast as white iron, then *malleabilized*; this refers to a heat treating process that imparts ductility to an otherwise brittle material.

White Cast Iron

This product results when carbon in solution in the molten iron does not form graphite on solidification but remains in the form of massive carbides (a compound of carbide and one or more metal elements such as iron). White iron is hard and has limited ductility and high compressive strength (squeezing-type load), with good retention of strength and hardness at elevated temperatures. However, in most instances white cast iron is used for its excellent resistance to wear and abrasion—due largely to the massive carbides in its microstructure (Fig. 3-41).

Gray Cast Iron

Given the right composition of iron and the right cooling rate on solidification, a substantial part of the carbon in gray iron separates out of the liquid to form flakes of graphite (Fig. 3-42). Gray iron is easy to machine at hardness levels conducive to good wear resistance. It also resists galling under conditions of poor and limited lubrication. Its damping capacity (ability to absorb vibrations, which are cyclical stresses) is rated outstanding, as is its resistance to thermal shock (a condition caused by the cycling of temperatures).

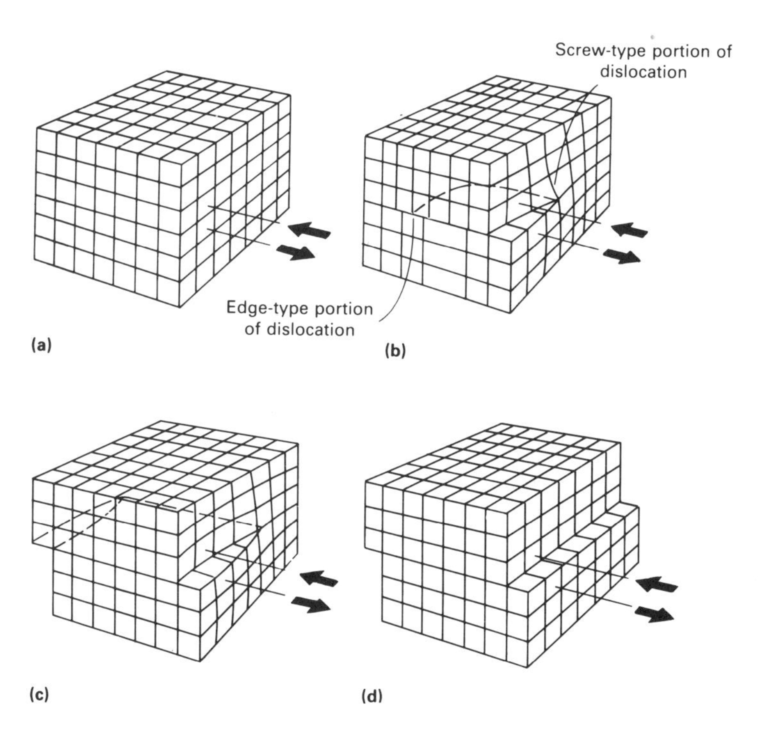

Fig. 3-36 Four stages of slip formation. (a) Crystal before displacement. (b) Crystal after some displacement. (c) Complete displacement across part of crystal. (d) Complete displacement across entire crystal

Ductile Cast Iron

Ductile iron, also known as nodular iron or spherulitic graphite cast iron, is similar to gray iron in composition. During the casting process, graphite is caused to nucleate (the start of a phase transformation) as spherical particles, rather than flakes. Nucleation is initiated by a very small addition of magnesium and cerium to the molten iron—a process called *inoculation*.

The main advantage of ductile iron over gray iron is its combination of strength and ductility. Compared to gray iron of similar strength, a ferritic ductile iron has an elongation up to 18%, while that of the gray iron is only 0.6%. Graphite nodules in ductile iron are shown as black objects in Fig. 3-43. They are surrounded by ferrite (white); the matrix (a phase in which another constituent is dispersed) is pearlite.

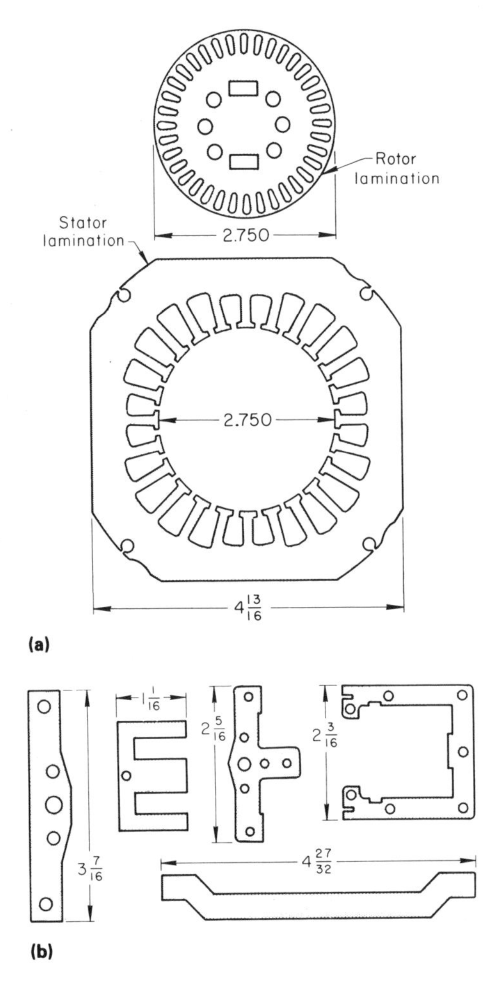

Fig. 3-37 Electrical steel sheet for rotating and nonrotating machinery

Malleable Cast Iron

Malleable iron also contains graphite nodules, but because they are the product of heat treating, rather than being formed from the melt, they are not spherical. Properties include considerable ductility and toughness. The choice between malleable iron and ductile iron hinges on economy and availability rather than properties. In certain applications, malleable iron has a clear edge; it is preferred for thin-section castings not feasible in ductile iron, and in applications that require a modulus of elasticity higher than that of ductile iron. The microstructure of pearlitic malleable iron is shown in Fig. 3-44.

Processing Advantages of Cast Irons

Casting often is preferred to forging or machining to make a desired shape. Liquid iron has high fluidity and easily fills intricate molds needed to duplicate intricate shapes. In addition, high-volume production favors casting. Casting does require a pattern and molds, but these costs can be amortized over a large number of parts.

The machinability of castings also can be used to advantage. A properly made gray or ductile iron is much more machinable than steel. Dispersed graphite particles facilitate the formation of machining chips and lubricate tools. Parts in the as-cast condition often meet mechanical property requirements, eliminating the need for additional processing. Also, as-cast parts may be in near-net-shape condition, meaning that secondary operations associated with wrought metal parts, such as machining, may be minimized or eliminated.

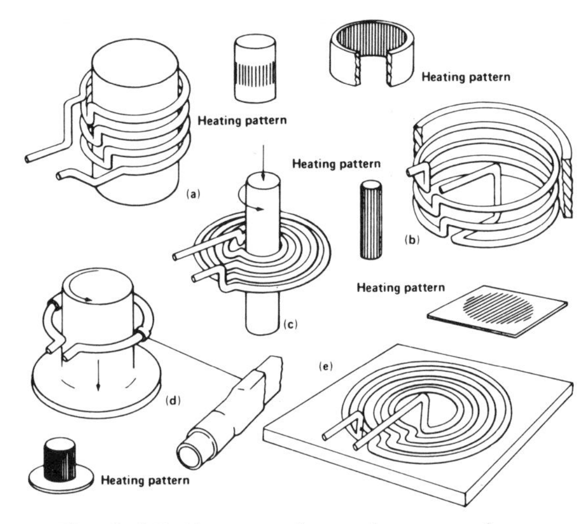

Fig. 3-38 Various induction heating coils

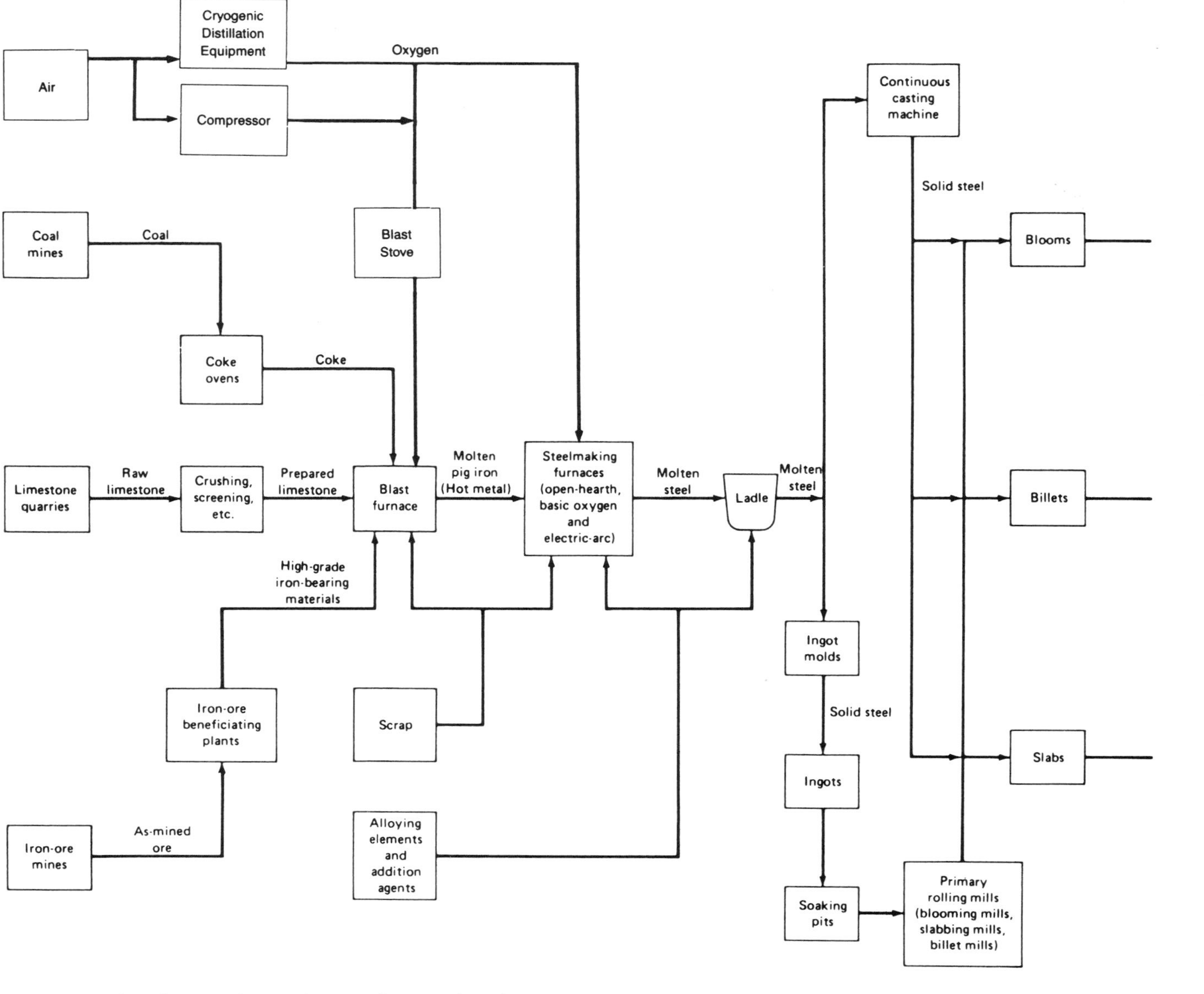

Fig. 3-39 Flow diagram for production of iron and steel

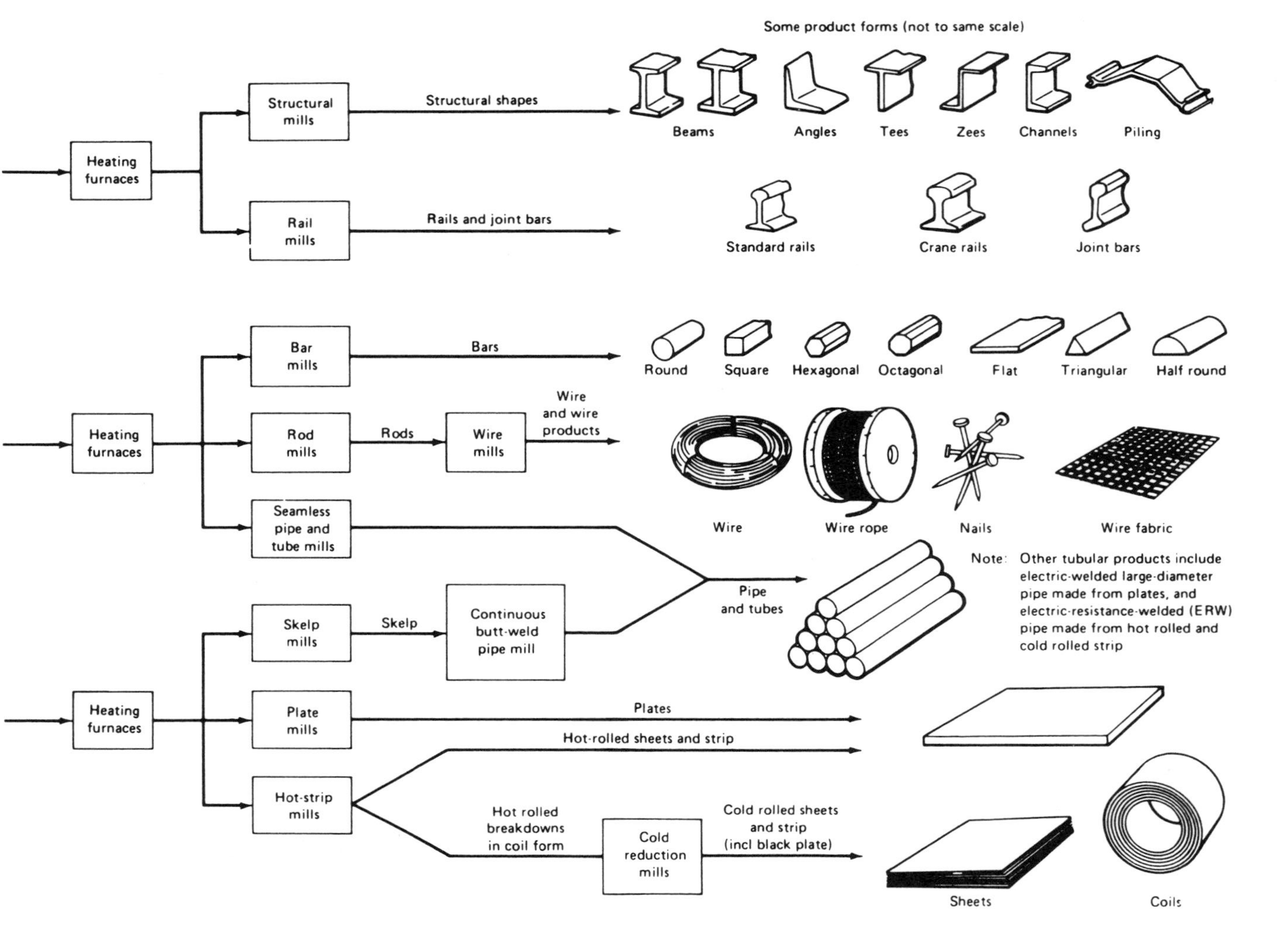

Fig. 3-40 Flow diagram showing process used in converting raw steel into mill products

Other advantages over steel:

- Because of the presence of graphite throughout their structure, cast irons have lower density.
- Thermal conductivity is higher, an advantage in applications such as automotive brakes. In braking, mechanical energy is converted to heat, which must be dissipated quickly.
- Wear properties are better.

Wear Resistance of Irons and Steels

Both metals often are used where wear can be caused by mating surfaces in contact with each other. In *abrasive wear,* metal is removed from a surface by harder (than metal) particles, generally nonmetallic. Such wear is a problem in earthmoving operations, in materials handling, and in the cement, coal, and mining industries. Types of abrasive wear include:

- Erosion and low-stress abrasion
- High-stress grinding abrasion
- Gouging abrasion

Erosion and *low-stress abrasion* are caused by light rubbing contact with sharp, abrasive particles. Screens, chute liners that handle sand, and parts exposed to airborne abrasives are subject to this type of wear. Tests show that an alloyed white iron (a high-carbon composition containing chromium and molybdenum) has much more resistance to abrasive wear than a low-carbon steel.

High-stress grinding abrasion involves the removal of relatively fine particles from a wearing surface. The punching action of two metal surfaces causes the abrasive to fragment. Unit compressive or shear stresses are very high.

Fig. 3-41 Microstructure of white iron

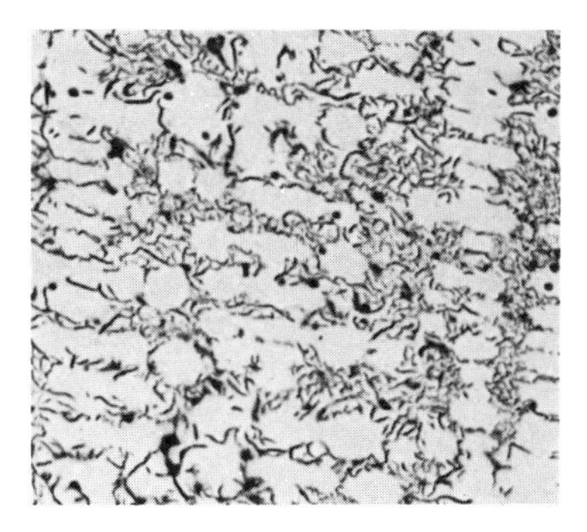

Fig. 3-42 Microstructure showing graphite distribution in gray iron

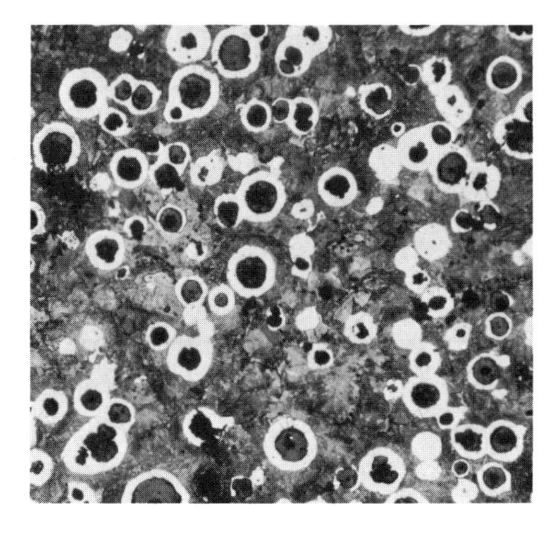

Fig. 3-43 Graphite nodules in as-cast ductile iron, surrounded by ferrite (white) in a matrix pearlite

Fig. 3-44 Microstructure of pearlitic malleable iron. Dark areas are graphite nodules.

Harder abrasives, such as quartz, can indent or remove steel very high in hardness. However, data on steel show that resistance to grinding abrasion increases with hardness and carbon content.

In *gouging abrasion,* relatively coarse particles are removed from a steel wearing surface in a manner similar to the removal of metal by machining or grinding with a coarse-grit grinding wheel. Such wear is experienced by large rock crushers and the teeth of power shovels handling large rocks. Test results for a variety of steels and cast irons are presented in Fig. 3-45, which shows the relationship between gouging wear and carbon content. The effect of both carbon content and microstructure is dra-

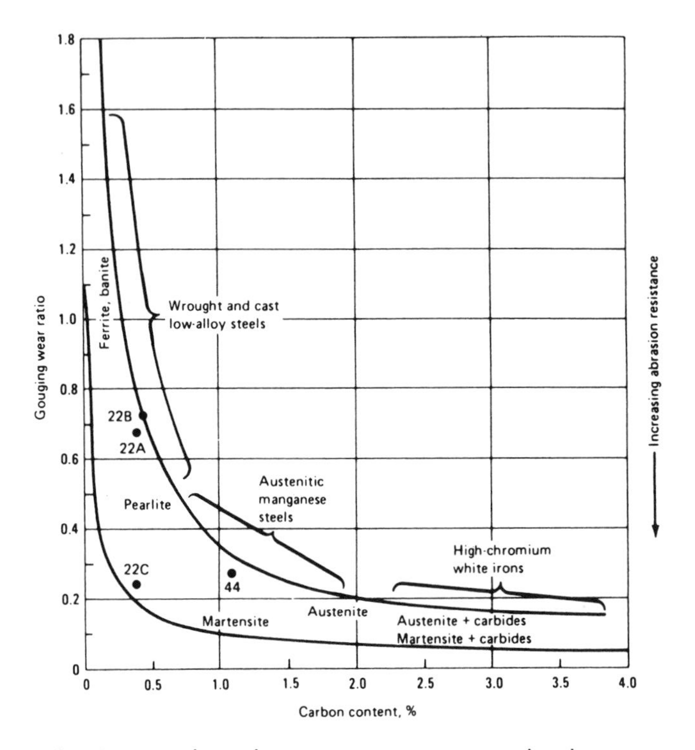

Fig. 3-45 Relation between gouging wear and carbon content of steels and cast irons

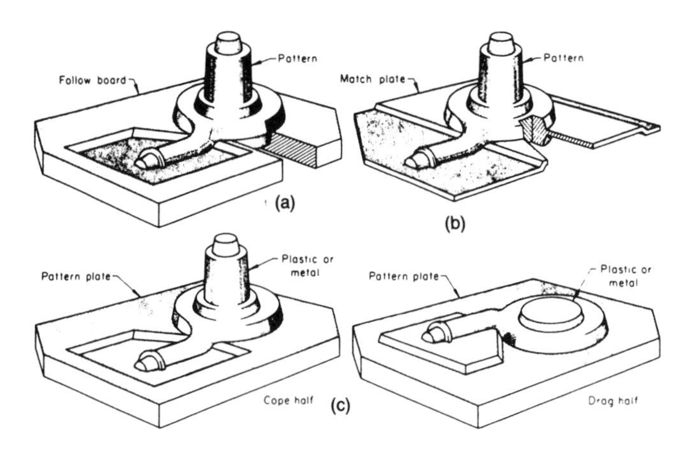

Fig. 3-46 Three types of patterns for making a water pump casting

matic. Hard martensitic structures are considerably more resistant to gouging abrasion than softer microstructures containing pearlite or *austenite* (a solid solution of one or more elements in fcc iron). Alloyed white cast irons performed better than steels.

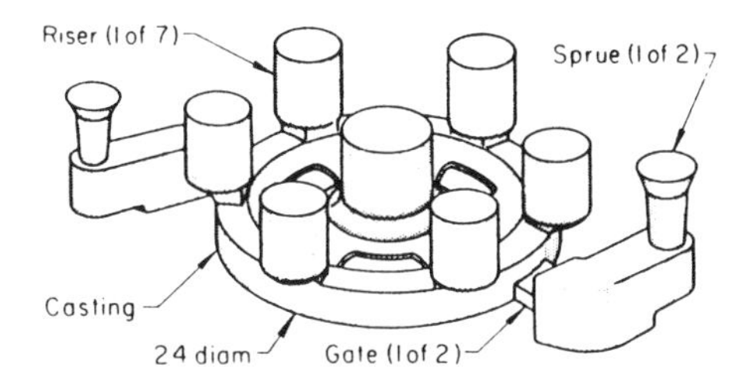

Fig. 3-47 Gating and feeding system for casting a gear blank

Producing Castings from Iron and Steel

Practices for casting iron and steel shapes in a foundry differ from those used in steelmaking in several ways. Patterns and molds are required for making cast shapes, along with typical technology involved in the melting of metals and their pouring into molds.

Design features of a pattern include a shrinkage allowance (to compensate for the shrinkage of liquid metal when cooling) and allowance for taper or draft on the vertical sides of a pattern to facilitate removal of the casting from the molding medium. Gating or other projections may be added to facilitate flow of the liquid metal. Patterns for making a water pump casting are shown in Fig. 3-46.

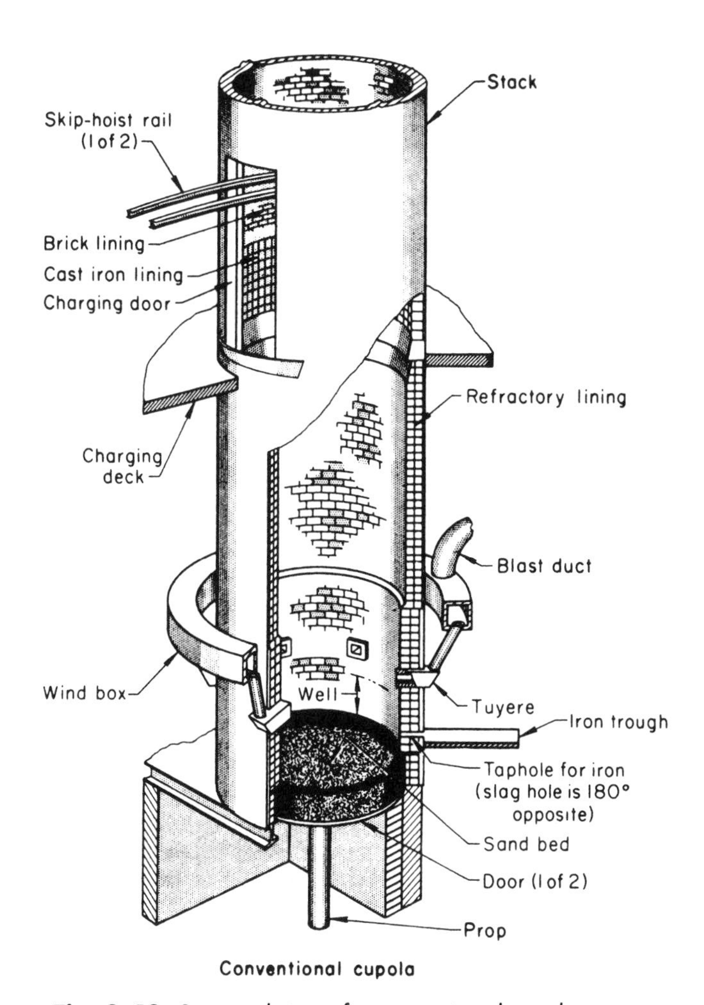

Fig. 3-48 Sectional view of a conventional cupola

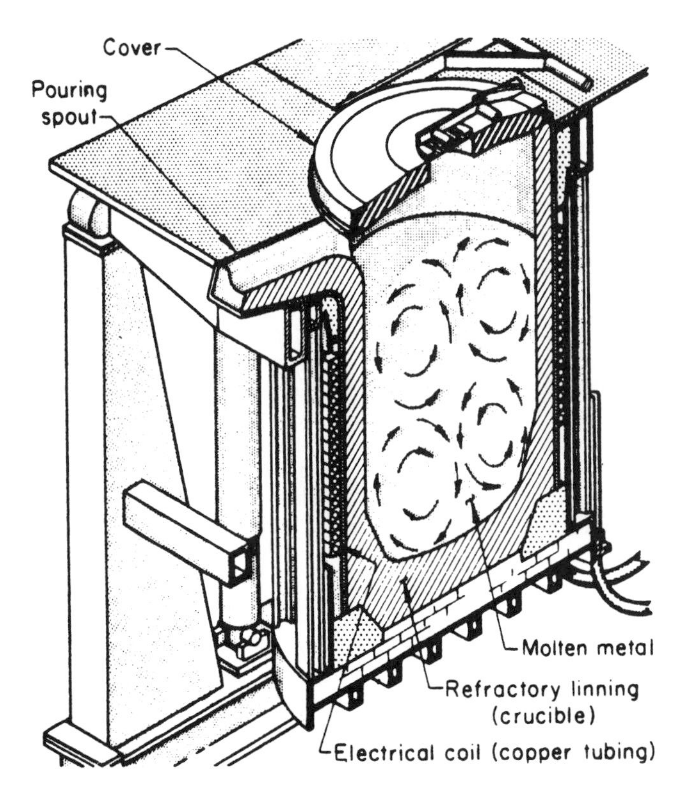

Fig. 3-49 Sectional view of a coreless induction furnace. Arrows show direction of stirring action.

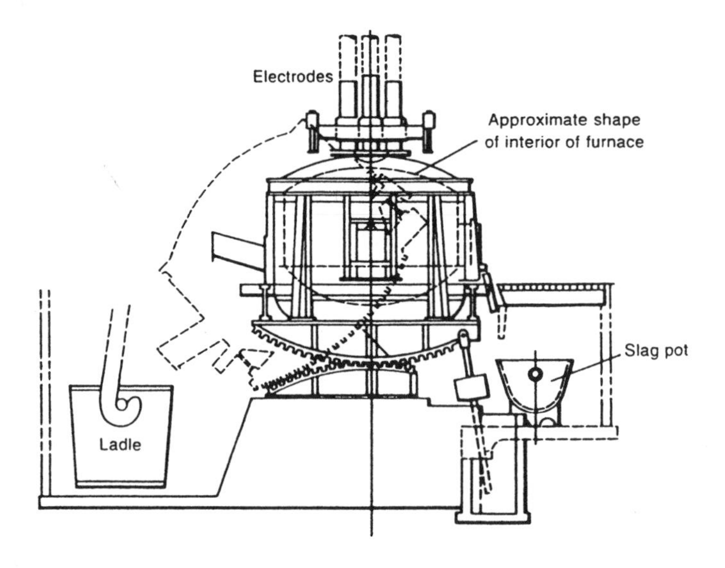

Fig. 3-50 Schematic of an electric-arc furnace

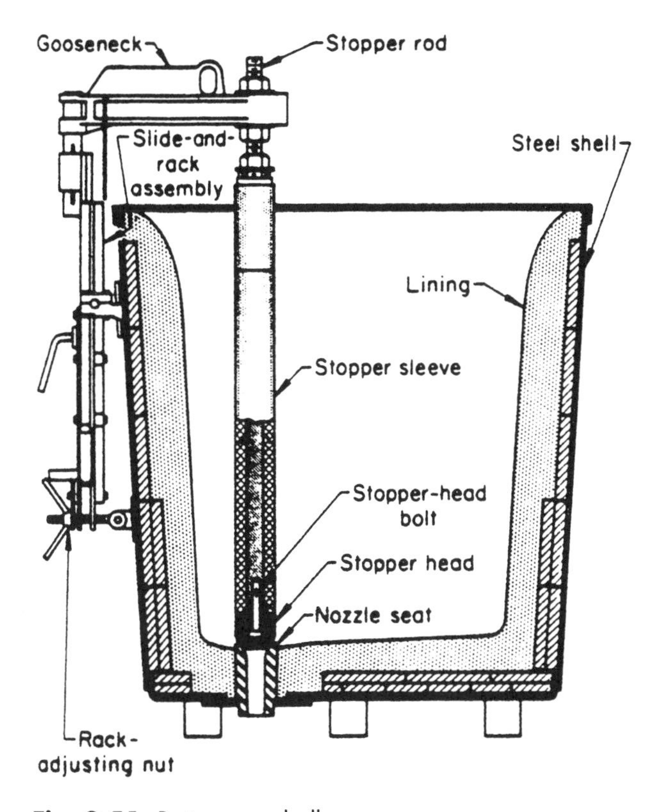

Fig. 3-51 Bottom-pour ladle

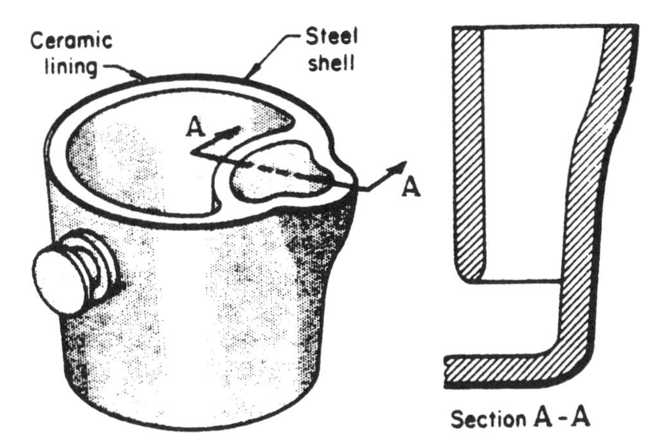

Fig. 3-52 Typical teapot-type ladle

Mold Features

The gating system is the complete assembly of sprues (where liquid metal enters the mold), runners, and gates in a mold through which metal flows to enter the casting cavity. Risers are reservoirs for excess metal to feed the casting as it shrinks during solidification. Figure 3-47 shows a gating and feeding system for casting a gear blank.

Molds for iron and steel usually are made of sand, combined with a suitable binder to hold it together. Sand is packed tightly about the pattern, so that when the pattern is removed, a cavity corresponding to the shape of the pattern remains. Molten metal poured into the cavity upon cooling develops a cast replica of the pattern. Remaining sand is easily broken up and removed.

Melting and Casting

Gray iron usually is melted in a cupola, induction furnace, or electric-arc furnace A cupola (Fig. 3-48) is a refractory-lined vertical steel shaft into which coke, flux (added to a melt to remove undesirable substances like sand, ash, or dirt), and metal are charged in alternating layers. Air for combustion enters through tuyeres near the bottom of the cupola. The taphole for iron is at the surface of the sand bed at the bottom of the cupola. An air supply, or "blast," burns the coke (made from bituminous coal, petroleum, or coal tar pitch), providing intense heat of combustion to meet the iron.

Metal at the surface of the coke bed trickles down through the hot coke and collects on the sand bottom below the tuyeres. The next layer of charge material moves down to replace the melted metal, and a fresh layer of coke replaces that which has burned. This process continues as long as the supply of air is maintained and the combination of coke and metallic charge is added. Molten slag (coke ash and nonmetallics in the charge) is also formed. It floats on the surface of the molten iron. Metal usually is tapped from the cupola to a transfer ladle or to a holding furnace (to maintain metal temperature) located at the front of the cupola. The use of cupolas is declining.

Induction furnaces are classed as coreless types (where a single crucible is surrounded by a water-cooled copper heating coil) or as core or channel furnaces (where molten metal forms a loop or channel around one leg of a transformer core). A coreless furnace is shown in Fig. 3-49. Mixing is excellent in this type of furnace. Alloying elements and fresh charges are absorbed rapidly.

Casting Gray Iron

Gray iron melted in a cupola is poured into a ladle or tilting forehearth at a metal temperature of about 1480 °C (2700 °F). Gating and feeding practice for gray iron is less critical than for other metals. This is because graphite precipitates during solidification, and the expansion of graphite compensates for normal solidification shrinkage.

Casting Ductile Iron

Ductile iron is similar to gray iron in terms of carbon and silicon content. Melting equipment, handling temperatures, and general chemistry also are similar. The important difference is that graphite separates during the nodule solidification of ductile iron in the form of spheroids instead of the flakes present in gray iron. This difference in structure is obtained by the addition of a few hundredths of a percent of magnesium. Because the presence of minute quantities of elements such as sulfur, lead, titanium, and aluminum in iron can interfere with or prevent the nodulizing effect, the molten iron must be purer than that used in gray iron. A small quantity of cerium added to the magnesium minimizes the effects that inhibit nodule formation.

Casting White Iron

Practice for melting and casting alloyed white cast iron is similar to that for steel (see next section). High bath temperatures are unnecessary. Castings thicker than 100 mm (4 in.) generally are poured at temperatures ranging from 1340 to 1400 °C (2450 to 2550 °F). Higher temperatures may be needed in pouring smaller castings. Control of the cooling rate is the most critical factor in making white cast irons. Cooling all the way down to room temperature is recommended to avoid cracking of the casting. A cooling rate that is too fast can cause the formation of hard martensite, which causes brittleness.

Casting Steel

Practice for melting and pouring steel for sand castings is similar to that for melting and pouring steel in ingots at the mill. At the mill, however, metal is destined for further processing by forging or rolling. Another difference is that higher tapping temperatures are used in the foundry.

Steel mills use both electric-arc and induction melting furnaces. The electric-arc furnace (Fig. 3-50) has a tilting mechanism for pouring off refined metal. The top and electrodes are raised and swung out of the way during charging.

Bottom-pour and teapot-type ladles generally are used. The bottom-pour ladle (Fig. 3-51) has an opening in the bottom that is fitted with a refractory nozzle. A stopper, suspended inside the ladle, pulls the stopper head up from its seat in the nozzle, allowing molten steel to flow from the ladle. The teapot ladle (Fig. 3-52) has a ceramic wall or baffle that separates the bowl of the ladle from its spout. As the ladle is tipped, hot metal flows from the bottom of the ladle, up the spout, and over the lip. Because metal poured is from the bottom of the ladle, it is free of slag and pieces of eroded refractory.

Chapter 4

Nonferrous Metals and Alloys: The Why Behind Where They Are Used

Compared with their steel counterparts, nonferrous metals and alloys are notable both in number and in diversity of properties. In addition, many nonferrous elements do double duty as important alloying agents for both steels and nonferrous metals.

If the count is confined to basic types of iron and steel, the total in this instance is five: iron, carbon steel, alloy steel, tool steel, and stainless steel. On the same basis, the total for nonferrous is 52.

The first segment of this chapter describes individual nonferrous metals and their alloys in brief summaries confined to representative properties and uses. Summaries are in alphabetical order, including the three nonferrous families: precious metals, rare earth metals, and refractory metals. The chapter concludes with more in-depth treatments of aluminum, copper, magnesium, titanium, lead, tin, and zinc.

Aluminum (Al)

Representative Properties. Aluminum is light in weight (about one-third that of steel), and some alloys have strength-to-weight ratios better than those of many steels and of other nonferrous metals. Resistance to corrosion in many atmospheric and chemical environments is excellent, due to the naturally occurring, tough oxide film on its surface. Aluminum is a good conductor of heat and is second only to copper as a conductor of electricity. Its relatively good ductility and low hardness make it easy to form and finish.

Representative Uses. Typical uses of commercially pure (99% minimum) aluminum (Fig. 4-1) include sheet metal work, foil, and chemical processing equipment. Aluminum-copper alloys can be heat treated to high strength and are used in the aircraft industry. Aluminum-silicon-magnesium alloys are used in moderate-strength structural parts, furniture, and bridge railing. Aluminum-zinc-magnesium alloys are noted for their heat-treated strength and are used in many aircraft structural applications.

Beryllium (Be)

Representative Properties. A high *modulus of elasticity* (elasticity up to the yield point, where change from elasticity to plasticity oc-

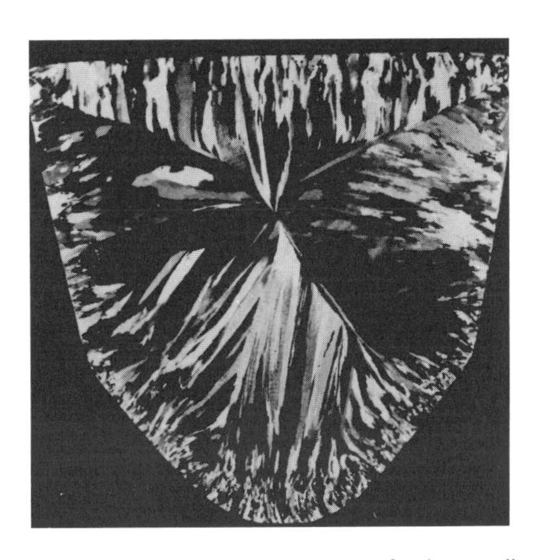

Fig. 4-1 Columnar structure of chemically pure aluminum ingot. 1.5×

curs) makes possible the design of light structures with thin sections that are quite rigid. The modulus is three times that of magnesium. Strength is maintained at temperatures up to about 595 °C (1100 °F).

Representative Uses. Applications include springs for aerospace equipment and fiber-optic waveguides. Beryllium is used as an alloying agent (Fig. 4-2).

Bismuth (Bi)

Representative Properties. Bismuth is brittle, easy to cast, and not readily formed by working methods. Its melting point of 183 °C (360 °F) is lower than that of tin-lead solder. Bismuth is one of the few metals to increase in volume upon solidification. It is *diamagnetic* (slightly repelled by magnetic fields) and has a low capture cross section for thermal neutrons.

Representative Uses. Low-melting-point alloys like bismuth-cadmium-tin-lead-indium alloy, which melts at 47 °C (117 °F), are called *fusible alloys*. One application is in sprinkler system triggering devices. A nuclear research application takes advantage of a combination of properties: High density gives bismuth excellent shielding properties for blocking the passage of gamma rays, while its low neutron cross section allows the passage of neutrons. Small additions of bismuth (0.1%) to iron and steel improve their machinability and mechanical properties.

Cobalt (Co)

Representative Properties. The more than 800 patented cobalt-containing alloys have a wealth of engineering properties, including resistance to oxidation and abrasion, resistance to the attack of many corrosive acids and chemicals, and electrical resistance.

Representative Uses. Cobalt-containing alloys are used in high-temperature alloys, spring alloys, bearing alloys, wear-resistant alloys, magnetic materials, tooling materials, and hot work steels. Selected applications include balls for ballpoint pens, high-temperature ball bearings, dental prostheses, searchlight reflectors, components for jet aircraft engines, components for nuclear submarines subject to severe wear in seawater, and permanent magnets for motors, radar, games, novelties, and door latches. Figure 4-3 shows the microstructure of a copper-cobalt alloy.

Copper (Cu)

Representative Properties. Copper and its alloys are known for their combination of excellent resistance to atmospheric and liquid corrosion, high thermal and electrical conductivity, and excellent hot and cold work properties. Brass and bronze are alloys of copper; their properties vary per alloy content. For example, high-leaded tin bronze offers resistance to corrosion, while high-strength yellow brass combines strength and resistance to wear and corrosion.

Representative Uses. Electrical equipment is among the applications for unalloyed copper, which is low in strength but high in electrical conductivity. High-leaded tin bronze is used in corrosion-resisting pumps and railway journal bearings. High-strength yellow brass is used in high-strength structural parts, gears subject to severe wear, and marine propellers, which require exceptional resistance to corrosion. The

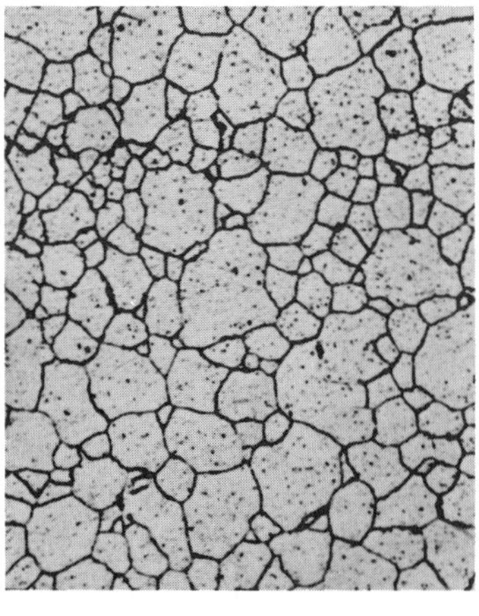

Fig. 4-2 Equiaxed grains of beryllium in copper have about the same dimensions in all directions. 300×

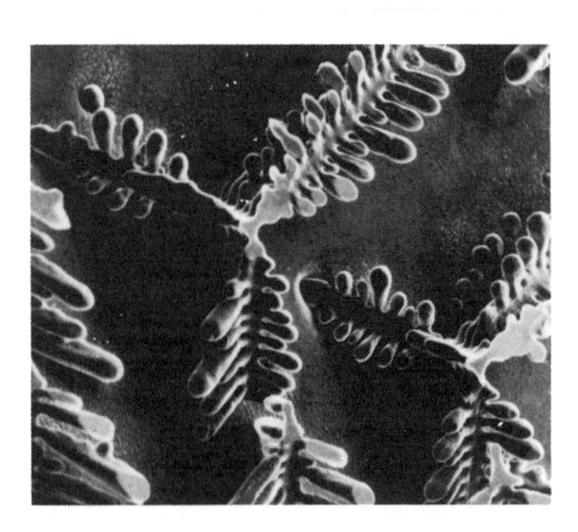

Fig. 4-3 Dendrites (crystals with a treelike branching pattern) in a copper-cobalt alloy casting. 150×

structure of a weld in copper bar is shown in Fig. 4-4.

Gallium (Ga)

Representative Properties. The chemical properties of gallium are between those of aluminum and indium. Its melting point is 29.78 °C (85.6 °F); service temperatures are increased by alloying. Like aluminum, gallium forms an oxide film on its surface when exposed to air. Like water, bismuth, and germanium, it expands on freezing. As an alloy with arsenic, gallium has a higher service temperature than germanium. It was first used in semiconductors and can operate at higher frequencies (number of cycles per unit of time) than silicon, which replaced germanium in military applications.

Representative Uses. Gallium can be used at high temperatures and high frequencies. Uses of gallium arsenide include semiconductors, tunnel diodes for FM transmitter circuits and amplifiers, and solar cells, such as those in satellites. Gallium-ammonium chloride is used in plating baths that deposit gallium onto whisker wires used in leads for transistors. Gold-platinum-gallium alloys are used in dental restoration.

Germanium (Ge)

Representative Properties. Germanium is a hard metallic element with a lower melting point (937 °C ± 1 °C, or 1718 °F ± 2 °F) than silicon, which means better fabrication properties. Below 600 °C (1110 °F), germanium has good resistance to oxidation; above that temperature, oxidation proceeds rapidly. Hydrochloric and sulfuric acids do not react with metallic germanium at room temperature, and it shows no evidence of toxicity.

Representative Uses. Germanium is primarily used for semiconductors in transistors, diodes, rectifiers, and infrared optics. It is also used in precious metal brazing alloys.

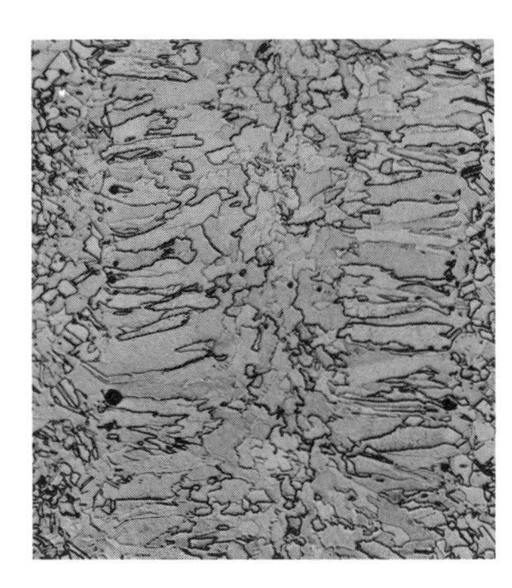

Fig. 4-4 Electron-beam-welded copper bar. Columnar grains in middle portion represent weld metal. Grain structure on either side is that of the bare metal (copper). Scattered black dots represent porosity. 35×

Hafnium (Hf)

Representative Properties. Hafnium is always associated with zirconium in nature. Both are similar in terms of chemical properties, such as resistance to corrosion in nuclear environments and high strength. Two differences: Hafnium is not as ductile or as easily worked.

Representative Uses. Like zirconium, hafnium is used in water-cooled nuclear reactors, but they serve different functions. Zirconium allows passage of thermal neutrons, while hafnium impedes their passage. It is also used in nuclear control rods as a neutron-absorbing material. However, the greatest usage of hafnium is as an alloying element for superalloys. The microstructure of hafnium plate is shown in Fig. 4-5.

Indium (In)

Representative Properties. Indium is one of the softest metals—easily scratched and highly plastic (can be deformed in compression almost without limit). It melts at 156.2 °C (313.1 °F), has low tensile strength and hardness, and is stable in dry air at room temperature. It is neither a toxin nor a skin irritant.

Representative Uses include semiconductor devices (where indium works well with germanium), bearings, and low-melting-point alloys. Low-melting-point applications include fusible safety plugs (an alloy containing 19.1% indium

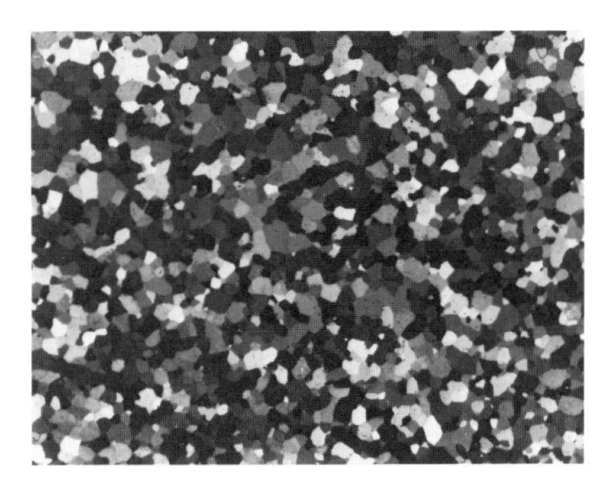

Fig. 4-5 Equiaxed grain structure of hot-rolled and annealed hafnium plate. 56×

melts at 16 °C, or 60 °F) and foundry patterns. An alloy containing 50% tin and 50% indium is commonly used to seal glass.

Lead (Pb)

Representative Properties. The limited structural properties of lead (such as low tensile strength and a low melting point of 327 °C, or 621 °F) are offset by a variety of desirable fabrication and service properties, ranging from corrosion and radiation protection castability, and coatability to use as an alloying element to improve machinability.

Representative Uses. The list includes: shielding agent against x-ray and gamma radiation; die-

Fig. 4-6 Large equiaxed and columnar grains in high-purity lead. Actual size

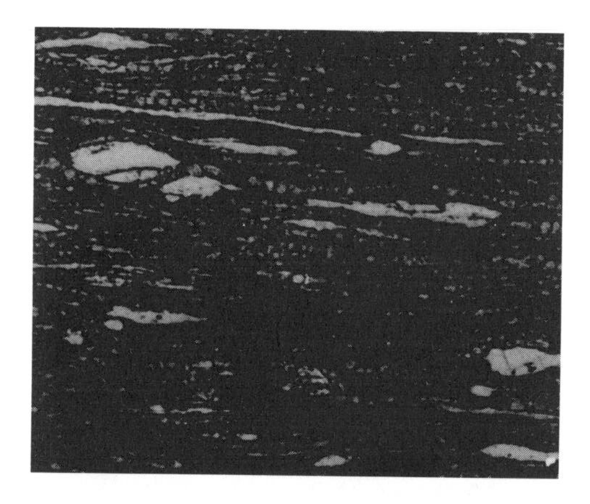

Fig. 4-7 Banded, hot-worked structure of a magnesium-zinc-zirconium alloy. Recrystallized grains are small. Light islands are deficient in zinc and zirconium. 250×

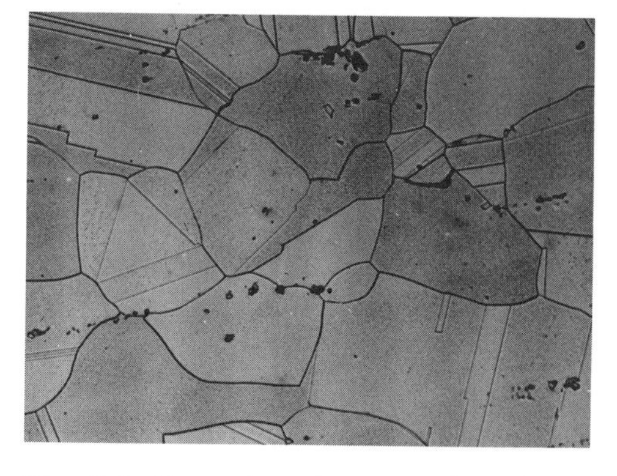

Fig. 4-8 Grain boundaries in forged Astroloy, a chromium-nickel-cobalt-molybdenum-titanium-aluminum-iron-carbon-boron-zirconium alloy. 100×

cast grids for batteries; coatings that prepare surfaces for soldering; spherical shot for shotgun shells; alloy bearings, such as babbitt; alloys that promote the machinability of both ferrous and nonferrous alloys; and solder (alloys containing 50% lead and 50% tin). Figure 4-6 shows the grain structure of high-purity lead.

Magnesium (Mg)

Representative Properties. Magnesium is the lightest of all structural metals; aluminum weighs 1.5 times more, iron four times more. Its machinability is excellent, and it can be cast by practically all commonly used methods. Sand or permanent mold castings have good strength, stiffness, and resistance to impact or shock loading. Electrical and thermal conductivities are relatively high.

Representative Uses. Unalloyed magnesium lacks the necessary properties for structural applications. Casting alloys include magnesium-aluminum-zinc types, which provide a combination of high yield strength, moderate elongation, and good pressure tightness; cast magnesium-rare earth-zirconium alloys, with or without zinc, are used at service temperatures ranging from 175 to 260 °C (350 to 500 °F). Commonly used sheet and plate alloys have good strength, toughness, formability, and weldability. The microstructure of a magnesium-zinc-zirconium alloy is shown in Fig. 4-7.

Manganese (Ma)

Representative Properties. Pure manganese is too brittle for structural applications, oxidizes easily, and rusts rapidly in moist air. Alloys have a number of desirable properties, including ductility, damping of vibrations, plus both high and low thermal expansion rates. Some alloys can be age hardened.

Representative Uses. Most commercial aluminum alloys contain manganese, usually in amounts of less than 1%, to increase recrystallization temperatures, slightly improve mechanical properties, and sometimes add resistance to corrosion. Other applications include specialty nonferrous alloys, such as manganese-copper and manganese-containing brasses, bronzes, and some nickel silvers. Virtually all steels contain small amounts of manganese, mainly to control and counteract the undesirable effects of sulfur and to slightly increase hardenability. Austenitic stainless steel contains iron-chromium-manganese and possibly nickel or nitrogen.

Nickel (Ni)

Representative Properties. Nickel and nickel-base alloys are vitally important to industry because of their ability to withstand a variety of severe operating conditions involving corrosive environments, high temperatures, high stresses, and combinations thereof.

Representative Uses. Nickel and its alloys have many applications, most of which are for resistance to corrosion and/or heat. Examples include aircraft gas turbines, steam turbine powerplants, turbochargers and valves in reciprocating engines, prosthetic devices, heat treating equipment, pollution control equipment, coal

Fig. 4-9 Gold-plated nickel-iron reed contact. Dark hood is copper overplate to protect the edge of gold (light object) during polishing. 250×

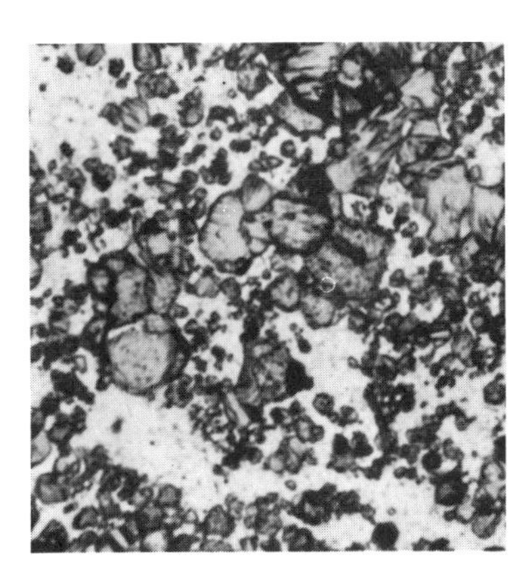

Fig. 4-10 Tungsten powder infiltrated with silver. Dark areas are tungsten; bright areas are silver. 500×

gasification and liquefaction systems, and components in pulp and paper mills. Grain boundaries in a nickel-base superalloy are shown in Fig. 4-8.

Precious Metals

Eight metals are in this family. Two of them—gold and silver—require little or no introduction. Not so well known are the six platinum group metals, which commonly occur together in nature: platinum, palladium, iridium, rhodium, ruthenium, and osmium.

Gold (Au)

Representative Properties. Gold is soft, ductile, and easy to deform. Chemically pure gold is 99.9% pure, and higher purity is available. Resistance to oxidation is its outstanding property; its color, yellow, is maintained in air or when heated. Gold is usually alloyed to increase hardness.

Representative Uses. The greatest use of gold is in jewelry and the arts. Industrial usage (mainly in electronics) ranks second, and dentistry is third. Monetary usage, historically the most prevalent application for gold, has dropped below 1%. Alloys such as gold-silver-copper can be made in a variety of colors, ranging from white to many shades of yellow. Figure 4-9 shows a gold-plated electrical contact.

Silver (Ag)

Representative Properties. Among metals, pure silver has the highest thermal and electrical conductivity. Next to gold, it is the most ductile. Silver is also one of the most common corrosion-resistant metals. It can be cold worked, extruded, rolled, swaged, and drawn.

Representative uses include electroplated ware, sterling ware, jewelry and arts, photography, mirrors, batteries, bearings, coins, and catalysts (which accelerate reactions in processing). Tungsten metal powder infiltrated with silver is shown in Fig. 4-10.

Platinum Group Metals

Platinum group metals are expensive and scarce, the main deposits are located in the Republic of South Africa, areas in the former Soviet Union, and Canada. These metals and their alloys typically are the only materials available to meet the requirements of advanced technology or special industrial applications. High material costs are counterbalanced by long, reliable service. Platinum is the most important and most used metal of the group; palladium ranks second.

Platinum (Pt)

Representative Properties. Platinum has a high melting point of 1770 °C (3216 °F). Remarkably resistant to corrosion and chemical attack, it retains its mechanical strength and resistance to oxidation in air even at elevated temperatures. Platinum has catalytic properties and is easily worked when heated.

Representative uses include spark plug electrodes (which require resistance to corrosion and erosion); high-temperature wiring (in which the base metal is clad with platinum for electri-

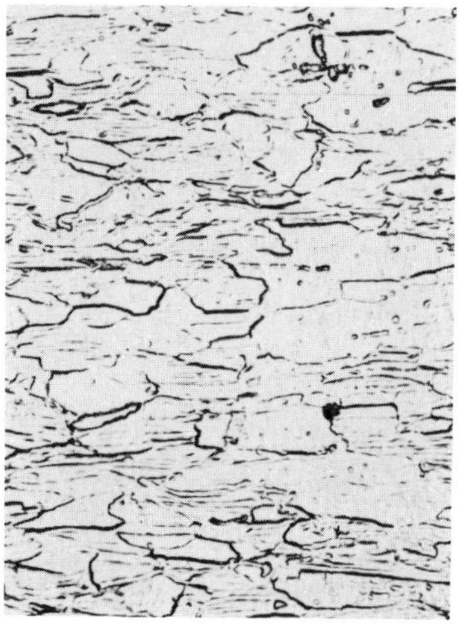

Fig. 4-11 Structure of platinum-ruthenium alloy strip cold rolled to 50% reduction. Grains are elongated in the direction of rolling. 500×

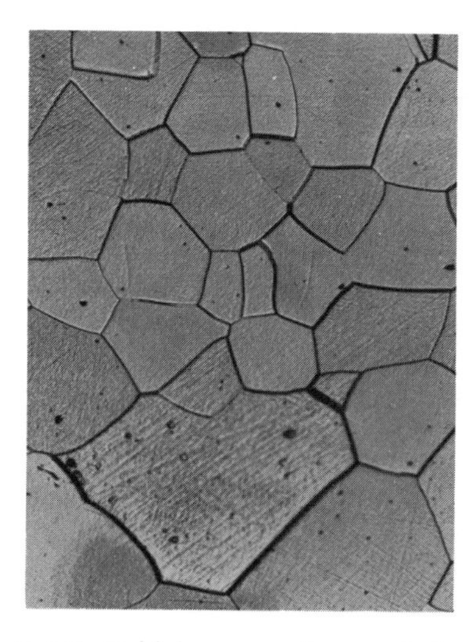

Fig. 4-12 Cold-drawn and annealed palladium wire. Grains are recrystallized; specks are impurities. 75×

cal conductivity and resistance to oxidation); brazing alloys for tungsten (due to ductility and high melting point); laboratory ware (high melting point combined with resistance to corrosion and heat); and fuel cell electrodes (for catalytic activity and resistance to corrosion). The structure of cold-rolled platinum-ruthenium alloy strip is shown in Fig. 4-11.

Palladium (Pd)

Representative Properties. Palladium has high electrical resistivity. It melts at 1522 °C (2826 °F), is very ductile, and is slightly harder than platinum. Palladium can be worked with conventional equipment, and as an alloying agent improves hardness and strength without detriment to the corrosion resistance of the parent metal.

Representative Uses. Palladium metal is used as a conductor material in printed circuits (due to its solderability and resistance to corrosion). In combination with alumina, the most commonly used ceramic material, palladium serves as a catalyst for removal of oxygen from hydrogen (because of its activity at low temperatures). Alloys and their uses include palladium-gold in spark-plug anodes, gold-palladium in thermocouples (for temperature stability), and palladium-platinum sensing elements in gas analysis (for catalytic action). The structure of cold-drawn palladium wire is shown in Fig. 4-12.

Iridium (Ir)

Representative Properties. The melting point of iridium (2455 °C, or 4449 °F) is higher than that of platinum. This hard, brittle material is the most corrosion-resistant metal known.

Representative uses include spark-plug anodes, neutron absorption (due to its high-absorption cross section), and as a gamma ray source (due to its radiation energy and moderate half-life). Iridium-platinum alloys are used in dopant contacts for transistor junctions (*dopant* modifies electrical characteristics); iridium often is added to platinum to improve its mechanical properties. Elongated grains in platinum-iridium strip are shown in Fig. 4-13.

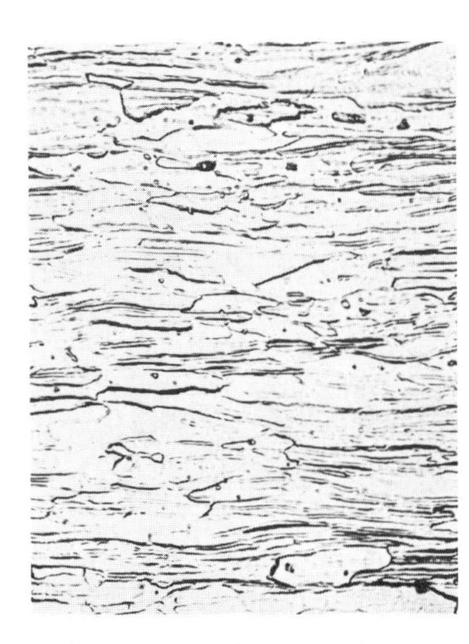

Fig. 4-13 Grains in platinum-iridium strip have been elongated and severely deformed by cold rolling. 500×

Rhodium (Rh)

Representative Properties. Rhodium is an important element in high-temperature applications up to 1650 °C (3000 °F). It is very hard and difficult to work.

Representative Uses. As an alloying agent, rhodium enhances mechanical properties and resistance to corrosion. Rhodium-platinum is used in rayon spinnerets (due to its resistance to corrosion, strength, and ductility), in glow plugs for jet engines (used to restart engines after flameouts), in heater windings for glass, ceramic, and ferrite research (for its resistance to oxidation and high melting point), and in thermocouple wire (for maintenance of accuracy up to high temperatures). Because of its electrical contact properties and resistance to wear, rhodium electroplate is applied on contact points for switches. Melted rhodium plating on an electrical switch is shown in Fig. 4-14.

Ruthenium (Ru)

Representative Properties. Ruthenium has high hardness and electrical resistivity. Its melt-

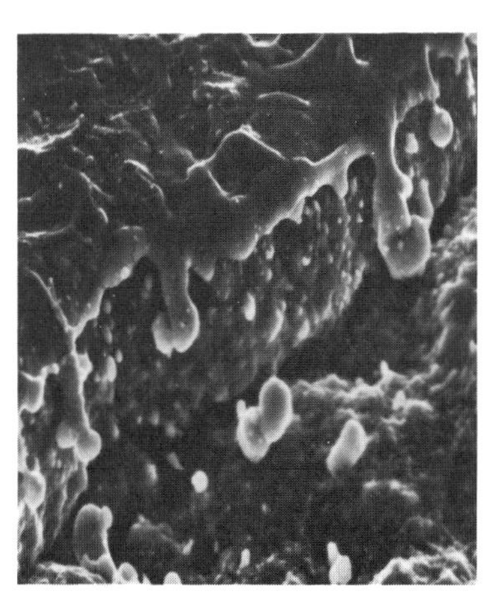

Fig. 4-14 Scanning electron micrograph of rhodium-plated reed switch showing melting of plate at contact point due to arcing. 3500×

ing point is 2500 ± 100 °C (4530 ± 180 °F). Some ruthenium-platinum alloys can be highly volatile and toxic. The metal is brittle and not workable in the pure state.

Representative Uses. Ruthenium metal is used in crucibles that handle molten bismuth. As an alloying agent for platinum, ruthenium improves mechanical properties and resistance to corrosion. This alloy is used in *resistors* (which measure and control electrical characteristics) and *potentiometers* (which measure electromotive forces).

Osmium (Os)

Representative Properties. Of all the platinum group metals, osmium has the highest specific gravity (weight per mass) and the highest melting point (2700 ± 200 °C, or 4900 ± 350 °F). When heated in air, it forms a highly volatile and poisonous material. It is practically unworkable.

Representative Uses. Osmium is used mainly as a catalyst.

Rare Earth Metals

The 17 rare earths make up a closely related group of highly reactive metals, which means they readily combine with oxygen at elevated temperatures to form very stable oxides. Reactive metals can become embrittled by interstitial absorption of oxygen, hydrogen, and nitrogen. Titanium, zirconium, and beryllium are other reactive metals.

The term *rare earth* is deceptive. Rare earths are neither rare nor earths. The name derives from oxide minerals in which they are discovered. The family tree has these members:

- Scandium
- Yttrium
- Lanthanum
- Cerium
- Praseodymium
- Neodymium
- Promethium
- Samarium
- Europium
- Gadolinium
- Terbium
- Dysprosium
- Holmium
- Erbium
- Thulium
- Ytterbium
- Lutetium

Mischmetal is an alloy of the cerium group and is the least expensive rare earth in alloy form.

Representative Properties. Some rare earths can be fabricated cold; all are poor conductors of electricity and are *paramagnetic* (have little mutual magnetic attraction). Below room temperature, some are strongly *ferromagnetic* (have strong mutual magnetic attraction). Some form a thin oxide film under ordinary atmospheric conditions.

As a family, the rare earths do not have properties that suggest usage as structural alloys. Scandium is similar to aluminum in density. Yttrium has a density similar to that of titanium and a melting point of 1522 °C (2772 °F), which allows passage of neutrons, and yttrium can be cold worked. Samarium forms a protective oxide film in air at temperatures up to about 595 °C (1100 °F).

Representative Uses. Rare earth metals are used in cored carbons for arc lighting. Mischmetal-iron alloys serve as flints for lighters. Glass is polished with cerium oxide.

Additions of mischmetal and various rare earth compounds improve the properties of cast irons and steels. An addition of 1% yttrium to 25% chromium steel increases the re-

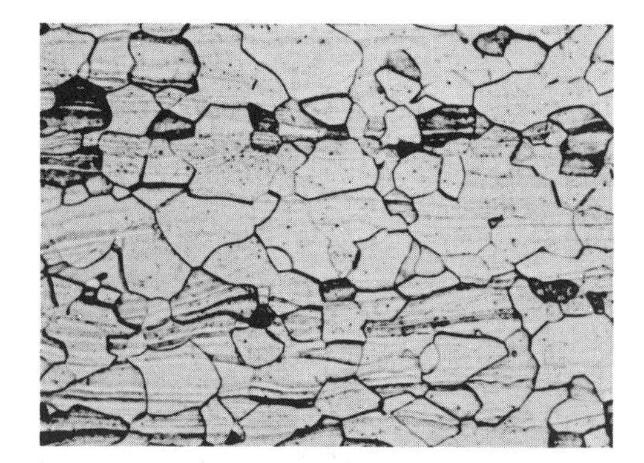

Fig. 4-15 Crystal structure of niobium alloy sheet after hot extrusion and warm rolling, with 50 to 75% reduction between anneals. 250×

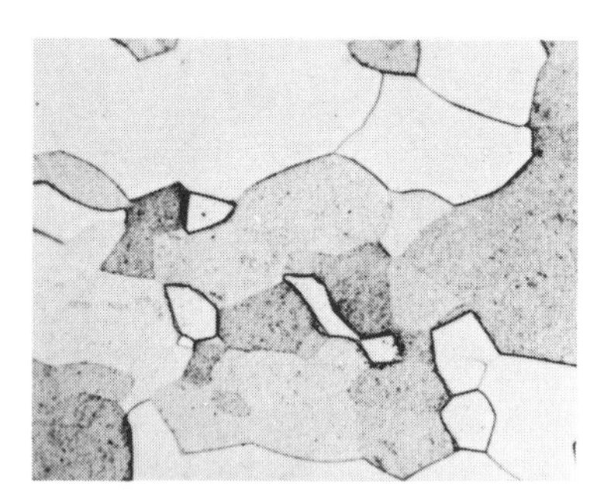

Fig. 4-16 Unalloyed tantalum sheet with recrystallized, equiaxed grains of different sizes after forging and 60% cold reduction. 250×

sistance of the alloy to oxidation at service temperatures ranging from 1095 to 1370 °C (2000 to 2500 °F).

Refractory Metals

This family has five members: niobium (sometimes called columbium), tantalum, molybdenum, tungsten, and rhenium. With the exception of two platinum group metals—osmium and indium—refractory metals have the highest melting temperatures of all metals. Tungsten has the distinction of having the highest melting temperature of all elements (3410 °C, or 6170 °F).

All refractory metals are readily degraded by oxidizing environments at relatively low temperatures, which has restricted their use in low-temperature or nonoxidizing, high-temperature environments. However, protective coating systems, primarily for niobium, have opened their application in high-temperature, oxidizing environments in space.

Usage at one time was limited to lamp filaments, electron tube grids, heating elements, and electrical contacts. Today, refractory metals also find application in the aerospace, electronic, nuclear, and chemical processing industries.

Niobium (Nb)

Representative Properties. In addition to the generic properties of refractory metals just described, niobium is used for its combination of relatively light weight and retention of high strength at service temperatures ranging from 980 to 1205 °C (1800 to 2200 °F). Niobium also impedes the passage of neutrons in nuclear applications.

Representative Uses. Most niobium is used in the production of high-strength low-alloy steels and stainless steels. Other uses include nose caps for hypersonic flight vehicles and superconductivity magnets for magnetic resonance imaging (MRI) machines, used in medical diagnosis. The crystal structure of niobium after warm rolling is shown in Fig. 4-15.

Tantalum (Ta)

Representative Properties. Tantalum has a high melting point of 2996 °C (5425 °F) and high resistance to corrosion, including corrosion in body fluids.

Representative Uses. The largest use for tantalum is as powder and anodes for electronic *capacitors* (devices that store electrical charges when voltage is applied). Another important use is in prosthetic devices and surgical staples. Figure 4-16 shows recrystallized tantalum sheet with equiaxed grains.

Molybdenum (Mo)

Representative Properties. Molybdenum has a high melting temperature of 2610 °C and (4730 °F), along with high stiffness, good creep strength, and retention of mechanical properties at high temperatures.

Representative Uses. Molybdenum is primarily used as an alloying agent for irons, steels, superalloys, and corrosion-resistant alloys to improve hardenability, toughness, abrasion resistance, strength, and resistance to creep at elevated temperatures. Elongated grains in a molybdenum-titanium alloy are shown in Fig. 4-17.

Tungsten (W)

Representative Properties. Tungsten, an extremely heavy metal, is an excellent alloying agent.

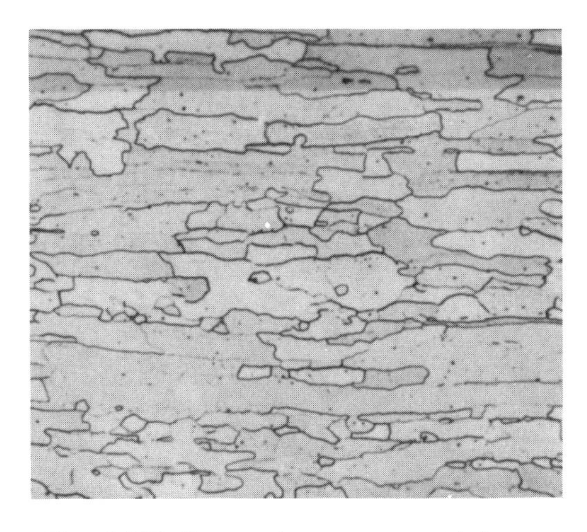

Fig. 4-17 Elongated grains in a molybdenum-titanium alloy after cold rolling and annealing. 200×

Fig. 4-18 Recrystallized, equiaxed structure of tungsten wire after annealing. 200×

Representative Uses. The major use of tungsten is in cemented carbide metal-cutting tools and wear-resistant materials. A *carbide* is a compound of carbon with one or more metallic elements. Hard carbide particles are bound together, or cemented, by a soft and ductile metal binder, such as cobalt or nickel. In addition to machining, tungsten carbide tools are used in mining and in oil and gas drilling. The weight of tungsten is utilized in applications such as counterweights and flywheels. The microstructure of tungsten wire after annealing is shown in Fig. 4-18.

Rhenium (Re)

Representative Properties. The properties of rhenium are generally similar to those of molybdenum and tungsten. Its melting temperature of 3180 °C (5756 °F) is the second highest among refractory metals. Pure rhenium combines great room-temperature ductility with good high-temperature strength. Alloying and catalytic properties are good.

Representative Uses. As an alloying agent, rhenium improves the ductility of tungsten and molybdenum. As a catalyst, it protects fine chemicals from impurities such as nitrogen, sulfur, and phosphorus. In processing petrochemicals, it is used to increase the octane rating of lead-free gasoline. Figure 4-19 shows the microstructure of molybdenum-rhenium foil after warm working, cold working, and annealing.

Superalloys

Representative Properties. Superalloys are based on either nickel, iron-nickel, or cobalt. Classified as heat-resistant alloys, superalloys provide a combination of mechanical strength and resistance to surface degradation that is unmatched by other metallic alloys. Alloying elements include chromium, molybdenum, vanadium, cobalt, and tungsten. For additional information, see the sections on nickel and cobalt alloys in this chapter.

Representative uses include turbine disks and blades, jet engine blades and parts, integrally cast turbine wheels, jet engine sheet parts, and high-temperature aerospace applications. Superalloys also find use in gas turbines, coal conversion plants, and the chemical processing industries. Figure 4-20 illustrates the effect of cold working and different annealing temperatures and times on the grain size of cobalt-base superalloy.

Tin (Sn)

Representative Properties. Tin, one of the first metals known to humans, is nontoxic, soft and pliable, and suitable for cold rolling. Tin resists corrosion, making it an ideal coating for other metals. Hardness is increased by alloying elements, such as copper, antimony, bismuth, cadmium, and silver. Tin has a low coefficient of friction.

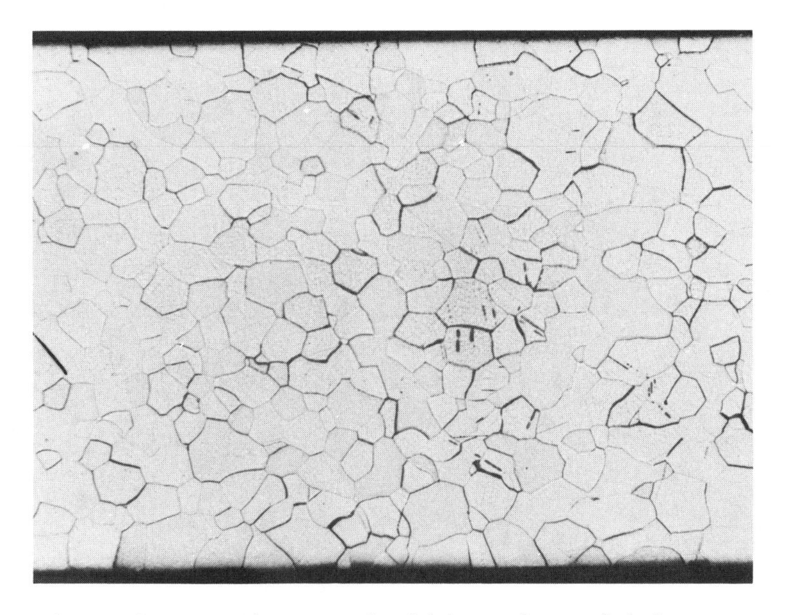

Fig. 4-19 Equiaxed structure of molybdenum-rhenium foil after warm working, cold working, and annealing. 200×

(a)

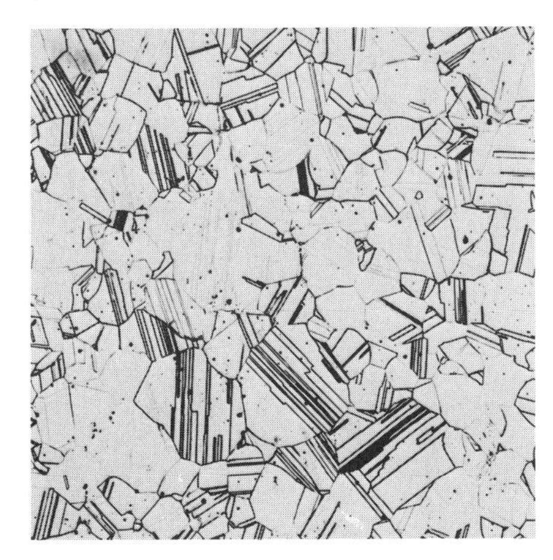

(b)

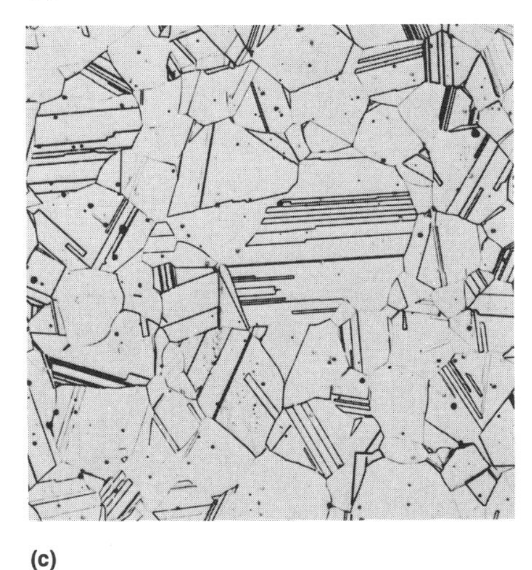

(c)

Fig. 4-20 Growth of grain size in cobalt-base superalloy (Haynes 25) with annealing at increasing temperatures and times following cold working (35% reduction each time). 100×

Representative uses include foil (for wrapping food), wire, pipe, collapsible tubing (e.g., for tooth paste), and cans. Alloys include pewter (a tin-base alloy containing antimony and silver) and bearing materials. The microstructure of a tin-antimony-lead-copper alloy is shown in Fig. 4-21.

Titanium (Ti)

Representative Properties. Titanium-base alloys fall into three categories:

- *Alpha alloys* are weldable, have good stability at temperatures up to 540 °C (1000 °F), and are strong and relatively tough at temperatures down to –423 °F = –253 °C. Compared with the beta and alpha-beta titanium alloys, alpha alloys are more difficult to form.
- *Beta alloys* are weldable, have good stability at temperatures up to 315 °C (600 °F), and are brittle at temperatures below –24 °C (–11 °F). At room temperature, they are very formable. Figure 4-22 shows precipitate in grains of Beta C titanium rod after cold drawing, solution treating, and aging.
- *Alpha-Beta* alloys may or may not be weldable, depending on the alloy. Stability is good up to a temperature of 425 °C (800 °F). These alloys are strong and relatively tough at temperatures down to –320 °F = –195 °C, and their resistance to corrosion is outstanding. In general, alpha-beta alloys are more formable than alpha alloys and beta alloys. A defect in an alpha-beta alloy billet is shown in Fig. 4-23.

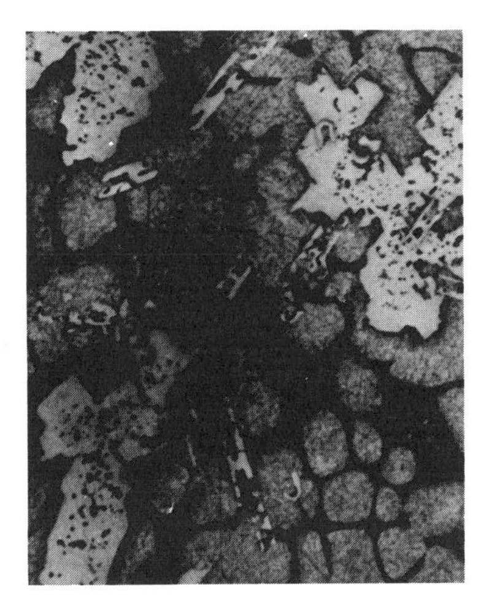

Fig. 4-21 Copper-tin needles and antimony-tin crystals (both light) in tin-rich solution. Alloy contains tin, antimony, lead, and copper. 150×

Representative Uses. Unalloyed titanium is used in airframes, chemical processing and distillation equipment, condenser tubing, marine and seawater parts, heat exchangers, cryogenic vessels, pulp and paper production equipment, and surgical implants. Alloyed titanium is used in chemical processing equipment; airframes; blades, disks, wheels, spacers, and fasteners for turbine engines; hydraulic tubing and fittings for aircraft; foil; space capsule components; pressure vessels; helicopter rotor hubs; downhole mining exploration equipment; and logging equipment.

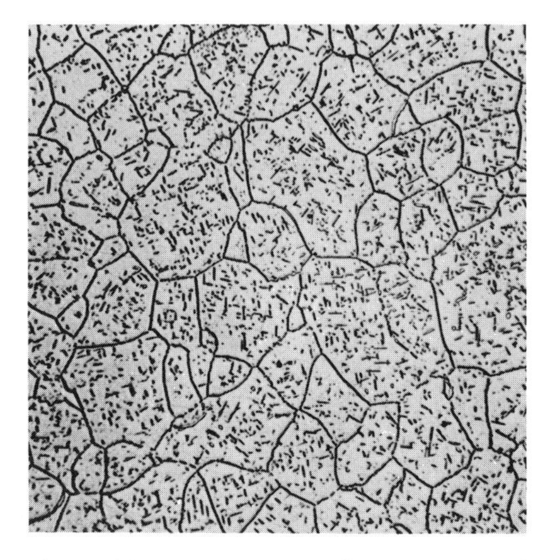

Fig. 4-22 Precipitate (tiny dots) in grains of Beta C rod (titanium-aluminum-vanadium-chromium-molybdenum-zinc alloy) after cold drawing, solution treating, and aging. 250×

Uranium (U)

Representative Properties. Uranium is a moderately strong and ductile metal that can be cast, formed, and welded by a variety of standard methods. It is used in nonnuclear applications largely because of its high density (68% greater than that of lead). Uranium and its alloys generally are considered difficult to machine, requiring special tools and conditions.

Depleted uranium is melted, fabricated, and machined following conventional metallurgical practice. Due to its mild radioactivity, chemical toxicity, and pyrophoric properties, special precautions are taken in processing. Its toxicity is similar to that of heavy metals. However, due to its pyrophoricity, powder or chips can self-ignite when exposed to oxygen. Inhalation of excessive amounts of dust or fumes can cause various health problems, including kidney damage.

Representative Uses. Uranium is generally selected over other very dense materials because it is easier to cast and fabricate than tungsten and less costly than gold or platinum. Typical nonnuclear applications of uranium and its alloys include radiation shields, counterweights, and armor-piercing ammunition, such as that for cannons mounted in tanks. As-cast unalloyed uranium billet with a coarse grain structure is shown in Fig. 4-24.

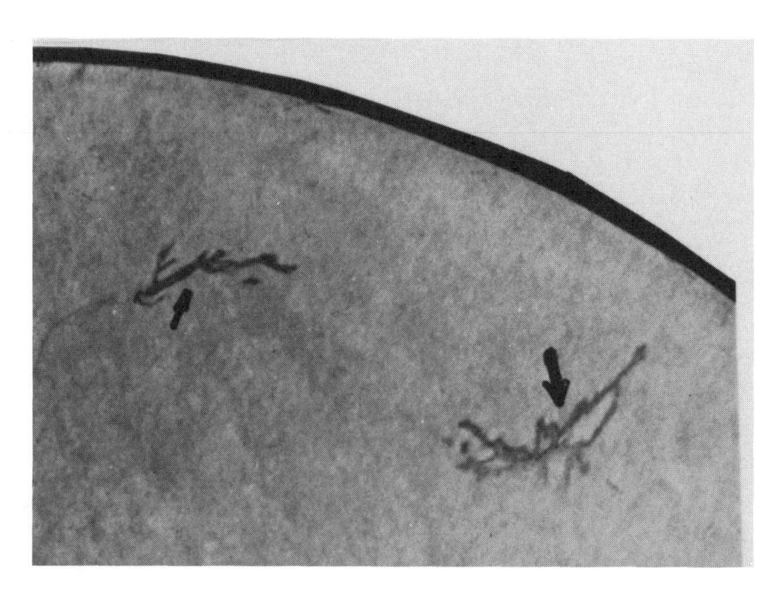

Fig. 4-23 Defect (see arrow) in alpha-beta titanium alloy (titanium-aluminum-vanadium) billet. Actual size

Vanadium (V)

Representative Properties. The melting point of vanadium is 1900 ± 25 °C (3450 ± 50 °F). Chemically active and closely related to niobium, tantalum, titanium, and chromium, vanadium is soft, ductile, and strong, and has good resistance to atmospheric corrosion and to hydrochloric and sulfuric acid environments. Its cold working properties are excellent. In hot working, vanadium must be heated in an inert gas atmosphere. It is suitable for drawing wire. As with more difficult-to-machine stainless steels, speeds must be low and feeds light to moderate. Welding is not difficult, but must be shielded from air with inert gas.

Representative Uses. Vanadium serves as cladding for fuel elements in nuclear reactors because it does not alloy with uranium. It does alloy with titanium and steel (remember the structural usage of vanadium alloy steel in Ford's Model T). Its thermal conductivity is rated good, and it also has fair thermal neutron cross section. Vanadium is an essential part of some titanium alloys (Fig. 4-25).

Zinc (Zn)

Representative Properties. Zinc has a relatively low melting point of 419 °C (787 °F), resists corrosion, is ductile and *malleable* (permits plastic deformation in compression without fracture), and is highly soluble in copper.

Representative Uses. As a coating, zinc provides corrosion protection for iron and steel. It is widely used in die casting parts for autos, household appliances, computer equipment, and builder hardware. Brass alloys are produced by mixing zinc (up to 35%) with copper. Rolled wrought zinc is used in products ranging from shells for dry batteries to gutters and downspouts. Zinc-rich dendrites are shown in Fig. 4-26.

Zirconium (Zr)

Representative Properties. Zirconium is associated with hafnium in nature. Its nuclear properties are excellent, and it is stable in severe corrosive environments. Zirconium is similar to stainless steel in ductility, has high strength, and is easy to fabricate with common shop equipment.

Representative Uses. The primary application of zirconium is in the construction of nuclear reactors, generally as cladding and for structural elements. The reactor grade must be free of hafnium, because hafnium absorbs thermal neutrons while zirconium allows their passage. Zirconium also is used in chemical processing equipment and in flash bulbs. The zirconium

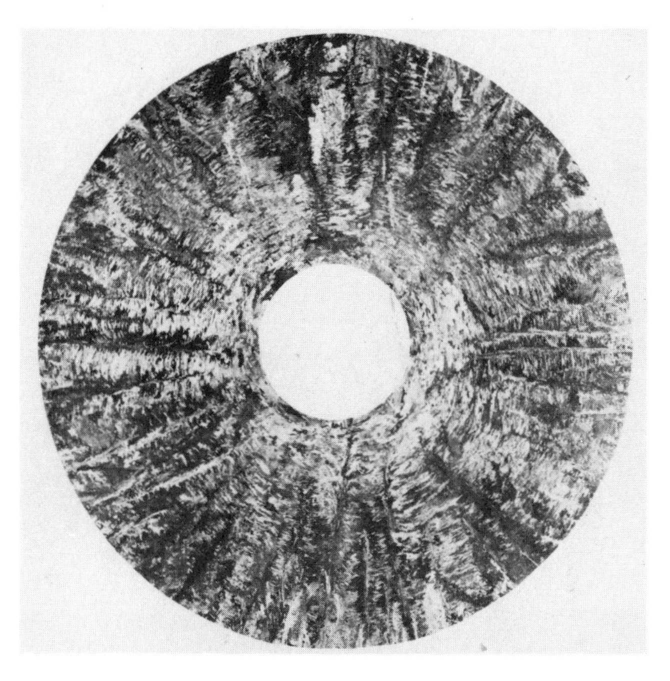

Fig. 4-24 As-cast unalloyed uranium billet with coarse grain structure. Half actual size

tubing in Fig. 4-27 shows signs of chemical corrosion.

Aluminum and Its Alloys

Aluminum is one of the most economical and structurally effective materials used for commercial and military equipment applications. Some of its attractive properties include:

- Good appearance
- Ease of fabrication
- Good corrosion resistance (especially in the newer 7000 series aluminum alloys, which perform very well, but are not completely immune, in environments that promote hydrogen embrittlement and intergranular corrosion)
- Low density and high strength-to-weight ratio
- Higher strength by alloying, heat treatment, or cold working, or a combination of all three
- High fracture toughness

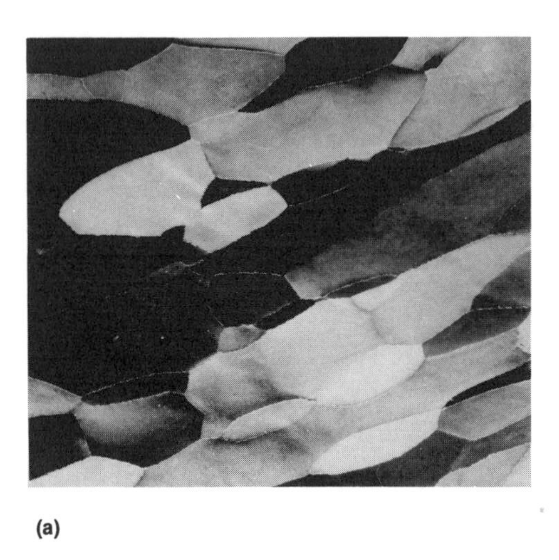

(a)

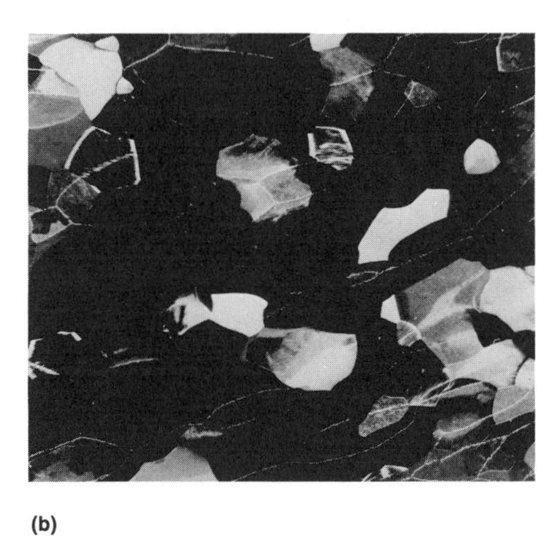

(b)

Fig. 4-25 Titanium-vanadium-iron-aluminum alloy. (a) Deformed grains. (b) Recrystallized grains. Magnification not known

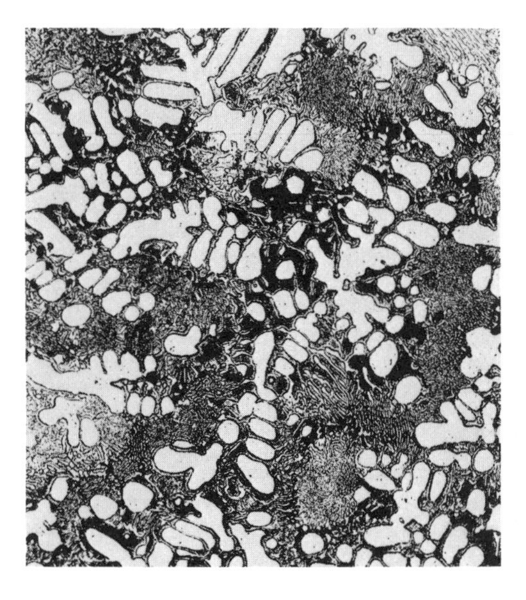

Fig. 4-26 Coarse zinc-rich dendrites (white) in a zinc-aluminum-copper-magnesium alloy

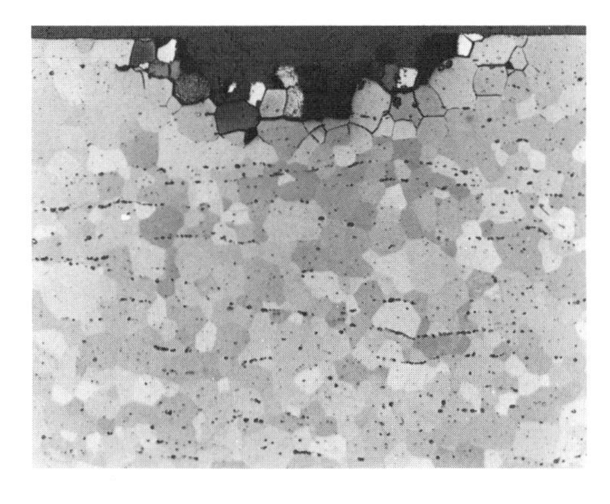

Fig. 4-27 Cold-worked and annealed zirconium tubing showing signs of intergranular chemical corrosion (black dots). 140×

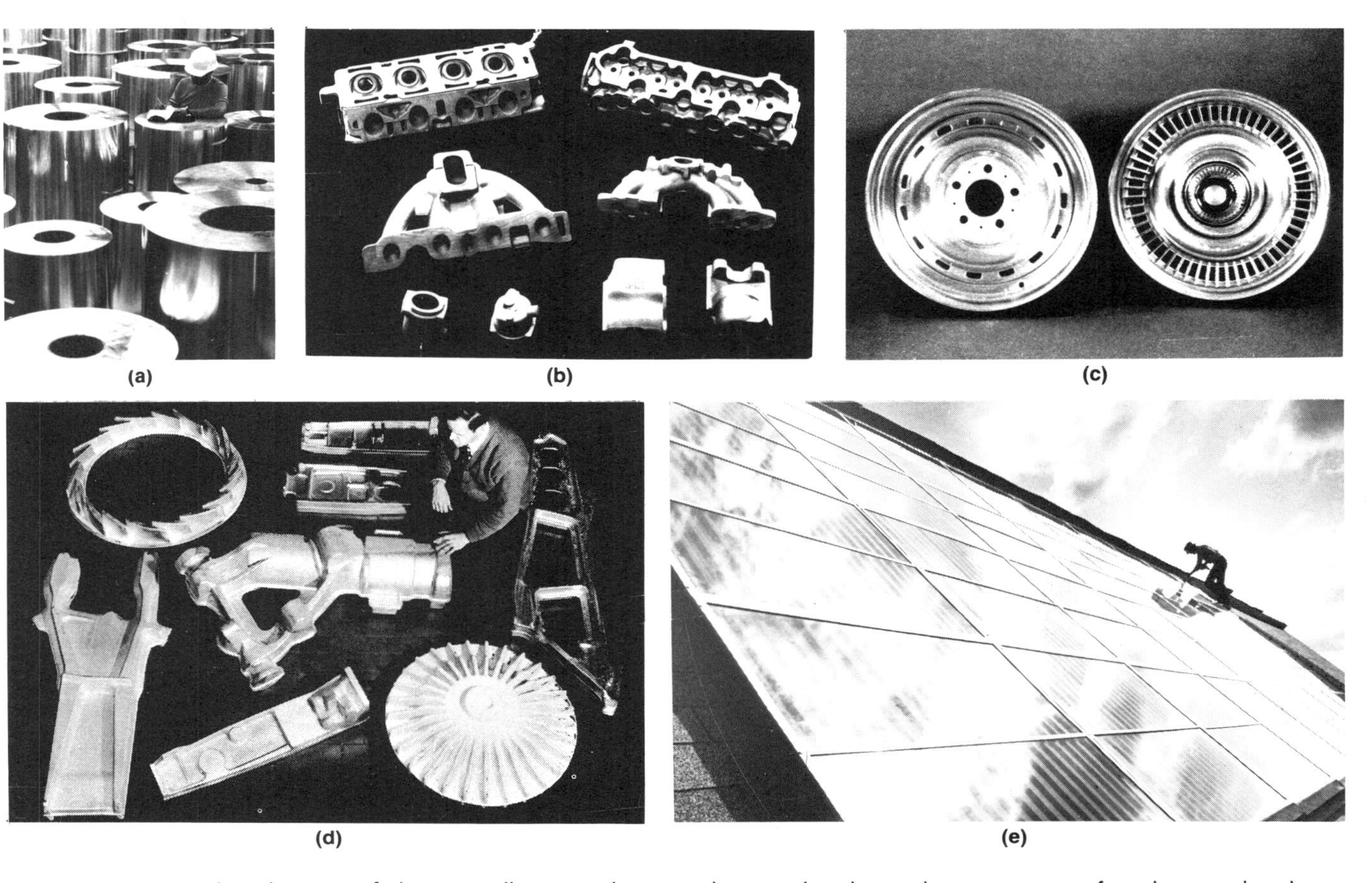

Fig. 4-28 Typical applications of aluminum alloys. (a) Aluminum sheet ready to be made into a variety of products, such as beverage cans, bakeware, and containers. (b) Castings (intake manifolds, cylinder heads, and calipers). (c) Production sheet aluminum wheel with and without wheel cover. (d) Forgings for aerospace applications. (e) Architectural applications, such as the window frames and solar collectors pictured

Representative Treatments. The principal strengthening treatment for aluminum is cold working; another is solution heat treatment and precipitation hardening. For example, additions of 4 to 8% zinc and 1 to 3% magnesium enhance the development of excellent precipitation-hardening characteristics in the 7000 series of alloys. In *tempering,* a metal that has been hardened is reheated to a specific temperature to decrease hardness and increase toughness. The most commonly used temper is T6, which provides good mechanical properties as well as enhanced machinability and corrosion resistance. When resistance to stress-corrosion cracking (scc) is required, the T7 temper is employed. Artificial aging times and temperatures that are higher than those used for maximum hardness produce the T7 tempers.

Some of the major objectives for improvement of aluminum alloys include enhancement of strength, toughness, and corrosion resistance. Wrought aluminum powder metallurgy (P/M) alloys have been developed to attain these goals. For example, the fracture toughness of P/M aluminum alloys has been observed to be superior to that of many 7000 series alloys produced by ingot metallurgy (I/M). Powder metallurgy forgings can be used in parts currently produced by conventional aluminum forging, including connecting rods, pistons, gears, and other automotive structures.

To enhance cost and strength-to-weight characteristics, AISI 434 stainless steel has been clad to aluminum alloy 5052 for use in aircraft firewalls. *Cladding* is essentially a combination of thin layers of different metals. The production of clad metal strip by continuous roll bonding of two or more metallic strips is based on solid-state welding.

Representative Applications. Major applications for aluminum are in aircraft. Alloy 7075-T73 has been used for large structural forgings on the McDonnell Douglas DC-10 because of its resistance to SCC and fatigue crack propagation, and because of its high fracture toughness. Alloy 6061 has been used in vertical takeoff and landing aircraft for low-pressure hydraulic tubing and rotor blades because of its good corrosion resistance. This alloy is also useful in master cylinder pistons, valve parts, and aircraft wheels. Typical applications are illustrated in Fig. 4-28.

Sources of Aluminum. The principal source of aluminum is the mineral bauxite, from which aluminum oxide is extracted and prepared for the smelter by crushing, grinding, chemical processing, and calcination. Primary aluminum contains iron and silicon as major impurities. Iron contents may vary from 0.05 to 0.6% and silicon from 0.04 to 0.03%. Additionally, very small amounts of many other elements are present as impurities.

Secondary Aluminum. Aluminum recovered from scrap (secondary aluminum) has been an important contributor to the total metal supply for many years. New scrap is defined as that generated by plants making end products; old scrap is that recovered from metal that has been previously used by consumers. In addition, considerable amounts of scrap generated at various stages of mill processing are recycled. The energy required to remelt secondary aluminum preparatory to fabrication for reuse is only 5% of that required to produce new aluminum. One type of shredding equipment is shown in Fig. 4-29.

Removal of Impurities. Smelter grade is refined to remove impurity elements that degrade electrical conductivity, bright finishing capabil-

Fig. 4-29 Equipment used to shred aluminum cans for recycling

ity, corrosion resistance, fabricability, or electrochemical characteristics. In addition, several million pounds of super-purity aluminum (99.9% minimum) are used annually, principally for manufacture of electrolytic capacitor foil and for increasing the purity of alloys used in bright-finish applications.

Copper and Its Alloys

Copper and its alloys constitute one of the major groups of commercial metals. They are widely used because of their excellent electrical and thermal conductivity, outstanding resistance to corrosion, and ease of fabrication, together with good strength and fatigue resistance. Generally they are nonmagnetic and can be readily joined by soldering and brazing, and many coppers and copper alloys can be welded by various gas, arc, and resistance methods. For decorative parts, standard alloys having specific colors are readily available. Copper alloys can be polished and buffed to almost any desired texture and luster. Examples of products manufactured from copper and its alloys are pictured in Fig. 4-30.

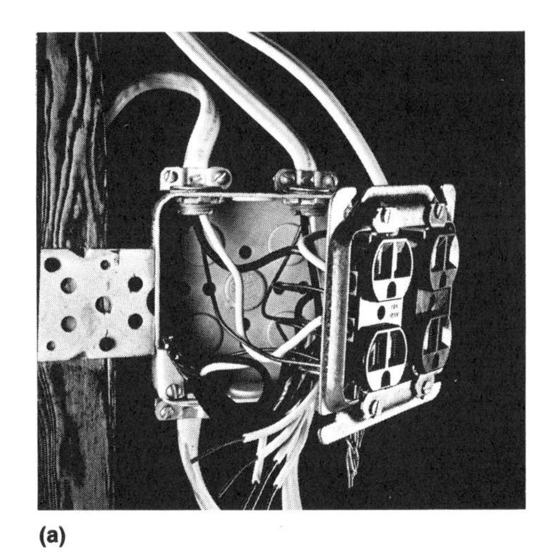

(a)

(c)

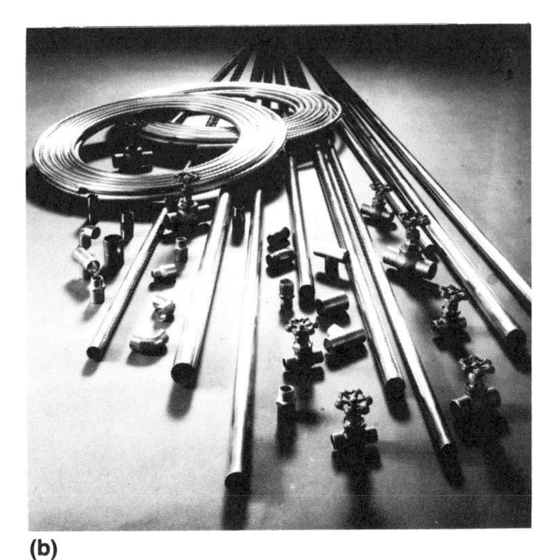

(b)

(d)

Fig. 4-30 Typical applications of copper and copper alloys. (a) Building wiring, which is the number one application of copper in the United States in terms of consumption. (b) Plumbing products rank a close second in terms of consumption of copper alloys. (c) This ship propeller is an example of a large copper-base alloy casting. (d) The Statue of Liberty is a familiar example of the architectural/decorative uses of copper.

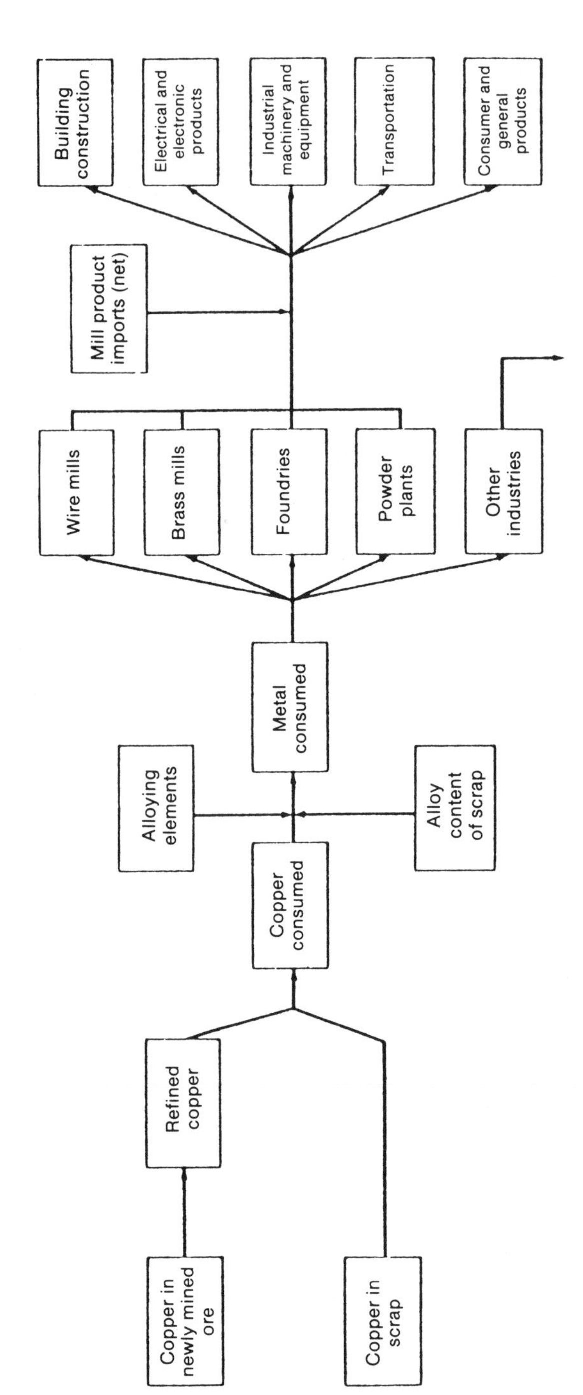

Fig. 4-31 Flow of copper from mine production through end use

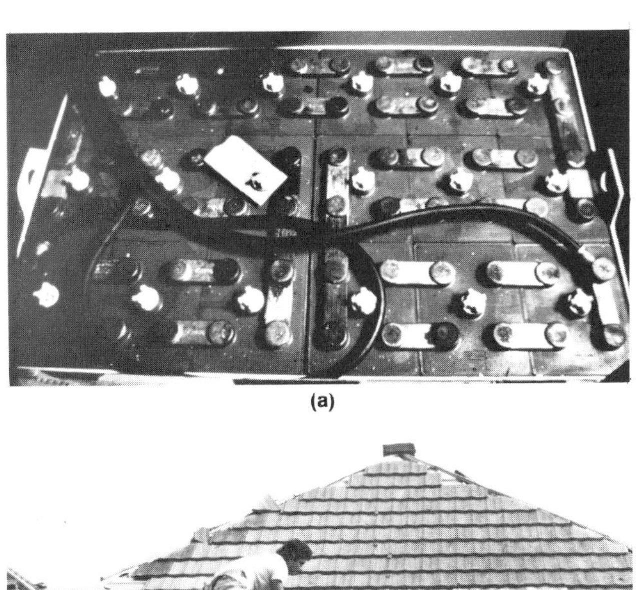

(a)

(b)

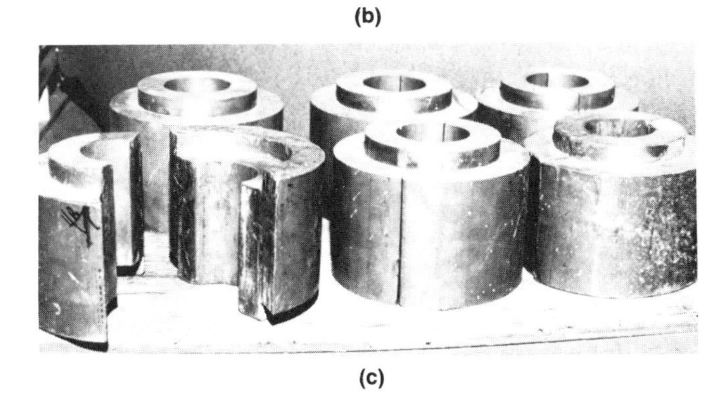

(c)

Fig. 4-32 Typical applications of lead alloys. (a) The lead acid battery is the portable power source of choice to fill automotive electrical needs, to power a variety of industrial vehicles, and to provide uninterruptible emergency voltage during power outages. (b) The construction industry uses lead sheet as a durable, malleable roofing and waterproofing material. (c) Cast lead components are deployed by the nuclear industry to shield against radiation and to clad cooling pipes in nuclear power plants.

Representative Uses. Pure copper is used extensively for cables and wires, electrical contacts, and a wide variety of other parts that are required to pass electrical current. Coppers and certain brasses, bronzes, and cupronickels are used extensively for automobile radiators, heat exchangers, home heating systems, panels for absorbing solar energy, and various other applications requiring rapid conduction of heat across or along a metal section. Because of their outstanding ability to resist corrosion, coppers, brasses, some bronzes, and cupronickels are used for pipes, valves, and fittings in systems carrying potable water, process water, or other aqueous fluids.

Representative Properties. Selection criteria include good resistance to corrosion, good electrical conductivity, good thermal conductivity, color, and ease of fabrication coupled with strength, resistance to fatigue, and ability to take a good finish.

Corrosion Resistance. Copper is a noble metal but, unlike gold and other precious metals, can be attacked by common reagents and environments.

Electrical and Thermal Conductivity. Copper and its alloys are relatively good conductors of electricity and heat. In fact, copper is used for these purposes more often than any other metal. Alloying invariably decreases electrical conductivity and, to a lesser extent, thermal conductivity. For this reason, coppers and high-copper alloys are preferred over copper alloys containing more than a few percent total alloy content when high electrical or thermal conductivity is required.

Color. Copper and certain copper alloys are used for decorative purposes alone, or when a particular color and finish are combined with a desirable mechanical or physical property of the alloy.

Ease of Fabrication. Copper and its alloys are generally capable of being shaped to the required form and dimensions by any of the common fabricating processes. Copper metals can be polished, textured, plated, or coated to provide a wide variety of functional or decorative surfaces.

Copper and copper alloys are readily assembled by any of the various mechanical or bonding processes commonly used to join metal components. Crimping, staking, riveting, and bolting are mechanical means of maintaining joint integrity. Soldering, brazing, and welding are the most widely used processes for joining copper metals. Selection of the best joining process is governed by service requirements, joint configuration, component thickness, and alloy composition(s).

Copper Alloys. The most common way to catalog copper and its alloys is to divide them into six families: coppers, dilute copper alloys, brasses, bronzes, copper nickels, and nickel silvers. The first family, the coppers, is essentially commercially pure copper, which ordinarily is soft and ductile and contains less than about 0.7% total impurities. The dilute copper alloys contain small amounts of various alloying elements that modify one or more of the basic properties of copper.

Representative Applications. Of the chief customer industries, the largest is building construction, which purchases large quantities of electrical wire, tubing, and parts for builder hardware and for electrical, plumbing, heating, and air-conditioning systems. Next are electrical and electronic products, including those for telecommunications, electronics, wiring devices, electric motors, and power utilities. The industrial machinery and equipment category includes industrial valves and fittings; industrial, chemical, and marine heat exchangers; and various other types of heavy equipment, off-road vehicles, and machine tools. Transportation applications include road vehicles, railroad equipment, and aircraft parts; automobile radiators and wiring harnesses are the most important products in this category. Finally, consumer and general products include electrical appliances, fasteners, ordnance, coinage, and jewelry.

In the three categories that account for the greatest usage of wrought copper—telecommunications, automotive, and plumbing and heating—a continuing effort has been made to conserve materials and to manufacture products more efficiently. Most often this effort involves redesign of components and is accomplished through reductions in material gage. In some instances, such as small motors for appliances and other devices, the trend is toward using more copper for each unit to increase the energy efficiency of the end product. Figure 4-31 shows the flow of copper from mine production through end use.

Lead and Its Alloys

The properties of lead—high density, low melting point, corrosion resistance, malleability, unusual electrical properties, and the ability to form useful alloys and chemical compounds—combined with its readily available forms and relative low cost, make it a unique material for solving a variety of problems.

Alloys of Lead. Lead forms a wide range of low-melting alloys and readily alloys with tin in all proportions, forming the tin-lead solders used widely in industry. Alloyed with antimony or calcium, lead is used as both castings and sheet in automotive and standby storage batteries. Bearing alloys include lead alloyed with

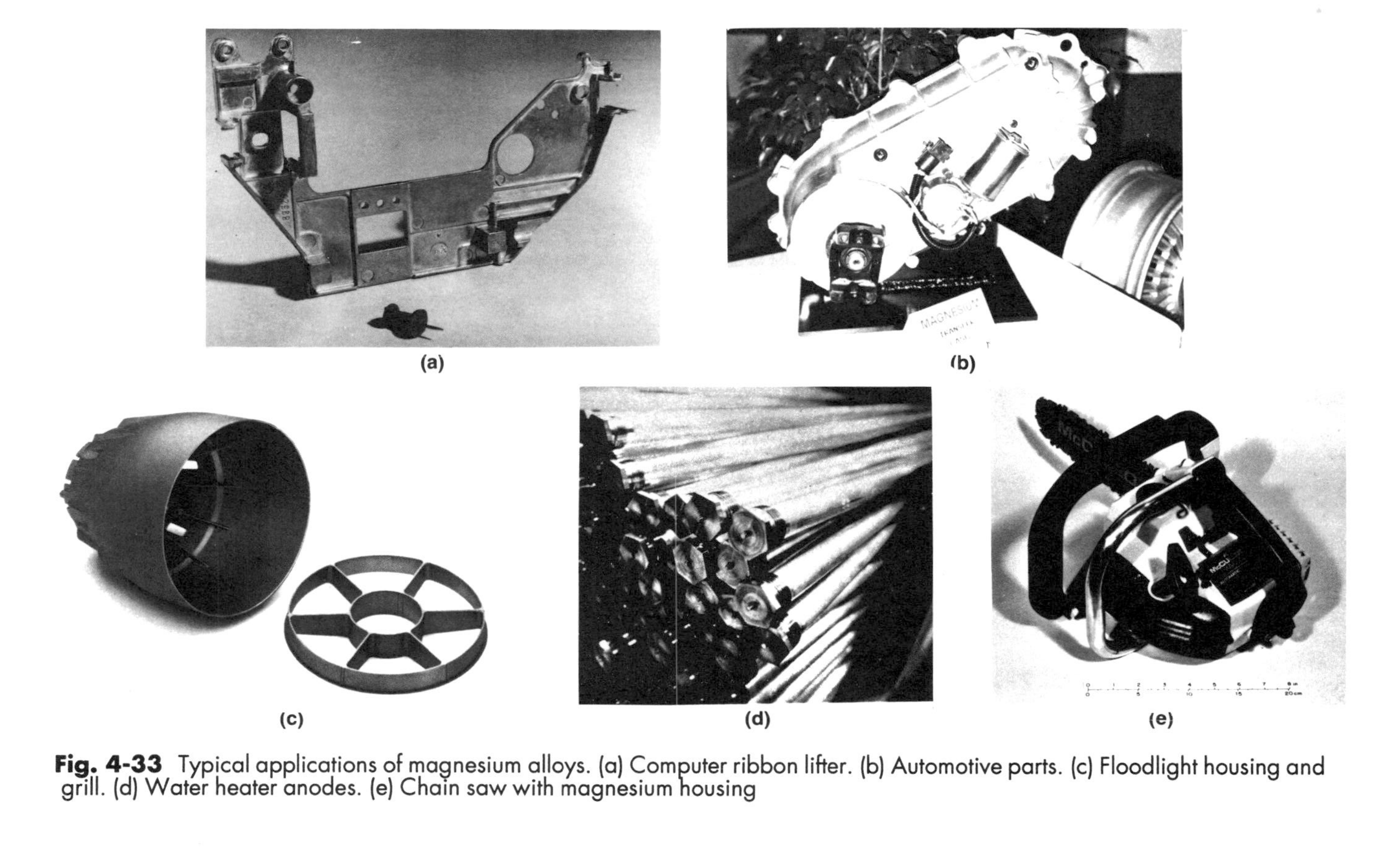

Fig. 4-33 Typical applications of magnesium alloys. (a) Computer ribbon lifter. (b) Automotive parts. (c) Floodlight housing and grill. (d) Water heater anodes. (e) Chain saw with magnesium housing

combinations of antimony, arsenic, and copper to yield suitable hardness and *embeddability* (the ability of a bearing material to embed harmful foreign particles and reduce their tendency to cause scoring or abrasion). Type metals are alloys containing antimony, tin, and arsenic, with excellent casting and hardness as required by the graphic trades. When lead is added to other alloys, such as steel, brass, and bronze, it promotes machinability, corrosion resistance, or other special properties.

In addition to its physical and mechanical properties, the chemical properties of lead, mainly corrosion resistance, account for many of its uses. Lead is durable under varying weather conditions, exposure to most types of soil and atmospheres (including marine and industrial), and the action of many corrosive chemicals. Its resistance to sulfuric acid is used advantageously in the manufacture of the acid and in the most common method of storing electricity—the storage battery. Typical applications of lead alloys are pictured in Fig. 4-32.

Forms of Lead. The malleability of lead (and most of its alloys) allows it to be rolled to any desired thickness. Lead is extruded in the forms of pipe, rod, wire, and practically any cross section, such as H-shape window frames, hollow stars, rectangular ducts, and so on. Common flux-cored solders and collapsible tubes are typical of the variety of lead and lead alloy extrusions.

The low melting point of lead (327 °C, or 621 °F) makes it one of the simplest metals to cast at about 370 °C (700 °F). It is used in massive counterweights, in sailboat keels, and as tiny die castings in instruments. Type metal is noted for its ability to produce fine detail, and storage battery metal grids are examples of commercial lead castings.

Lead shot is produced by taking advantage of the fact that the surface tension of lead is such that when molten lead is poured through a sieve and allowed to free-fall, it forms perfect spheres before solidifying. The size of the shot is controlled by the sieve size.

Lead powder, particles, and flakes are produced, usually by atomization, in diameters of 4 microns and up. The particles impart useful properties when added to grease and pipe-joint compounds.

Lead wool, a loose rope of fibers, is produced by passing molten lead through a fine sieve and allowing it to solidify. When forced into a crevice under considerable force, the fibrous rope cold welds into a homogeneous mass, forming a solid metal seal. This caulking process is useful where temperature or explosion hazards prohibit the use of flame heating. The gaps between lead shielding sheets in nuclear submarines are often filled and caulked with lead wool.

Tin-Lead Solder Alloys. Tin-lead alloys are the most widely used of all solders. They have the advantage of a low melting range, which makes them ideal for joining most metals by convenient heating methods with little or no damage to heat-sensitive parts.

A pure metal always melts at a single temperature. Most solder alloys melt over a range of temperatures. The highest temperature at which a solder is completely solid is called the *solidus*. The temperature at which it is completely molten is the *liquidus*. Between these temperatures, part of the solder is molten and part is solid; thus, the solder has a pasty consistency.

Care should be taken in specifying the correct solder for the job, since each alloy is unique with regard to its composition and, in general, its properties. When referring to tin-lead solders, the tin content is customarily given first—for example, 40/60 refers to 40% tin and 60% lead by weight.

Solders containing less than 5% tin are used for sealing precoated containers, for coating and joining metals, and for applications where service temperatures exceed 120 °C (250 °F). At those temperatures, strength is taken care of by design, and the solder functions primarily as a seal. The 10/90, 15/85, and 20/80 solders are used for sealing cellular automobile radiators and for filling seams and dents in automobile bodies.

General-purpose solders are 40/60 and 50/50. Soldering of automobile radiator cores, plumbing, electrical, and electronic connections, roofing seams, and heating units are but a few of the typical uses for these solders.

The 60/40 and 63/37 alloys are used where components are heat sensitive and where minimum heat should be used to make a solder joint. These alloys also provide the greatest ease and speed of joining. Electronic devices, computers, and communications equipment are typical products using these solders.

For the electronics industry, silver is added to tin-lead solders to reduce the dissolution of silver from silver alloy coatings. Silver may also be added to improve creep resistance. Tin-silver-lead alloys exhibit good tensile, creep, and shear strengths. Some are used for higher-temperature bonds in sequential soldering operations. Fatigue properties are increased by the addition of silver to the solder. The 1Sn-1.5Ag-97.5Pb solder is used in cryogenic equipment because it does not embrittle at low temperatures.

Sources of Lead. Lead occurs in widespread deposits throughout the world. Although not considered a rare metal, it makes up only 0.0016% of the earth's crust. Lead ores do, however, tend to be concentrated in pockets, which makes mining of such ores economical.

Magnesium and Its Alloys

Magnesium is the lightest of the commercially available metals and has a high strength-to-weight ratio when alloyed with other elements; typical materials alloyed with magnesium are aluminum and zinc. Because magnesium is about two-thirds the weight of aluminum, magnesium alloys have higher strength-to-weight ratios than many aluminum alloys, making it useful in aerospace structures.

Important properties of magnesium and its alloys include:

- Resistance to alkalis, oils, solvents, some acids, and most organic chemicals under ordinary atmospheric conditions
- Ease of machinability and fabrication by most metalworking processes, such as welding
- Good static fatigue and ductility characteristics
- Good toughness, impact resistance, and thermal diffusivity

Representative Applications. Magnesium and magnesium alloys are used in a wide variety of applications. Structural applications include industrial, materials handling, commercial, and aerospace equipment. In industrial machinery, such as textile and printing machines, magnesium alloys are used for parts that operate at high speeds and thus must be lightweight to minimize inertial forces. Materials-handling equipment includes dockboards, grain shovels, and gravity conveyors. Commercial applications include luggage and ladders. Good strength and stiffness at both room and elevated temperatures combined with light weight make magnesium alloys especially valuable for aerospace applications.

Magnesium is also employed in various nonstructural applications. It is used as an alloying element in alloys of aluminum, zinc, lead, and other nonferrous metals. Typical products manufactured from magnesium alloys are pictured in Fig. 4-33.

Alloying Elements. Although the mechanical properties of magnesium are relatively low, additions of alloying elements such as aluminum, zinc, manganese, and, for special purposes, tin, cerium, thorium, beryllium, zirconium, and lithium greatly improve these properties. Copper, iron, and nickel are considered impurities and must be kept to a minimum to ensure the best corrosion resistance.

Representative Properties. Corrosion resistance to ordinary outdoor atmosphere has been substantially improved through the use of higher-purity magnesium. Although commercial alloys are reasonably stable in inland atmosphere, protective treatment such as anodizing (a coating) and painting are frequently employed. Salt-bearing atmospheres (as in seacoast locations) are definitely corrosive to magnesium alloys, requiring the use of protective chemical treatment and painting.

Galvanic corrosion is a problem when magnesium and a dissimilar metal such as steel are in contact with each other. In the presence of water, which conducts electricity, the magnesium/steel combination functions like a battery, causing corrosion. One solution is to use metals that are chemically compatible with magnesium, such as aluminum.

A caveat: Magnesium and magnesium alloys are prohibited in components that come into contact with fuel. Water, which is an electrolyte, can accumulate in fuel systems, generating an aggressive corrosion reaction if magnesium or magnesium alloys are present.

Formability of magnesium alloys is good. They can be readily fabricated by most metalworking processes.

Its excellent machinability allows greater speeds to be used for magnesium than for any other structural material. Many magnesium alloys have a machinability rating at least five times that of free-machining brass. Low cutting pressures and high thermal conductivity, with rapid dissipation of heat, produces long tool life, high dimensional accuracy, and excellent surface finish. Tools must be kept sharp, and high rake angles are desirable. Magnesium chips and dust, if properly cared for and not allowed to accumulate, are not a serious fire hazard.

Weldability of alloys is good, and joints have high efficiency. Common welding methods used for joining magnesium include gas tungsten-arc, gas metal-arc, and electrical resistance processes. In addition, brazing, soldering, adhesive bonding, bolting, and riveting may be used for assembly.

Heat treatment followed by water quenching of cast magnesium alloys may cause cracking if the quenching temperature is too high or if the water is too cold. Certain alloys can be aged or heat treated, but strengthening is not attained to the same degree as in aluminum alloys.

Source of Magnesium. The chief source of magnesium is from seawater. Extraction of magnesium from seawater is by electrolytic reduction of magnesium chloride.

(a)

(b)

Fig. 4-34 Typical aerospace applications of titanium alloys. (a) The largest closed-impression die forging of titanium ever produced, a main landing gear support beam for the Boeing 747. Four of these forgings are for a 747, each measuring 620 cm (245 in.) long and weighing approximately 1600 kg (3500 lb). (b) Helicopter rotor hub manufactured from a titanium alloy (Ti-6Al-4V) weighing 250 kg (550 lb) as shipped, with a diameter of 100 cm (40 in.)

Titanium and Its Alloys

Titanium and its alloys are used primarily in two areas of application where the unique characteristics of these metals justify their selection: corrosion-resistant service and strength-efficient structures. For these two diverse areas, selection criteria differ markedly. Corrosion applications normally utilize low-strength unalloyed titanium mill products fabricated into tanks, heat exchangers, or reactor vessels for chemical processing, desalination, or power generation plants. In contrast, high-performance applications typically utilize high-strength titanium alloys in a very selective manner depending on factors such as thermal environment, loading parameters, available product forms, fabrication characteristics, and inspection and/or reliability requirements. As a result of their specialized usage, alloys for high-performance applications normally are processed to more stringent and costly requirements than "unalloyed" titanium for corrosion service.

Historically, titanium alloys have been used instead of iron or nickel alloys in aerospace applications because titanium saves weight in highly loaded components that operate at low to moderately elevated temperatures. Many titanium alloys have been tailored to have optimum tensile, compressive, and/or creep strength at selected temperatures and, at the same time, to have sufficient workability to be fabricated into mill products suitable for specific applications.

Representative Properties. Titanium is a low-density element (approximately 60% of the density of steel) and can be highly strengthened by alloying and deformation processing. Titanium is nonmagnetic and has good heat-transfer properties. Its coefficient of thermal expansion is somewhat lower than that of steels and less than half that of aluminum. Titanium and its alloys have melting points higher than those of steels, but maximum useful temperatures for structural applications generally range from 425 to 540 °C (800 to 1000 °F).

Titanium has the ability to *passivate* its surface (to change from a chemically active to a much less reactive state) and provide a high degree of immunity to attack by most mineral acids and chlorides. Titanium is nontoxic and generally biologically compatible with human tissues and bones.

The combination of high strength, stiffness, good toughness, low density, and good corrosion resistance provided by various titanium alloys at very low to moderately elevated temperatures allows weight savings in aerospace structures and other high-performance applications. Excellent corrosion resistance and biocompatability, coupled with good strength, make titanium and its alloys useful in chemical and petrochemical applications, marine environments, and biomaterials applications. Two aerospace applications are pictured in Fig. 4-34.

Sources of Titanium. Titanium is the fourth most abundant structural metal in the crust of the earth, after aluminum, iron, and magnesium. Development of its alloys and processing technologies started only in the late 1940s. Difficulty in extracting the metal from ores, its high reactivity in the molten state, its forging complexity, machining difficulties, and sensitivity to segregation and inclusions necessitated the development of special processing techniques.

Tin and Its Alloys

The largest single application of tin is in the manufacture of tinplate (steel sheet coated with tin), which accounts for about 40% of total world tin consumption. Since 1940, the traditional hot-dip method of making tinplate has been largely replaced by electrodeposition of tin on continuous strips of rolled steel.

More than 90% of the world production of tinplate is used for containers (Fig. 4-35a). Traditional tinplate cans are made of three pieces of tin-coated steel: two ends and a body with a soldered sideseam. Innovations in can manufacture have produced two-piece cans made by drawing and ironing. Tinplate cans find their most important use in packaging of food and beverage products, but also are used for holding paint, motor oil, disinfectants, detergents, and polishes. Other applications of tinplate include fabrication of signs, filters, batteries, toys, gaskets, and containers for pharmaceuticals, cosmetics, fuels, tobacco, and numerous other commodities.

Electroplating accounts for one of the major uses of tin and tin chemicals. Tin is used in anodes, and tin chemicals are used in formulating various electrolytes for coating a variety of substrates.

Hot-Dip Coatings. Coating of steel with lead-tin alloys produces a material called *terneplate.* Terneplate is easily formed and easily soldered and is used as a roofing and weather-sealing material and in construction of automotive gasoline tanks, signs, radiator header tanks, brackets, chassis, and covers for electronic equipment and sheathing for cable and pipe.

Hot-dip tin coatings are used on wire for component leads as well as food handling and processing equipment. In addition, hot-dip tin coatings are used to provide the bonding layer for babbitting of bearings shells.

Solders account for the second largest use of tin (after tinplate) (Fig. 4-35b). Tin is an important constituent in solders because it wets and adheres to many common base metals at temperatures considerably below their melting points. Tin is alloyed with lead to produce solders with melting points lower than those of either tin or lead. Small amounts of various metals, notably antimony and silver, are added to tin-lead solders to increase their strength. These solders can be used for joints that are subjected to high or even subzero service temperatures. Both solder compositions and applications of joining by soldering are many and varied.

Alloys For Organ Pipes. Tin-lead alloys are used in the manufacture of organ pipes. These materials commonly are named *spotted metal* because they develop large nucleated crystals or "spots" when solidified as strip on casting tables. The pipes that produce the range of tones in organs generally are made of alloys with tin contents varying from 20 to 90% according to the tone required. Broad tones generally are produced by alloys rich in lead; as tin content increases, the tone becomes brighter. Cold-rolled tin-copper-antimony alloys (95% tin) also have been used in the manufacture of pipes, and adoption of these alloys has improved the efficiency and speed of fabrication of finished pipes. This composition provides for a bright surface that is more tarnish resistant than the tin-lead alloys.

Pewter is a tin-base white metal containing antimony and copper. Originally, pewter was

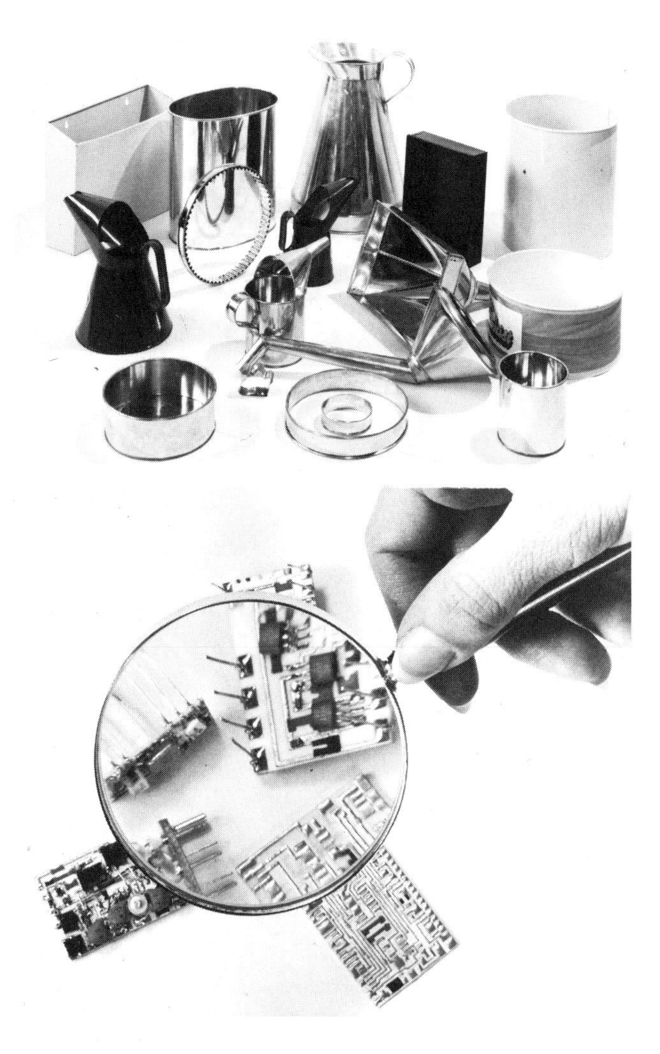

Fig. 4-35 Two principal applications for tin. (a) Tinplate for products such as standard liquid measures and containers. (b) Solders, in this case precision tinning of conductor paths on thick-film circuits

defined as an alloy of tin and lead, but to avoid toxicity and dullness of finish, lead is excluded from modern pewter. These modern compositions contain 1 to 8% antimony and 0.25 to 3.0% copper. Pewter casting alloys usually are lower in copper than pewters used for spinning hollowwares and have greater fluidity at casting temperatures.

Pewter is malleable and ductile and is easily spun or formed into intricate designs and shapes. Pewter parts do not require annealing during fabrication. Much of the costume jewelry

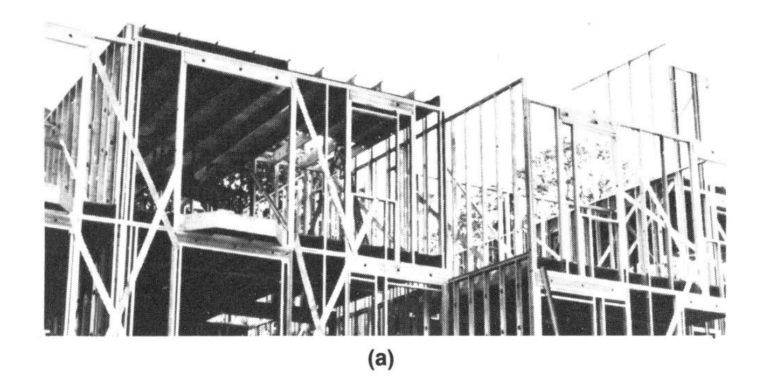

(a)

(b)

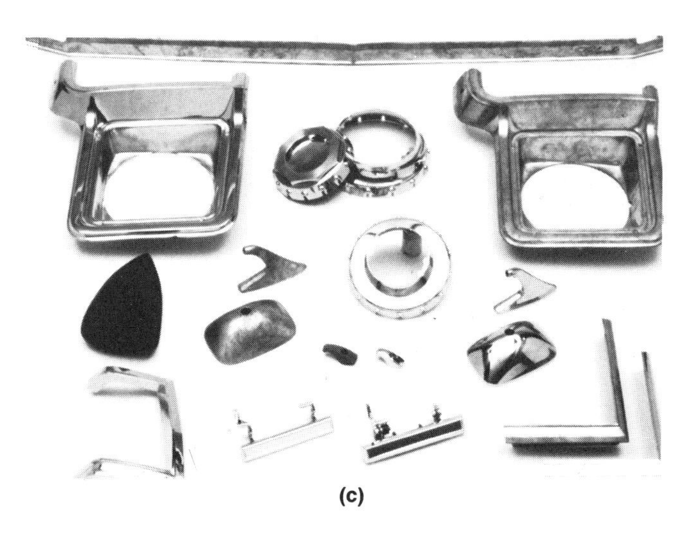

(c)

Fig. 4-36 Typical applications of zinc. (a) Galvanized steel framing for residential and light commercial construction produced on hot-dip continuous galvanizing lines that submerge sheet steel into anticorrosive baths of zinc or zinc alloys. (b) Thicker zinc coatings are applied by hot-dip galvanizing after fabrication on components such as highway guard rails. (c) Zinc die casting is used to mass-produce high-quality parts quickly and efficiently.

produced today is made of pewter alloys centrifugally cast in rubber or silicone molds.

Bearing Materials. Tin has a low coefficient of friction, which is the first consideration in its use as a bearing material. Tin is a structurally weak metal, and when used in bearing applications it is alloyed with copper and antimony for increased hardness, tensile strength, and fatigue resistance. Normally, the quantity of lead in these alloys, called tin-base *babbitts,* is limited to 0.35 to 0.5% to avoid formation of a low-melting-point constituent, which would significantly reduce strength properties at operating temperatures.

Sources of Tin. Primary ore deposits can contain very low percentages of tin (0.01%, for example); therefore, large amounts of soil or rock must be worked to provide recoverable amounts of tin minerals. Unlike ores of other metals, cassiterite is very resistant to chemical and mechanical weathering, but extended erosion of primary lodes by air and water has resulted in deposition of the ore as *eluvial* (formed from rock) and *alluvial* (sedimentary) deposits.

Production of tin ore generally is centered in areas far distant from centers of consumption. The leading tin-producing countries (excluding the former Soviet Union and China) are, in descending order, Malaysia, Bolivia, Indonesia, Thailand, Australia, Zaire, and Nigeria. These countries supply more than 85% of total world production.

Cassiterite, a naturally occurring oxide of tin, is by far the most economically important tin mineral. The bulk of the world's tin ore is obtained from low-grade placer deposits of cassiterite derived from primary ore bodies or veins associated with granites or rocks of granitic composition.

Zinc and Its Alloys

In industrial usage, zinc ranks fourth, behind iron, aluminum, and copper. Slab zinc and zinc oxide are the primary materials in the vast majority of applications.

Zinc has five principal areas of application: in coatings and anodes for corrosion protection of irons and steels; in zinc casting alloys; as an alloying element in copper, aluminum, magnesium, and other alloys; in wrought zinc alloys; and in zinc chemicals. In the corrosion protection category, hot-dip or continuous galvanizing accounts for the majority of zinc consumption. Almost all the zinc used in zinc casting alloys is employed in die casting compositions. Among zinc-containing alloys, copper-base alloys such as brasses are the largest zinc consumers. Rolled zinc is the principal form in which wrought zinc alloys are supplied, although drawn zinc wire for metallizing is increasingly being used. In the zinc chemical category, zinc oxide is the major compound.

Zinc oxide is employed in rubber, paints, ceramics, chemicals, agriculture, photocopying, floor coverings, and coated fabrics and textiles. The largest use for zinc oxide is in rubber products, where it is employed as an activator for the accelerators used to speed the vulcanization process. It is used as a pigment in paints and ceramics, as a soil nutrient in agriculture, and as a stabilizer in plastics, and it provides the photosensitive character for coated papers used in some photocopying techniques. Examples of typical products using zinc or zinc alloys are pictured in Fig. 4-36.

Corrosion Protection. Almost half the zinc consumed in the world is used for protective coatings on irons, mild steels, and low-alloy steels. Zinc corrodes at much lower rates than do steels in atmospheric exposure and will corrode sacrificially when the coated steel is exposed, such as at scratches or cut ends. Zinc anodes are also used to provide galvanic sacrificial protection in underwater and underground applications.

Zinc Coatings. The specific coating techniques by which zinc is applied to provide corrosion protection include postfabrication hot-dip galvanizing, continuous hot-dip galvanizing, electrogalvanizing, electroplating, metallizing, zinc dust/zinc oxide painting, and mechanical plating/sherardizing.

Sources of Zinc. Zincite, a native zinc oxide, usually occurs in massive or granular form. Zinc is a bluish white crystalline element that is brittle when cold and malleable at 110 to 210 °C (230 to 410 °F).

Chapter 5

Heat Treatment of Steel

Chapters 3 and 4 outlined the properties of ferrous and nonferrous metals and their alloys and the impact of those properties on the selection of candidate materials for a given application. In this chapter and the next, the spotlight is on how for a given application the metallurgist is able to manipulate the properties of these metals and alloys through heat treatment.

Some of the Basics

Steel products are available in a wide variety of mechanical properties, ranging from very soft and ductile to extremely hard and wear resistant. Such properties are made to order by changing the microstructure of steel through a number of heat treating processes.

Heat treatment is defined as controlled heating and cooling of a solid metal or alloy by methods designed to obtain specific properties by changing the microstructure. Notice in particular that heat treating takes place below the melting point of the metal and that the changes in microstructure take place inside the solid metal. Further, changes in microstructure are due to the movement of atoms within crystal lattices in response to heating or cooling over a period of time. Temperatures at which these changes take place are called *transformation temperatures*. The ability to tailor properties by heat treatment has contributed greatly to the usefulness of metals and their alloys in a potpourri of applications, ranging from sheet metal for cars and aircraft to razor blades and hardware steels.

Almost all metals respond to some form of heat treatment in the broadest sense of the definition, but the responses of individual metals and alloys are by no means the same. Almost any metal can be softened or annealed after cold working—another method of altering properties—by means of a suitable heating and cooling cycle, but only a few alloys can be strengthened or hardened by heat treatment.

Practically all steels can be strengthened by heat treatment. Many alloys of aluminum, copper, nickel, magnesium, and titanium can also be strengthened to different degrees by heat treatment, but not to the same degree or by the same techniques used to treat steel. Changing the properties of these nonferrous alloys will be discussed in Chapter 6.

Heat treating processes include annealing, normalizing, tempering, stress relieving, and surface hardening. *Quenching,* or cooling from a higher temperature, is an integral part of the heat treating process. Cooling media include water, oil, salt baths, and water solutions of polymers. Parts also may be cooled in air, with or without a fan, depending on requirements.

Brief Description of Processes

Annealing is generally used to soften steels or to bring about required changes in properties, such as machinability, mechanical or electrical properties, or dimensional stability. When the sole purpose is to relieve stresses in a part, the process is called *stress-relief annealing*.

Normalizing is typically used to refine the grain structure (i.e., reduce the grain size) of steels that have been subjected to high temperatures during forging or other hot work operations that increase grain size. Steel is heated above a given temperature, then cooled in air.

Tempering. A steel that has been hardened is reheated to reduce hardness and to increase toughness so that it has resistance to impact, which means that the steel is able to bend without breaking. In broader terms, the process is used to manipulate properties. Aluminum alloys, for example, have a number of different tempers. For example, the T2 temper applies to parts that are cold worked to improve their strength after cooling from an elevated-temperature shaping process.

Stress Relieving. As in tempering, this process is carried out by heating below the transformation temperature of a steel. Its purpose is to relieve residual stresses put into a part during forming, rolling, casting, machining, welding, or straightening. Stress relieving is a time- and temperature-dependent operation, followed by cooling at a relatively slow rate in order to avoid the creation of new stresses.

Case hardening is a generic term used for several processes that put a hard, wear-resistant surface layer on a part. The area below the surface, known as the *core,* is softer and/or tougher, by comparison. Commonly used processes include carbonizing, carbonitriding, induction hardening, flame hardening, nitriding, and nitrocarburizing. Both heating and controlled cooling are involved.

Iron + Carbon = Steel

Iron, by itself, has limited structural usefulness. However, in combination with carbon it makes steel. Steel is called the world's unique structural material.

Fundamentally, all steels are alloys of iron, carbon, manganese, silicon, and phosphorus. Sulfur and phosphorus are present as residual elements in the raw materials used in steelmaking. A typical analysis or composition of plain carbon steel is 0.03 to 1.6% carbon, 0.25% silicon, 0.5% manganese, 0.04% sulfur, and 0.05% phosphorus. Other alloy steels, by comparison, contain other elements—ranging from chromium and aluminum to nickel and molybdenum.

Of particular importance is the fact that the carbon atom is much smaller than the iron atom, and thus is able to fit within the interstices or open spaces between iron atoms, as shown by the small dark sphere or carbon atom in the face-centered cubic (fcc) lattice in Fig. 5-1. Ferrous metallurgists are involved primarily with the fcc and body-centered cubic (bcc) structures (Fig. 5-2). The bcc structure has one atom at each corner of the cube and one in the center (the dark sphere in this instance). Heat spurs the movement of atoms, and the atoms are rearranged within the grain (lattice) when steel or iron is heated through the temperature at which changes in crystal structure occur. These temperatures are commonly known as *transformation temperatures.*

The shifting of atoms is referred to as an *allotropic* change. Iron is an allotropic element, which means it can exist in more than one crystalline form, depending on temperature. Allotropy is one of the most important facts in ferrous metallurgy. The science of heat treating steel depends largely on the allotropy of iron and differences in the size of iron and carbon atoms.

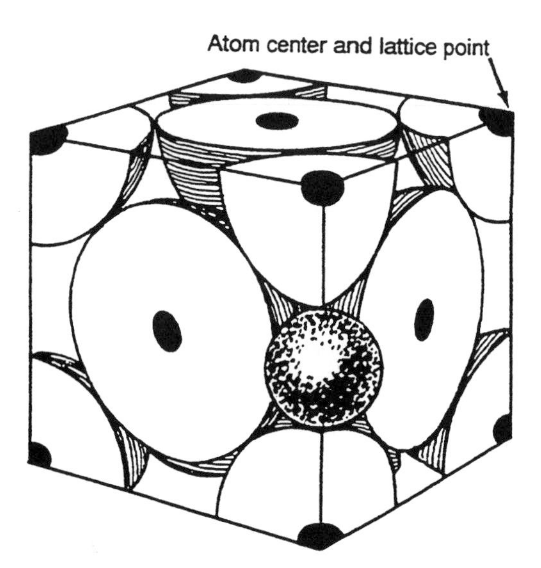

Fig. 5-1 Face-centered cubic structure

Allotropic Changes in Pure Iron

Changes in the structure of iron that take place as the metal transforms from liquid to solid are shown schematically in Fig. 5-3. At 1540 °C (2800 °F) molten iron starts to solidify (called *freezing*), and no change in temperature occurs until the iron is completely solid, which accounts for the horizontal jog in the cooling curve (point A in Fig. 5-3). Note that after the iron is solid, its temperature falls again and continues to do so at a uniform rate until 1395 °C (2540 °F) is reached.

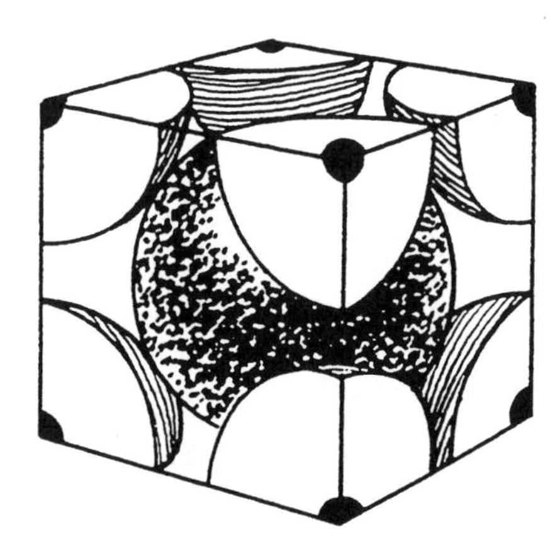

Fig. 5-2 Body-centered cubic structure

Between 1540 and 1395 °C (2800 and 2540 °F) the iron that is now solid is known as *delta iron* or *delta ferrite,* and has a bcc structure. The horizontal jog at point B at 1395 °C (2540 °F) indicates that a change takes place in the solid iron. Delta iron changes or transforms to *gamma iron* (fcc).

When the transformation is complete at 1395 °C (2540 °F), the temperature falls again at a uniform rate until it reaches 911 °C (1672 °F) at point C, and the temperature again remains constant briefly. At this temperature, gamma iron (fcc) changes to *alpha iron,* which has a bcc structure. This change is of great significance in the heat treating of steel. The change in shading at 770 °C (1420 °F) at point D is of no great significance. What happens at this point is a change from nonmagnetic iron to magnetic iron.

All the changes or transformations illustrated in Fig. 5-3 represent what happens during very slow cooling. During slow heating, the same changes would take place in the reverse order and at the same pace—a rate of change described as *near equilibrium,* or a condition approaching "no change." Commercial practice is much faster.

More on Carbon and Phases

Carbon is almost insoluble in alpha iron at temperatures below 911 °C (1672 °F), because the interstitial positions available in the bcc lattice are not large enough to accommodate it.

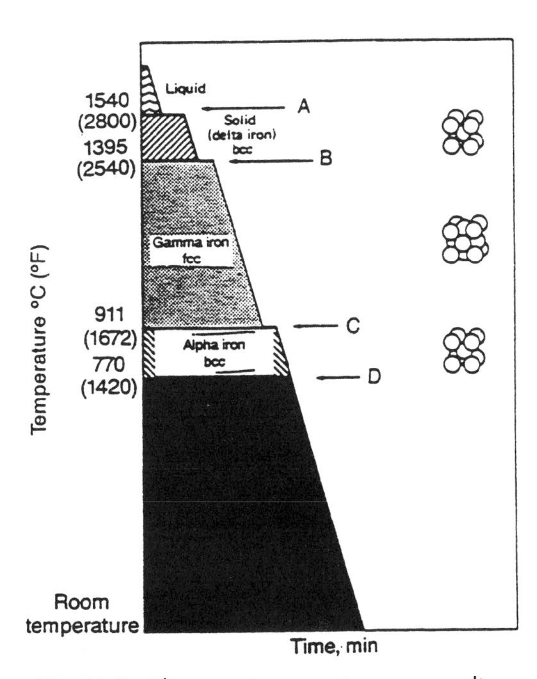

Fig. 5-3 Changes in pure iron on cooling from molten state to room temperature

However, carbon is quite soluble in gamma iron, because the interstices in the fcc lattice are considerably more accommodating (larger) to carbon.

The process by which iron changes from one atomic arrangement to another when heated through 911 °C (1672 °F) is called a *phase transformation.* Transformations of this type occur not only in pure iron but also in many of its alloys. Each alloy composition transforms at its own characteristic temperature. Phase transformations make possible the variety of properties that can be obtained through carefully selected heat treatments.

Phases in iron-carbon alloys are molten alloy, or gamma phase, also known as austenite; alpha or delta phase, also known as ferrite; cementite; and graphite. These phases are discussed in the following section.

Iron-Carbon Phase Diagram

Figure 5-4 is a simplified iron-carbon phase diagram, also known as an iron-cementite phase diagram. Selected temperatures run vertically, on the left. Selected percentages of carbon are shown horizontally, left to right, at the bottom.

Some of these terms were used in Fig. 5-3: delta iron (bcc), gamma iron (fcc), and alpha iron (bcc). In Fig. 5-4, however, gamma iron is called *austenite,* a new term. Other new terms are *ferrite,* which is bcc alpha iron, and *cementite,* which is a chemical compound of iron and carbon, also known as carbide or Fe_3C. Not shown in the phase diagram are three other new terms—*pearlite, bainite,* and *martensite.* These will be discussed later in this chapter.

First, a guided tour of points of interest in the phase diagram, starting at the bottom of Fig. 5-4 and reading from left to right. Note that the carbon ranges from 0.008% on the extreme left to 6.69% on the extreme right. Also note that the carbon range for steel runs from 0 to 2.0%; and the carbon range for cast iron runs from 0.77 to 6.69%.

Temperatures, listed vertically on the left, range from room temperature to 1540 °C (2800 °F), which is the melting point of steel. Keep in mind that from room temperature to 1040 °C (1900 °F) the steel is solid. Melting starts (mushy stage first) at 1050 °C (1920 °F). The top temperature for heat treating is generally 1040 °C (1900 °F).

At the far left, the long skinny area extending from room temperature to 911 °C (1672 °F) shows the limited solubility of carbon in alpha ferrite at this range of temperatures: 0.008% at room temperature to a maximum of 0.025% at 725 °C (1340 °F). A photomicrograph of commercially pure iron is shown in Fig. 5-5.

The large area to the right of alpha ferrite is identified as alpha ferrite and cementite. The temperature range is room temperature to 725 °C (1340 °F); the carbon range is 0.008 to 6.69%. Alternate layers of ferrite and cementite are shown in Fig. 5-6.

Back to the left is a triangle labeled austenite and ferrite. Here the temperature range is 725 to 911 °C (1340 to 1670 °F); the carbon range is 0.025 to 0.77%.

The large area labeled gamma austenite starts at 0.77% carbon and 725 °C (1340 °F) and extends laterally to 2.0% carbon at about 1150 °C (2100 °F), then swings back to the left to a temperature of 1490 °C (2715 °F), which is approaching the melting point. The area down and to the left is located at a temperature of 1395 °C (2540 °F). Carbon content is close to zero.

The two triangles located to the top left of the austenite area are identified as delta iron, and delta iron and austenite. Both are above the heat treating temperature and near the temperature at which steel begins to transform from solid to liquid. In both instances, carbon content is less than 0.5%.

In the remaining areas of the phase diagram, steel is in the interim stages of changing from

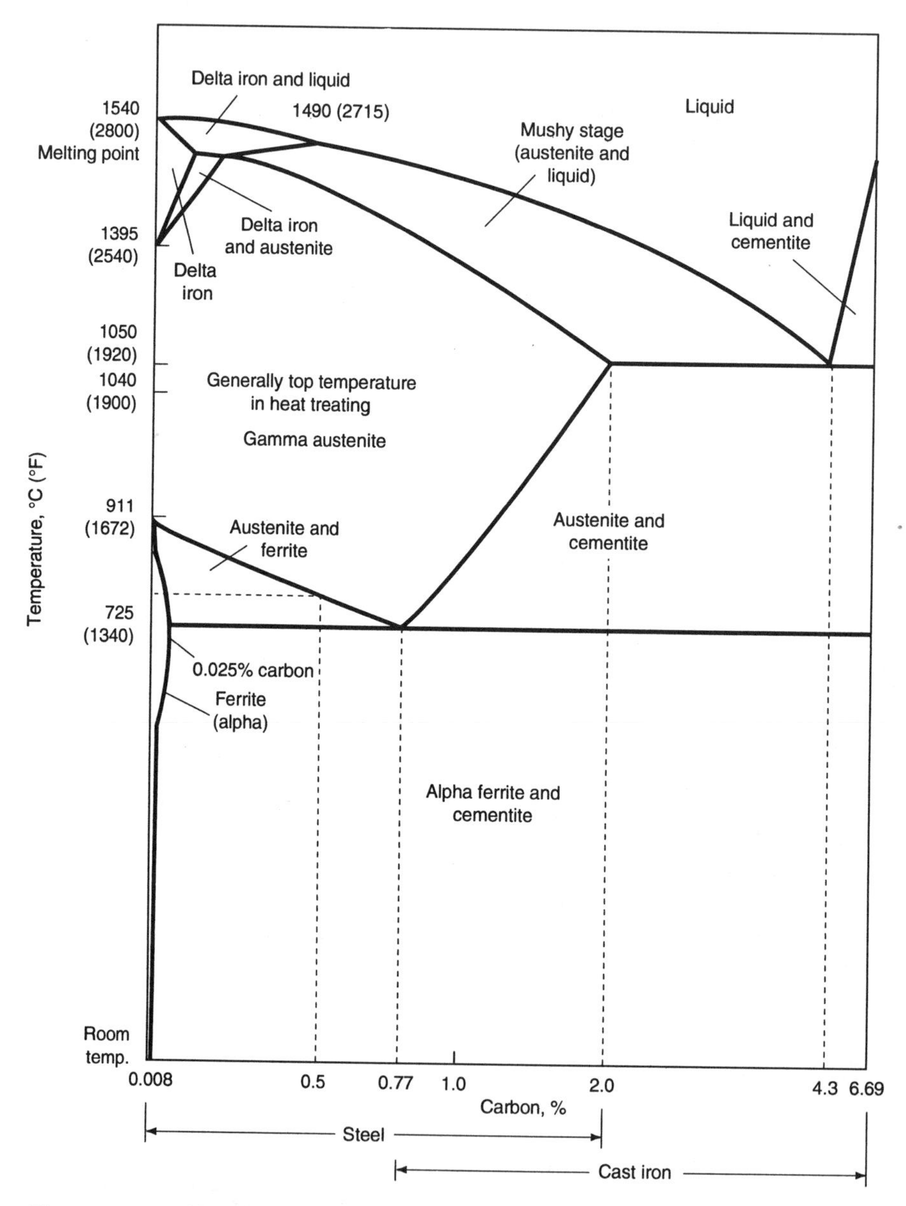

Fig. 5-4 Simplified iron-carbon phase diagram

solid to liquid, and finally to 100% liquid. In the transition condition, phases coexist (from right to left, cementite-liquid, austenite-liquid, and delta iron-liquid). Transition temperatures range from about 1150 to 1540 °C (2100 to 2800 °F). Note that in the carbon range of 4.3% the transition from solid to liquid is completed at about 1150 °C (2100 °F), while in the 0% carbon range the transition is completed at 1540 °C (2800 °F).

A caveat: The data in Fig. 5-4 are based on near-equilibrium conditions and do not represent commercial practice.

Transformation of Austenite to Pearlite

Suppose that a steel containing 0.5% carbon is heated to 815 °C (1500 °F) (check this carbon content and temperature in Fig. 5-4). At this point 0.5% carbon is dissolved in the interstices, or open spaces between iron atoms, in the gamma phase (fcc crystal).

Next, say slow cooling is started when the temperature of the solid metal drops below about 790 °C (1450 °F) and the alloy starts to transform to the alpha phase (bcc). As the temperature continues to fall, more ferrite forms and austenite increases in carbon content just above 725 °C (1340 °F). Austenite now contains about 0.77% carbon (see Fig. 5-4) and transforms to pearlite.

The formation of pearlite from austenite is a nucleation and growth process. It takes time for the atoms to form a nucleus and subsequently grow. When austenite is cooled below 725 °C (1340 °F), it is unstable and has a very strong tendency to change or transform into pearlite. The more cooling proceeds, the stronger the tendency to form pearlite.

Transformation to Bainite

At temperatures below 565 °C (1050 °F), the mobility of carbon atoms in austenite slows, retarding the transformation of austenite to pearlite. When the point is reached where not enough carbon atoms are available to support the transformation to pearlite, a new product, called *bainite,* is formed. This structure varies from a fine mixture of ferrite and cementite to needles or plates of ferrite, and cementite is present but not visible. The plate-like structure of bainite is shown in Fig. 5-7.

Transformation to Martensite

After an iron-carbon alloy is transformed from the gamma phase to the alpha phase (austenite to ferrite) and is cooled rapidly to below 275 °C (530 °F), another transformation product, called *martensite,* is formed. In this process, carbon atoms are trapped in the ferrite lattice because they do not have time to escape. As a result, the lattice is deformed. As in the case of aging aluminum, this produces a significant increase in strength. This is the major hardening and strengthening mechanism for steel. The structure is a new one, called *body-centered tetragonal (bct)* (Fig. 5-8). Lath martensite in a low-carbon steel is shown in Fig. 5-9.

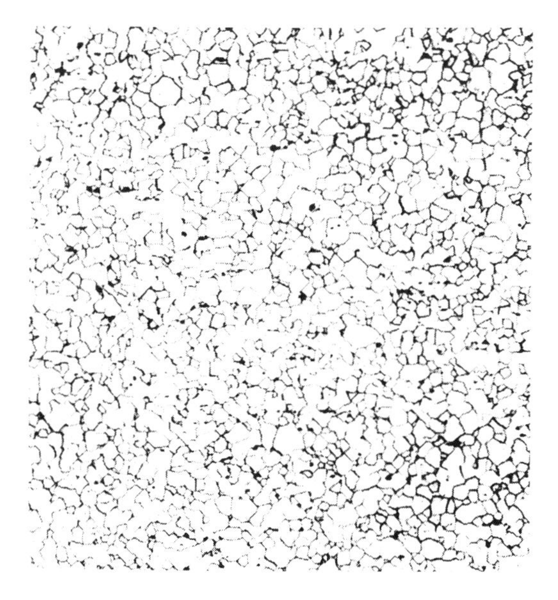

Fig. 5-5 Micrograph of commercially pure iron. 100×

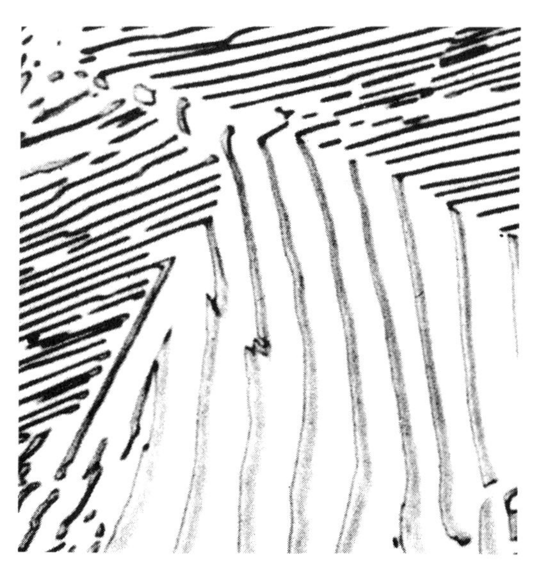

Fig. 5-6 Microstructure of steel, showing alternate layers of ferrite (lighter areas) and cementite (darker areas). 2500×

With steady cooling, the formation of martensite continues until the reactions tops at a temperature called M_f, or martensite finished. If cooling is interrupted, transformation stops.

Properties of Transformation Products

As just demonstrated, steel transforms from austenite to pearlite, to bainite, or to martensite. Hardness, toughness, ductility, and strength depend on carbon content and transformation products, especially the latter.

Hardness can be obtained in steels by controlling the amounts of pearlite, bainite, ferrite, cementite, and martensite. Each contributes to overall hardness, in proportion to the amount present.

Rate of Cooling

Cooling rates at a given starting temperature and type of quenching medium depend on the cross section (mass) of a part. Consider two bars with the same composition but different thicknesses. The thinner bar will always cool faster than the thicker one; this is because it takes more time for both the surface and the center of the thicker part to cool.

Summary, Plus A New Topic

Development of martensite as the as-quenched structure is usually the primary objective in heat treating. There are two main classes of heat treating processes: those that do not involve intentional changes in surface chemistry, and those that do involve such changes. The former group consists of processes that have been discussed in the first part of this chapter: normalizing, annealing, austenitizing for hardening, tempering, and stress relieving.

Normalizing. Steels containing 0.77 to 2.0% carbon are heated above the temperature at which the solution of cementite in austenite is completed, usually about 40 °C (70 °F) above this temperature. Cooling is in air to a temperature well below the temperature transition range.

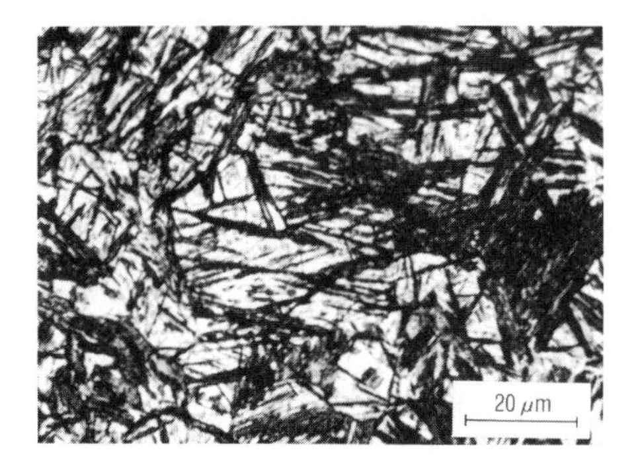

Fig. 5-7 Platelike microstructure of bainite (dark plates) formed in 4150 steel

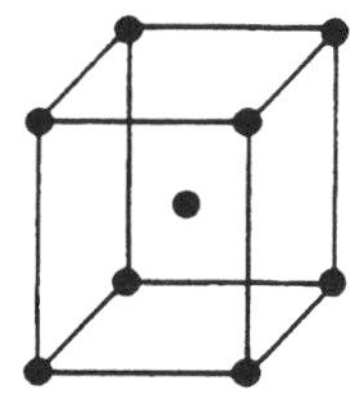

Fig. 5-8 Body-centered tetragonal structure

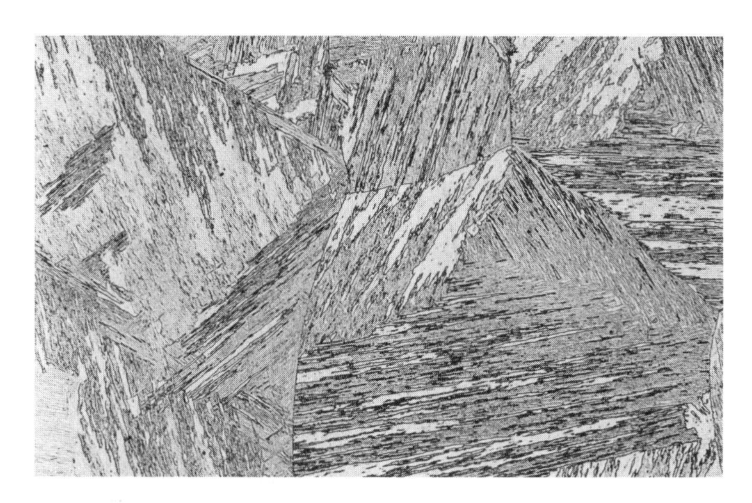

Fig. 5-9 Lath martensite in a low-carbon alloy steel. Etched with 2% nital. 100×

Normalizing is typically used as a conditioning treatment, especially in refining the grain structures of steels that have been subjected to high temperatures during forging or other hot working operations. Normalizing is sometimes followed by a second heat treating operation, such as austenitizing, annealing, or tempering.

Annealing. Softening of steels is the main application. Alternative uses include making changes in other properties, such as machinability, mechanical or electrical properties, or dimensional stability.

The process is called *stress-relief annealing* when the sole purpose is to relieve stresses. This operation is carried out below the temperature at which austenite begins to form during heating. Another application, *in-process annealing,* is used mainly to reduce hardness caused by work hardening, to allow further cold working of a steel.

Austenitizing. Steel is heated at least 40 °C (70 °F) above the temperature at which the transformation of ferrite to austenite is completed. Austenitizing is the first part of a two-step process, the second step being normalizing, annealing, or quench hardening. Also, austenite is a phase that usually exists only at elevated temperatures.

Tempering. All tempering operations are carried out at temperatures below which austenite begins to form during heating. The purpose is to adjust various combinations of mechanical properties.

Stress Relieving. As in tempering, stress relieving is always performed at the temperature below which a steel begins to form austenite during heating. This is a time-dependent operation which is continued long enough to reduce residual stresses to an acceptable level. Temperatures can vary from ambient to about 700 °C (1300 °F). Relatively high temperatures are needed for total relief.

Case Hardening Processes

In these processes—which include carburizing, carbonitriding, cyaniding, nitriding, and nitrocarburizing—changes in surface composition are deliberate. Though all will be described, only the most widely used of the case hardening processes—carburizing—will be treated in some detail.

Case hardening develops hard, wear-resistant surfaces on parts—hence the term *case.* Below the case, parts have softer and/or tougher cores. *Toughness* means resistance to impact; in other words, a part will bend but will not break.

Two approaches are used in case hardening. In the first, the steel already has enough carbon to produce the required hardness on heating and cooling, so the treatment is confined to areas of a part that require hardening. In the second, a steel that is low in hardenability is used. Generally, surface compositions are altered accordingly by processes that include shell hardening, flame hardening, and induction hardening.

Carburizing. In carburizing, austenitized metal is brought into contact with an environment containing carbon in a sufficient amount to cause its absorption into the surface of the part. This process is done at elevated temperatures, generally in the range of 845 to 955 °C (1550 to 1750 °F). The environment may be gaseous, for gas carburizing; a liquid salt bath, for liquid carburizing; or with all surfaces of the part covered with a solid carbonaceous compound, for pack carburizing.

In all cases, the objective is to start with a relatively low-carbon steel (e.g., 0.20% carbon) and increase the carbon content at the surface of the part. The result will be a high-carbon, hardenable steel on the outside—with carbon content gradually decreasing with depth, or distance from the surface.

Gas carburizing is the most widely used carburizing process. It is faster than liquid or pack carburizing and cases are deeper and higher in carbon content. Natural gas, which is largely methane, is the most common source of carbon. But the amount used is limited, because natural gas is too rich to be used directly. The gas is diluted in the carrier gas, which is noncarburizing or nearly so. Liquid hydrocarbons are also used as sources of carbon and are usually proprietary compounds. The term *carrier gas* is descriptive as it is the vehicle used to transport the carburizing gas into the work areas.

Gas carburizing is effective because at a temperature of about 925 °C (1700 °F) the surface of a steel is extremely active; if the carbon potential of the furnace environment is higher than that of the steel, the process works. The maximum rate at which carbon can be added to a steel surface is limited by the rate of diffusion of carbon in austenite. The rate is temperature dependent. At 925 °C (1700 °F), for example, the rate is about 40% greater than it is at 870 °C (1600 °F).

Control of the carbon potential of a given carburizing gas is complex. Control systems are available, however. All an operator need do is set the instrument for the desired surface carbon content, and the gas mixture is controlled accordingly.

The term *effective case* is widely used in heat treating. It is defined as the point at which hardness drops below 50 HRC ("R" for the Rockwell hardness test, and "C" for the type of test). Depth of any carburized case is a function of time and temperature.

A number of different types of conventional heat treating furnaces are used for gas carburizing, including pit, rotary, box, and continuous types. Furnaces are discussed later in this chapter.

Liquid Carburizing. In this process, ferrous metals are held in a molten salt bath at temperatures above their transformation temperatures. The salt decomposes and releases carbon, and sometimes nitrogen, which diffuses into the surfaces of a part so that a high hardness can be developed in quenching. Many liquid carburizing baths contain cyanide, which introduces both carbon and nitrogen into the process. Because of health and safety concerns, noncyanide carburizing is becoming more prevalent.

Parts that have been liquid carburized must be washed after quenching to remove adherent salt. Because of problems associated with salt removal, liquid carburizing is not recommended for parts with small holes, threads, or recessed areas that are difficult to clean.

The process has two advantages:

- Selective carburizing is possible because only those areas requiring surface hardening need to be immersed in the salt bath.
- Parts different in size and shape and with different case depth requirements can be carburized simultaneously.

Typical liquid carburizing furnaces are shown in Fig. 5-10.

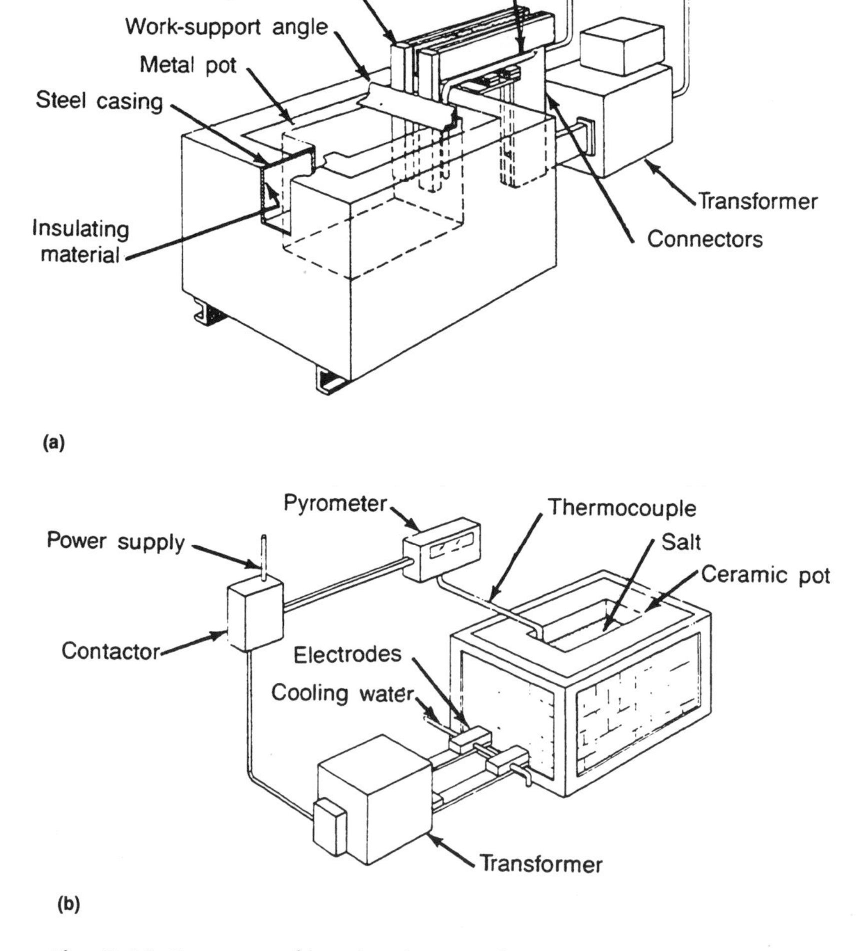

Fig. 5-10 Two types of liquid carburizing furnaces. (a) Immersed electrodes. (b) Submerged electrodes

Pack Carburizing. Carbon monoxide present in a solid carburizing compound decomposes into carbon and carbon dioxide at the surface of the metal. Carbon is absorbed into the surface, and the carbon dioxide reacts immediately with the carbonaceous material in the carburizing compound to produce fresh carbon monoxide. Carburizing continues as long as enough carbon is present to react with the excess carbon dioxide.

Fig. 5-11 Cross section of three sizes of water-hardening tool steel (W1) after heating to 800 °C (1475 °F) and quenching in brine. Black rings indicate hardened zones (cases) (65 HRC). Cores range from 38 to 43 HRC.

A prepared atmosphere is not required. Parts are placed in steel or heat-resistant alloy boxes surrounded by solid carburizing compound, usually a proprietary mixture. Boxes are sealed with fire clay and heated to the carburizing temperature, which usually is in the range of 815 to 955 °C (1500 to 1750 °F).

In terms of energy requirements, the process is highly inefficient. All materials involved, both boxes and compound, must be heated, and quenching is a problem because of the system design. Except for a few specialized applications, pack carburizing has been largely replaced by other methods.

Carbonitriding is a modified gas carburizing process. Ammonia (commonly 10%) is introduced into the gas carburizing atmosphere to add nitrogen to the carburized case as it is being produced. Ammonia dissociates to form nitrogen at the surface of the part, and it diffuses into the steel simultaneously with carbon. This results in a case with higher hardenability than a

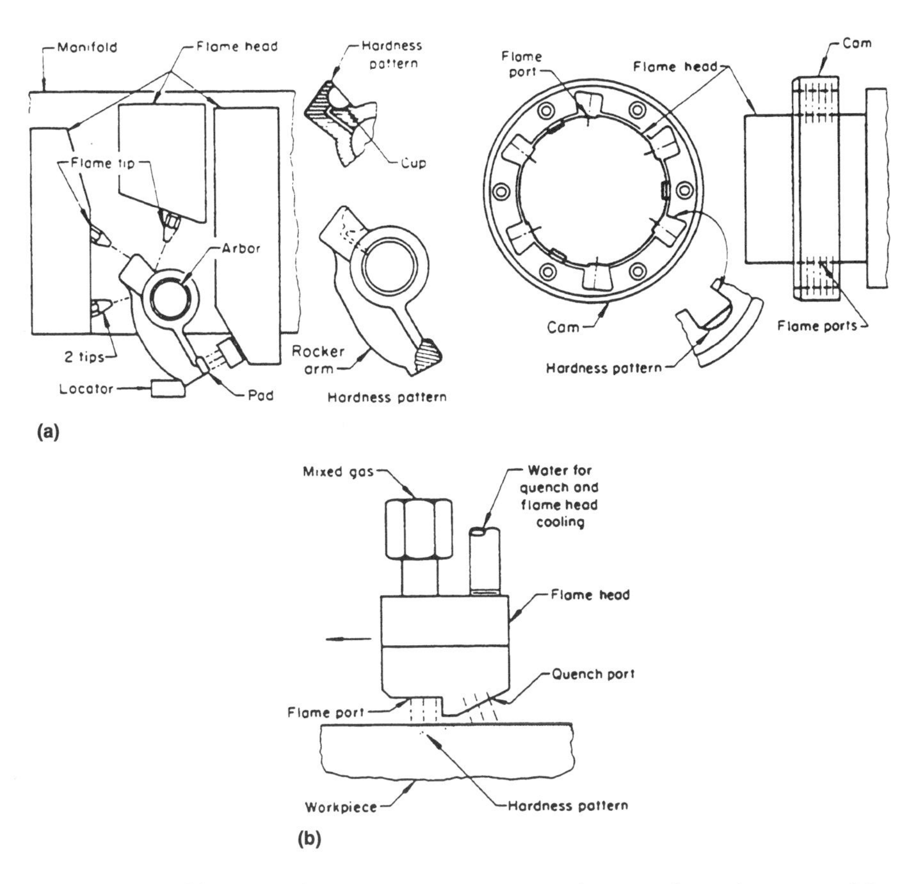

Fig. 5-12 Methods of flame hardening. (a) Spot (stationary) heating of a rocker arm and the internal lobes of a cam. (b) Progressive method

carburized case. A typical application would be a mass production operation for putting a hard, wear-resistant, relatively thin case on a variety of small hardware items. Gas carburizing furnaces are generally used, with some form of conveyor system added.

Nitriding. In the conventional nitriding process, nitrogen is introduced into the surface of a part by holding the metal in contact with a nitrogenous material at a temperature below the temperature at which austenite begins to form. Ammonia dissociates into its components: nitrogen and hydrogen. Nitrogen is very active at the moment of decomposition; it dissolves into the austenite and puts a case with high hardness on the part. Quenching is not required to get this result.

Nitrided cases are much harder than carburized cases and have good resistance to wear, fatigue, and corrosion. Also, nitrided cases have greater resistance to softening on heating than carburized cases.

An alternative process involves salt bath or liquid nitriding performed in molten salt at about the same temperatures as those used in gas nitriding (510 to 565 °C, or 950 to 1050 °F).

Fig. 5-13 Inductor with separate internal chambers for flow of quenchant and cooling water

The case hardening medium is molten cyanide. A typical salt bath is a mixture of sodium and potassium salts. Conventional salt bath furnaces may be used.

Nitrocarburizing is a relatively new proprietary process available under several different names. Selling points include the formation of an antiscuffing compound surface layer and improved resistance to fatigue. The gas used is a blend of carbon monoxide, hydrogen, and nitrogen, along with small amounts of carbon dioxide, water vapor, and methane. Operating temperatures are near 570 °C (1060 °F). Treatment usually takes 1 to 5 hours.

Localized Hardening Processes. One way to produce a part with a hard surface and a relatively soft core without altering composition is to use a steel with low hardenability (in other words, one that does not have through-hardening capability). Other techniques include shell hardening, flame hardening, and induction hardening.

Low Hardenability Process. In this process, heating is not preferential (the entire cross section of a part is heated). The steel transforms from the austenitic state to softer structures, such as ferrite or pearlite, so rapidly that only the outerlayer is cooled quickly enough to form martensite. Figure 5-11 shows that only a narrow band of full hardness is obtained after heating and quenching. Note that depth of hardness is greatest on the section smallest in diameter. A more severe quench produces martensite to a greater depth, but increases the likelihood of distortion and cracking.

Shell Hardening. The part is immersed in a high-conductivity heating medium such as molten lead or a salt bath. Immersion time is only long enough to completely heat the outer surfaces; quenching follows. The core is never heated to a temperature high enough to support the formation of austenite. The shell is martensitic, while the core is essentially pearlitic. Cold work dies are a typical application.

Flame Hardening. Any steel with the required carbon potential or ability to dissolve carbon is a candidate for this process. Equipment varies from a hand-held torch to sophisticated mechanized systems. Representative equipment is shown in Fig. 5-12.

The oldest process involves the hardening of selected areas of a part by heating its surface with an oxyacetylene flame, which supplies a very intense and concentrated heat. The hardened layer may vary from a very thin skin or case to one with a depth of 6 mm (0.25 in.). Depth of hardening is controlled by the amount of time heat is applied. With short times, a thin skin is made austenitic and hardened. With longer times, the case is thicker.

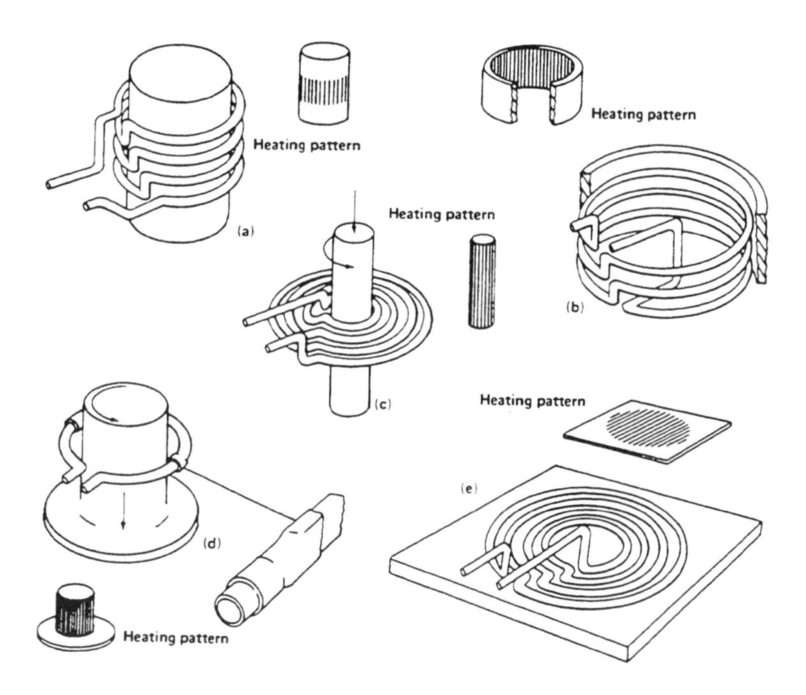

Fig. 5-14 Typical work coils for induction heating

More often than not, parts are spray quenched with water for two reasons: (1) Spraying with oil presents a fire hazard, and (2) the amount of water must be controlled closely to reduce the possibility of cracking parts—a likely result with immersion quenching. All flame-hardened parts must be tempered after hardening. Stresses due to hardening are relieved, and the hard case gains a degree of toughness.

Induction hardening is a surface hardening process in which heating by electromagnetic induction is confined to the surface layer of a part. Alteration of composition is not required, and the method can be used in normalizing, annealing, heating, hardening, and tempering.

The metallurgical principles involved are essentially the same as those for flame hardening. Both processes are characterized by rapid heating and close control of quenching. Any part that is electrically conductive is a candidate. Basic equipment consists of a power supply, controls, a means of matching power supply within the load, an inductor, quenching equipment, and a system for holding and positioning parts. An inductor is shown in Fig. 5-13, and typical work coils for induction heating are shown in Fig. 5-14.

Heat Treating Equipment

Heat treating equipment includes furnaces, fixtures and holding devices, quenching systems, and atmosphere- and temperature-control systems.

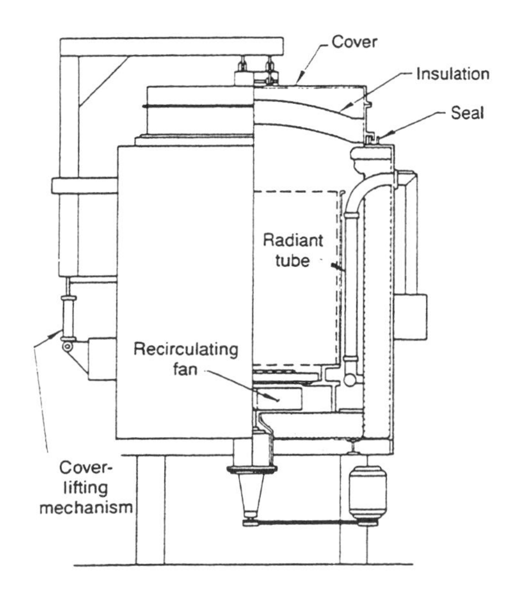

Fig. 5-15 Pit-type batch furnace. Dashed lines outline location of workload.

Furnace Sizes and Designs

Furnaces range in size from small models that sit on a bench and can accommodate only a few ounces of metal, to large car-bottom furnaces that handle hundreds of tons of parts in a single load. Furnaces may be directly fired with fuel, with the parts being treated exposed to combustion gases, or they may be indirectly fired, with the parts separated from combustion gases. Electric resistance systems provide a second type of heating.

A batch furnace is loaded with a charge, then closed for a preestablished heating cycle. After the cycle is completed, parts may be cooled in the furnace after the heat is turned off, or a method such as quenching in water or oil may be used.

The batch furnace shown in Fig. 5-15 is a pit-type, top-loading unit that can be adapted to almost any heat treating operation. As the furnace is indirectly fired, the work area is surrounded by gas-heated radiant tubes. Circulation of heat is assisted by a bottom fan.

Car-bottom furnaces like the one shown in Fig. 5-16 are used for heat treating large castings, forgings, and weldments. A rail car serves as the bottom or hearth of this furnace. The hearth is insulated to prevent heat from reaching wheel roller surfaces. Such furnaces usually are built at floor level and frequently equipped with a lift door. Work to be processed is placed on the hearth; the car is moved into the furnace on rails, and the door is closed.

Continuous furnaces are basically of the "in-one-end-and-out-the-other" types, and are ordinarily used for continuous volume production of relatively small parts. They can accommodate a variety of heat treatments.

Roller Hearth Furnace. In the roller hearth furnace shown in Fig. 5-17, work is conveyed through the furnace by means of rollers, the ends of which project through the walls of the furnace to external air- or water-cooled bearings. Usually, rolls are power driven by a common source through a chain-and-sprocket mechanism. Work is placed on revolving rollers at the charge end and carried through and out of the furnace by the friction between the work surface and the revolving roll surfaces.

Moving Hearth Furnace. In the moving hearth furnace, another type of continuous unit, a conveyor belt carries parts through the furnace. Belts may be constructed of woven wire mesh, flat cast alloy links, or a more open flat design.

Pusher furnaces, another continuous type, are similar in many respects to the belt type but are designed to carry higher-volume loads. Parts are

Fig. 5-16 Car-bottom furnace for homogenizing cylindrical parts

placed on the hearth and pushed ahead periodically by a mechanical ram operating at the charging end. Parts may be loaded into sturdy, cast alloy baskets, tracks, or followers that ride on skid rails, tracks, or rollers built into the floor of the furnace.

Liquid Bath Furnaces. The concept here differs from that of heating in gaseous atmospheres. Heating in a molten metal, usually lead, is an age-old practice. A pot-type furnace containing molten lead provides an effective means of heating steel parts. However, special precautions must be taken because of the toxicity of molten lead fumes.

Molten salt baths are an alternative. A variety of these salts are available, including barium chloride, sodium chloride, potassium chloride, and calcium chloride. Some salts change the surface chemistry of the steel.

Figure 5-18(a) illustrates a common type of fuel-fired pot furnace that can be used for either molten metal (usually lead) or molten salt. This type of furnace is used in heating metals to about 900 °C (1650 °F). The furnace shown in Fig. 5-18(b) is similar, but is heated by electrical resistance.

Both types of furnaces are versatile, but are best suited for limited production of small parts. One shortcoming is that to achieve acceptable pot life, it is necessary to make pots from expensive nickel-chromium alloys.

Fluidized-bed furnaces consist of a bed of mobile inert particles, usually aluminum oxide, suspended by the combustion of a fuel-air mixture flowing upward through the bed. Components being treated are immersed in the fluidized bed, which acts like a liquid, and are heated by it. Heat-transfer rates are several times greater than those available in conventional furnaces.

There are two types of furnaces: (1) internally fired types for high-temperature applications (760 to 1205 °C, or 1400 to 2200 °F) and, (2) externally fired furnaces for controlled-atmosphere heating. In the internally fired bed (Fig. 5-19a), fuel and air are mixed in controlled proportions and passed through a porous ceramic plate, above which particles are fluidized in the gas stream. In the externally fired bed (Fig. 5-19b), an excess air burner fires into a plenum chamber, above which the fluidized bed is supported by a porous metallic plate. The bed is fluidized

Fig. 5-17 Roller hearth annealing furnace for processing such products as steel wire coil, wire rod coil, and parts

by the flow of controlled-atmosphere gases. External heating is required so that products of combustion do not contaminate the fluid bed.

Vacuum Furnaces. Heated chambers in this type of furnace are evacuated to a very low level, depending on the material and process by which metal is being treated. In most operations, evacuation is to about 1 torr, the unit for measurement of vacuum. Under these conditions the amount of original air remaining in the work chamber is about 0.1% or less. Some heat treating operations involving highly alloyed materials require a vacuum of less than 1 torr.

Side and end sectional views of a cold-wall vacuum furnace are presented in Fig. 5-20. The three-chamber furnace includes a loading vestibule (left portion of side view), heating chamber (center portion of side view), and an elevator quenching arrangement (right portion of side view). Vacuum furnaces are heated by electric resistors, frequently of the graphite type, as shown in Fig. 5-20.

Temperature-Control Systems

Temperature control in heat treating is of particular importance for management of quality. Tolerances can be as tight as ±5 °C (±10 °F). The general range is ±10 °C (±20 °F).

In any temperature-control system, three steps must be executed (Fig. 5-21). First, temperature must be sensed by some device that responds to changes in temperature. Second, the amount of change must be indicated or recorded, then controlled. Third, controller output must be transmitted to the final element, a compound of the process itself. Final elements relay controller output and initiate corrective changes in the process.

Thermocouples are the most widely used sensors for determining the higher temperatures involved in most heat treating operations. Commonly used type K thermocouples have a base composition of 90% nickel and 9% chromium. The melting point of the alloy is 1350 °C (2460 °F), and the maximum service temperature is 1260 °C (2300 °F).

Radiation pyrometers also are used to sense temperatures in heat treating furnaces (Fig. 5-22). These extremely accurate devices sense changes in temperature based on changes of wavelength of light; for example, the change in the color of metal from red to white with increasing temperature is sensed by changes from longer to shorter wavelengths of light. Differences in wavelength are sensed without contact with the metal whose temperature is being measured.

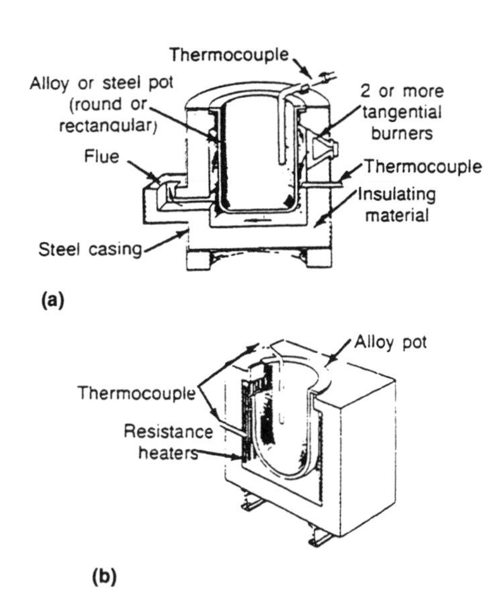

Fig. 5-18 Principal types of externally heated salt-bath furnaces. (a) Gas-fired or oil-fired. (b) Resistance heated

Furnace Atmospheres

In many heat treating operations, parts must be heated under conditions that prevent scaling and/or loss of carbon. As steel is heated, its surfaces become more active chemically. For example, oxidation of carbon steels commonly begins at about 425 °C (800 °F). As temperatures exceed about 650 °C (1200 °F), the rate of oxidation becomes exponential as temperatures increase. Also, the steel can lose carbon (called *decarburization*) even at moderate temperatures. The steel reacts with the furnace atmosphere so that the amount of carbon near the surface is substantially reduced—usually an undesirable condition.

Generally, furnace atmospheres serve one of two requirements: (1) They allow heat treating of parts with clean surfaces without changing surface conditions, or (2) they permit a controlled surface change to be made, as in certain case hardening operations. Controlled changes in surface chemistry can also be made in molten salts or in fluidized beds. The main types of gaseous atmospheres, based on increasing cost, are:

- Natural (conventional air)
- Derived from products of combustion
- Exothermic (generated by a chemical reaction in which heat is liberated)
- Endothermic (generated by a chemical reaction in which heat is absorbed)
- Nitrogen base
- Vacuum
- Dissociated ammonia

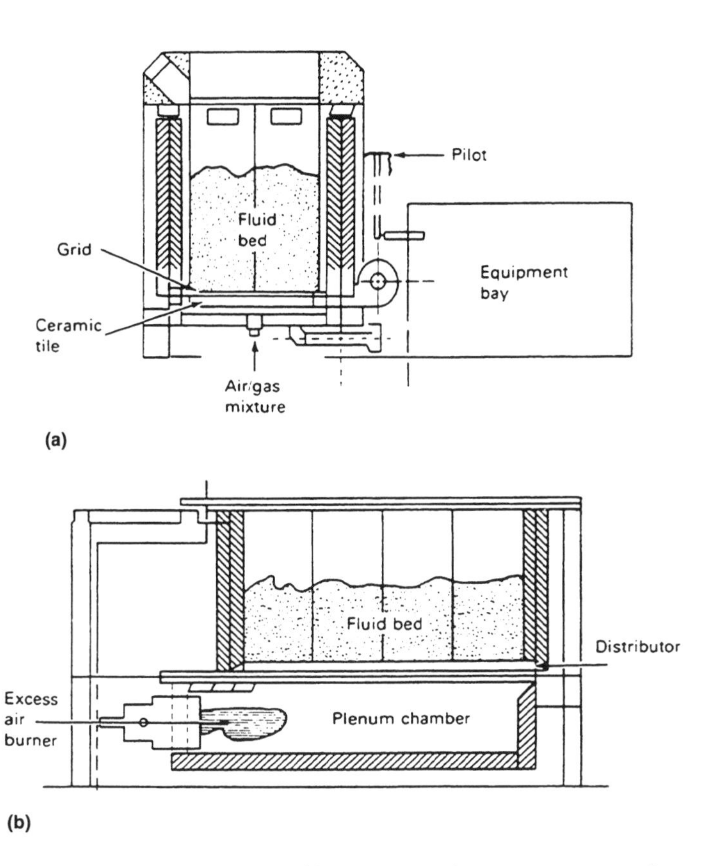

Fig. 5-19 Sectional views of fluidized-bed furnaces. (a) Internally fired. (b) Externally fired

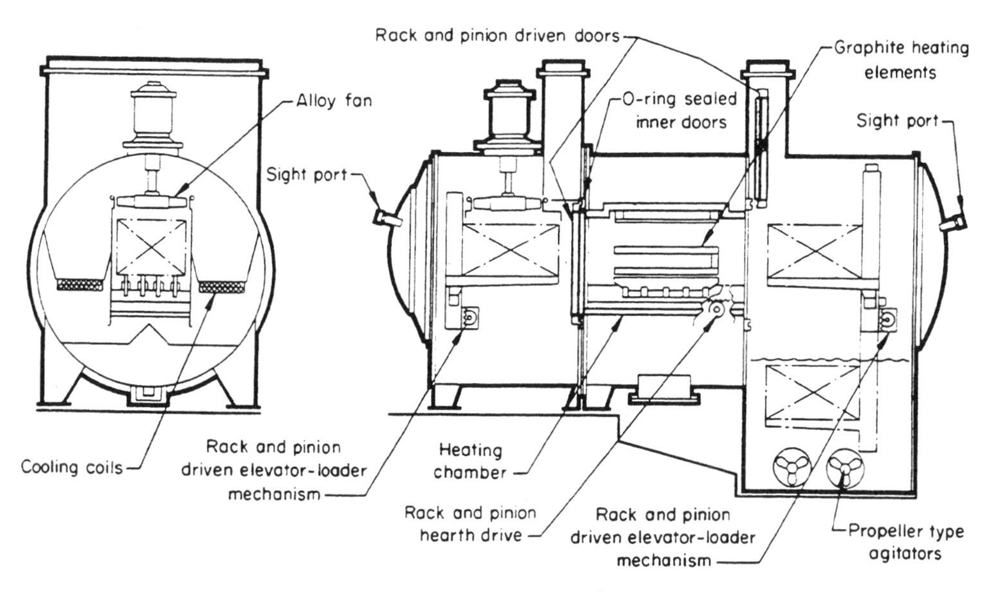

Fig. 5-20 Three-chamber cold-wall vacuum oil-quench furnace

- Dry hydrogen
- Argon (from bottles)

Natural Atmospheres. The air we breathe, termed a natural atmosphere, contains about 79% nitrogen, 20% oxygen, and 1% moisture, by volume. Such atmospheres exist in any heat treating furnace when the work chamber contains no products of combustion, or when specially prepared atmospheres are not involved. Natural atmospheres are suitable in many heat treating applications, including those where parts are to be machined after heat treating.

Atmospheres derived from products of combustion in direct-fired furnaces automatically provide protection a notch above that provided by natural atmospheres. When fuels are mixed with air and burned at the ideal ratio, some reaction with steel surfaces occurs. Ideal ratios for air-gas mixtures vary with the fuel. For example, when natural gas (methane) is burned, the ratio is about 10 parts air to 1 part gas, by volume. By comparison, when the gas is propane the ratio is 23 parts air to 1 part gas. Excess air in the mixture causes loose scale to form on the surfaces of parts. Excess gas (fuel) causes the formation of a tight, adherent oxide on part surfaces. In addition, all mixtures contain a certain amount of water vapor, which can cause the decarburization of carbon and alloy steels.

Exothermic atmospheres are widely used prepared atmospheres. They are produced by the

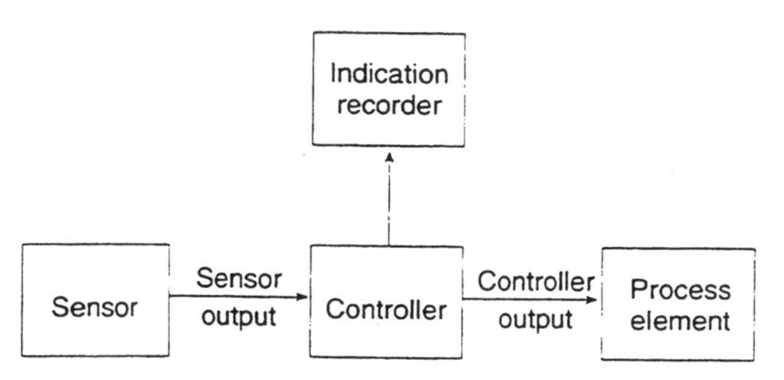

Fig. 5-21 Relationships among steps in temperature-control system

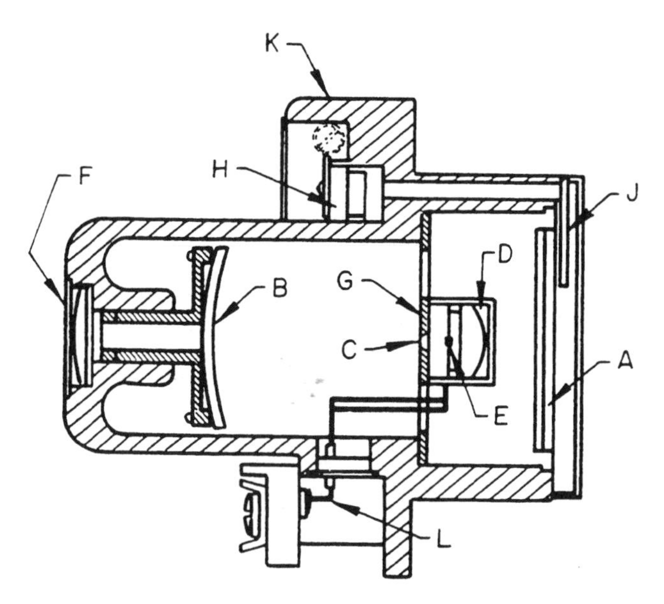

Fig. 5-22 Total radiation pyrometer. A, window; B, mirror; C, opening to thermopile; D, mirror; E, thermopile; F, lens; G, receiving disk; H, screw and gear adjustment for shutter; J, shutter (adjusted at factory only); K, pyrometer housing; L, electrical lead wires to temperature-indicating instrument

combustion of a hydrocarbon fuel such as natural gas or propane. The air-fuel mixture is closely controlled at a ratio of about 6 to 1.

Endothermic atmospheres are produced in generators that use air and hydrocarbon gas, such as methane or propane, as fuel. The ratio of air to gas is about 2.5 to 1.

Nitrogen-base atmospheres are used in a variety of heat treating operations. Because nitrogen is noncorrosive, special materials of construction are not required for most commercial nitrogen-base systems.

Vacuum atmospheres are quite flexible in terms of temperature and number of heat treating operations that can be performed. They are often considered a prepared atmosphere, even though this is not the case. The function is served by a very low-level atmospheric pressure.

Dissociated ammonia is produced from anhydrous ammonia and is a prepared atmosphere suitable for a variety of heat treating operations. However, it is very expensive.

Dry hydrogen is 98 to 99.9% pure in its commercially available form. Caution in handling is required, due to its explosive potential. Dry hydrogen is used in annealing stainless steels, low-carbon steels, and electrical steels.

Argon (from bottles) provides an excellent protective atmosphere for virtually any heat treating operation if furnaces are airtight, or if gastight furnace retorts are used. Its use is limited by high cost.

Quenching Media and Systems

Quenching is the term for rapid cooling of a metal or alloy from an elevated temperature. Parts are typically quenched by immersing them in quenching media such as water, oil, or a polymer-based water solution. The quenchant may be agitated to remove a gaseous surface film that retards cooling. Cooling may also be supplied by fan cooling or by still air. Molten salts are also used as quenching media.

Steels are quenched to control the transformation of austenite to form desired microstructures, which are identified by name in the continuous cooling transformation (CCT) dia-

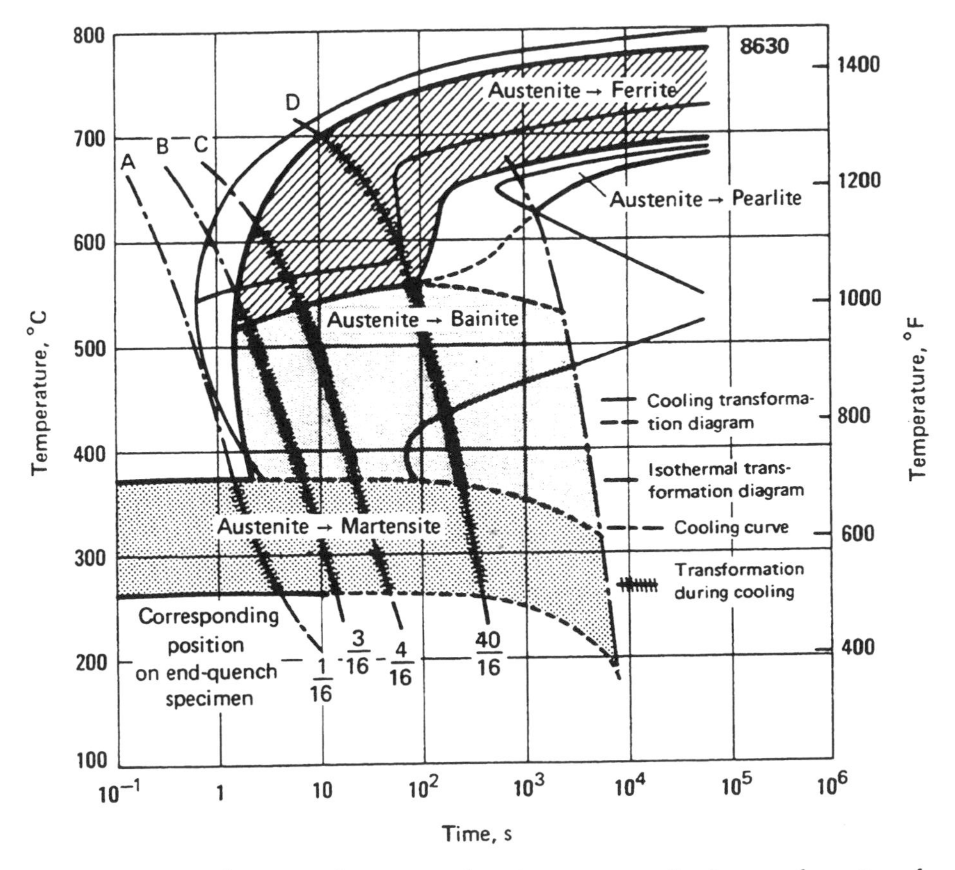

Fig. 5-23 Transformation diagrams and cooling curves, indicating transformation of austenite to other constituents as a function of cooling rate

gram in Fig. 5-23. Note that this diagram is different in some respects from the iron-carbon phase diagram in Fig. 5-4 but is similar in others. Temperatures, for example, are listed vertically, while time (shown in seconds) is listed horizontally, in comparison with the listing of carbon content in the iron-carbon phase diagram. Further, the subject of the CCT diagram is cooling and the steels involved are different: in this instance, an alloy steel (AISI 8630) versus an iron-carbon composition in the phase diagram. Differences in composition explain differences, for example, in the temperature marking the start of formation.

Martensite is the as-quenched microstructure usually desired in quenched carbon and low-alloy steels, as indicated by curve A in Fig. 5-23. Along cooling curves B, C, and D, some transformation to bainite and ferrite occurs, with a corresponding decrease in the amount of austenite formed and hardness developed.

Cooling Rates. To obtain the maximum amount of martensite, the cooling rate must be fast enough to avoid the nose of the time-temperature curve of the steel being quenched, as indicated by curve A in Fig. 5-23. The actual cooling rate required depends on the hardenability of the steel. The entire cross section of a part must cool at this rate, or faster, to obtain the maximum amount of martensite.

Under ideal conditions, water provides a cooling rate of about 280 °C (500 °F) per second at the surface of a steel cylinder 13 mm (0.5 in.) in diameter and 100 mm (4 in.) long. This rate decreases rapidly below the surface of a part, which means, for example, that for carbon steel only light sections with a high ratio of surface area to volume can be fully hardened throughout the cross section.

Quenching Media. Many systems are available, including:

- Water
- Water solutions of brine
- Water solutions of caustics
- Water solutions of polymers
- Oils
- Molten salts
- Molten metals
- Gases, still or moving
- Mists or sprays of water, for example
- Water-cooled dies that hold parts to minimize distortion

Water is the fastest of the quenching media. Oil provides a slower rate of cooling, and the cooling rate of water solutions of polymers is between those of water and oil.

Quenching Systems. The equipment required varies broadly. In small shops where only a few parts are heat treated daily, equipment may be as simple as a steel barrel containing water or brine. Heating or cooling facilities are not needed. If agitation of the quenching medium is required, the operator merely moves the part around in the coolant by hand.

In facilities operating continuously, on the other hand, the quenching system may contain all or most of the following components, regardless of the quenching medium used:

- Tanks or quenching machines
- Agitation equipment
- Fixtures that hold parts being quenched
- Cooling systems
- Storage or supply tanks
- Heaters
- Pumps

Chapter 6

Tailoring the Properties of Nonferrous Alloys

Aluminum-base and copper-base alloys are among the nonferrous alloys that are hardened and strengthened by solution treating and aging, also known as *precipitation hardening*. This route to property improvement is closed to the many nonaging nonferrous alloys, which are upgraded by such processes as cold working and/or other heat treating processes (e.g., annealing, stress relieving, and recrystallization annealing). For example, an alpha titanium alloy that contains 3% aluminum and 2.5% vanadium generally is used in the cold-worked and stress-relieved condition. Treatment of a wrought magnesium alloy, AZ61A, consists of annealing, stress relieving, and recrystallization annealing.

Solution annealing, also known as solution treatment, consists of heating an alloy to a selected temperature and cooling at a prescribed rate, usually very rapidly. The metal becomes softer, rather than harder, and is ready for either a hardening treatment (if it is hardenable) or for cold forming to a desired shape.

Precipitation hardening is also known as age hardening, or aging. When the process involves reheating a solution-annealed alloy for a specified time at a temperature that is lower than the solution treating temperature, it is referred to as *articial aging*. Strengthening is the end result. Some alloys age harden at room temperature with the passage of time after solution treatment, and this process is known as *natural aging*. Hardening is accompanied by an increase in strength and a reduction in ductility. A number of aluminum, copper, nickel, iron, and titanium alloys are precipitation hardenable.

Cold working usually takes place at room temperature. Also referred to as *strain hardening* or *work hardening,* this is one of the basic methods of hardening and strengthening metals. Hardness, yield strength, and tensile strength are improved. For example, tensile strength can be almost doubled, while yield strength can be increased by a factor of three to five. These improvements take place during plastic deformation. Deformation is permanent; the metal does not return to its original shape when the load causing deformation is removed.

Cold work has both positive and negative effects on properties. Higher hardness and strength are accompanied by reduced ductility, as shown in Fig. 6-1. Elongation is a measure of ductility. In shaping metals by cold working, there is a limit to the amount of plastic deformation possible without fracture. Annealing restores the metal to a structural condition similar

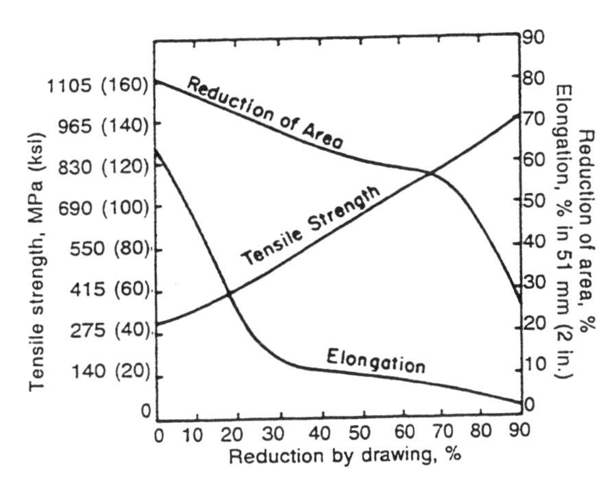

Fig. 6-1 Effect of cold work on tensile properties of yellow brass (66% copper and 34% zinc)

to that prior to deformation, making further cold working possible.

When a cold-worked, or strain-hardened, metal is heated to a given temperature for a specific time, a softening process called *recovery* is initiated. This is known as *stress-relief annealing*. The microstructure of a brass sheet that has been cold worked and stress relieved is shown in Fig. 6-2. The value given for elongation is 31%. The alloy was heated for an hour at 260 °C (500 °F).

Treatment at higher temperatures results in a change in microstructure. New grains form in the cold-worked matrix and grow until an entirely new set of grains is formed: a process called *recrystallization annealing*. Grains grow until all the strained material disappears. The microstructure of a cold-worked brass sheet after recrystallization annealing is shown in Fig. 6-3. The alloy was treated at 425 °C (800 °F) for an hour. The value for elongation in this instance is 51%.

The microstructure of the brass sheet after recrystallization and grain growth at a higher temperature of 705 °C (1300 °F) is shown in Fig. 6-4. In this instance, the value for elongation has jumped to 73%. Note that the same alloy (70% copper and 30% zinc) is involved in Fig. 6-2 to 6-4. Gains in elongation in each instance are accompanied by reductions in yield strength, tensile strength, and hardness with each increase in annealing temperature.

Mechanical properties	
Yield strength (0.2% offset)	395 MPa (57 ksi)
Tensile strength	485 MPa (70.5 ksi)
Elongation in 50 mm (2 in.)	31%
Hardness	78 HRB

Fig. 6-2 Microstructure of brass sheet (70% copper and 30% zinc) after cold working and stress relieving

Precipitation Hardening

Precipitation hardening is a two-step process: (1) solution treating followed by (2) age hardening. The desired improvement in hardness and strength—and possibly other properties of inter-

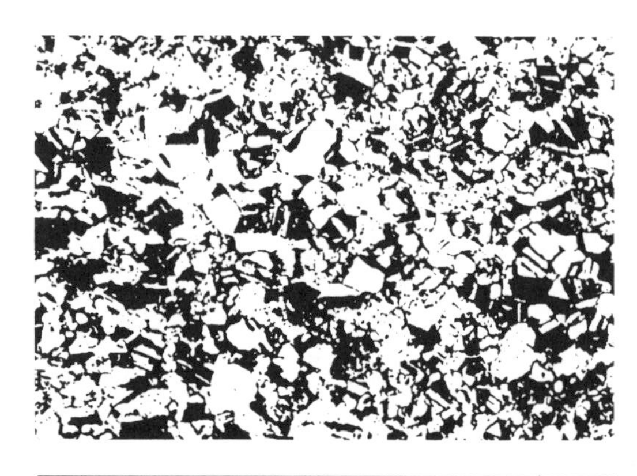

Mechanical properties	
Yield strength (0.20% offset)	170 MPa (24.8 ksi)
Tensile strength	380 MPa (55 ksi)
Elongation in 50 mm (2 in.)	51%
Hardness	41 HRB

Fig. 6-3 Microstructure of cold-worked brass sheet after recrystallization annealing

Mechanical properties	
Yield strength (0.20% offset)	72 MPa (10.4 ksi)
Tensile strength	310 MPa (45 ksi)
Elongation in 50 mm (2 in.)	73%
Hardness	51 HRB

Fig. 6-4 Microstructure of cold-worked brass sheet after recrystallization and grain growth

est, such as electrical conductivity or machinability—is achieved either by (1) *artificial aging,* which accelerates the process by rapid cooling of an alloy from the solution treating temperature, or (2) *natural aging* by slow cooling from the solution treating temperature back to room temperature over a period of several days.

Solution Treatment Basics. A prerequisite to precipitation hardening is the ability to heat an alloy to a temperature range at which all the solute is dissolved by the solvent, resulting in a single-phase structure, such as a combination of aluminum (the solvent) and copper (the solute). In the example presented in Fig. 6-5, A is the solute and B is the solvent. In this instance, the maximum solubility of B in A is 10%, which is reached at the T_2 temperature. By holding at this temperature for sufficient time, a single-phase alpha structure (combining A and B) is formed. The atoms of A and B occupy positions in a common crystal lattice structure. This structure is retained at ambient temperature by cooling rapidly (i.e., water cooling) from the alpha range to prevent a precipitate of B from forming. The structure is supersaturated with solute B, and for this reason is unstable.

Age Hardening Basics. The solution-treated alloy, when held for a period of time at room temperature or somewhat higher, must develop a coherent precipitate in order to cause age hardening. This is called the *aging* treatment. Aging increases hardness and yield and tensile strengths, but usually reduces ductility.

In artificial aging, precipitation is initiated by the application of heat, which accelerates the diffusion of atoms. With the single-phase alloy in an unstable condition, even a slight increase in temperature will energize the B atoms enough to start the precipitation process. At the beginning, the precipitate particles are so small they cannot be seen even under an optical microscope, but their presence dramatically strengthens the A-B alloy by pinning or impeding dislocation movement.

Up to a point, the time required to reach maximum hardness is reduced by increasing temperature. For example, the time-temperature relationship for an aluminum-copper alloy is shown in Fig. 6-6. Note that a satisfactory hardness with aging is obtained at 205 °C (400 °F). With higher aging temperatures, maximum hardness is reached in a shorter time. However, the hardness reached at higher aging temperatures is not as high as the hardness level achieved with lower aging temperatures.

In natural aging, alloys harden at room temperature. After the quenching step in solution treatment, hardness increases slowly over a matter of days or even weeks until it reaches a maximum. As in artificial aging, hardness is increased by strains put in the lattice structures by the trapped precipitate.

Coherency Theory. The alloy is in a supersaturated condition after solution treatment but before aging, and ordinarily the precipitate and parent solid solution have totally different lattice arrangements. However, during the initial stages of precipitation, the submicroscopic precipitate particles are forced to conform to the parent lattice structure, as shown in Fig. 6-7(a) and (b). Noncoherent structures are shown in Fig.6-7(c) and (d).

As long as the precipitating crystals are small, they can adjust themselves by distortion to fit the lattice of the parent solid solution. Accommodation to the parent phase is called coher-

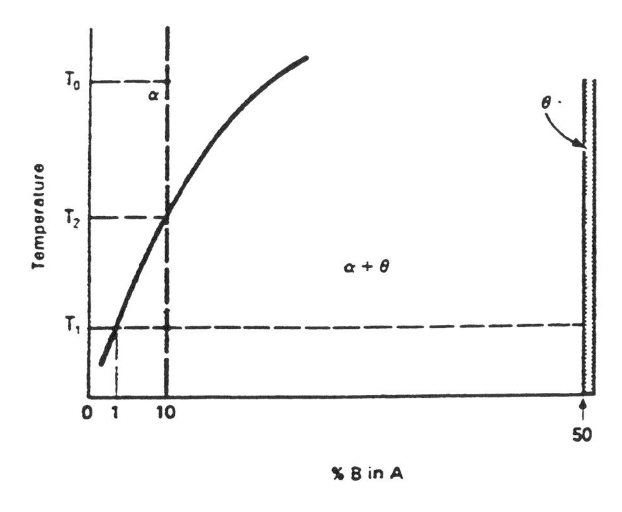

Fig. 6-5 Hypothetical phase diagram of system A-B

ency, and the stress developed by the distortion of the parent phase is believed to be a contributing cause of this type of hardening (Fig. 6-7a and b). Coherency is lost after the precipitate particle reaches a certain size and forms its own crystal structure, and the severe lattice stresses disappear (Fig. 6-7c and d).

The coherency theory explains what happens to the atomic lattice structure of an alloy during each of the three stages of aging:

- Solute atoms diffuse through the solvent lattice and segregate to form a few very small particles. This stage may be described as a grouping or rearranging of solute atoms so that they may begin to form a precipitate. Result: a slight increase in hardness.
- The rearranging and migration of atoms bring about the formation of many small regions attempting to form a separate precipitating phase. However, the precipitate (solute) remains coherent with the parent (solvent) lattice structure. Significant gains in hardness and strength result.
- Coherency is lost after the precipitate particle reaches a certain size, and the solute forms it own crystal structure (Fig. 6-7c and d). Loss of coherency reduces lattice stresses and results in decreasing hardness and strength. Hardness continues to drop as the precipitate particles continue to grow.

Heat Treating of Aluminum Alloys

One group of aluminum alloys can be hardened only by cold work; a second group is hard-

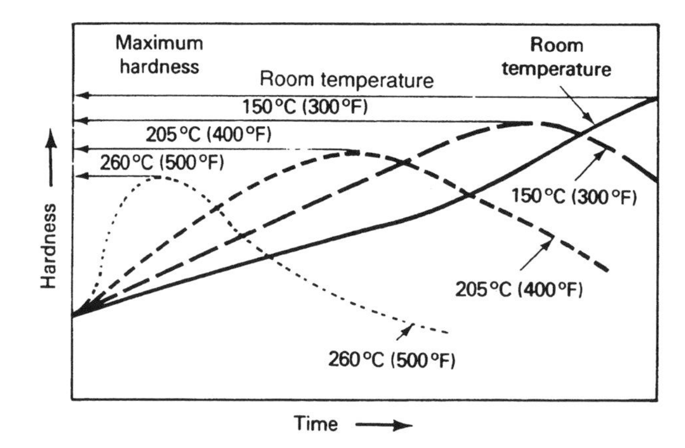

Fig. 6-6 Effect of time and temperature on the aging properties of an aluminum-copper alloy

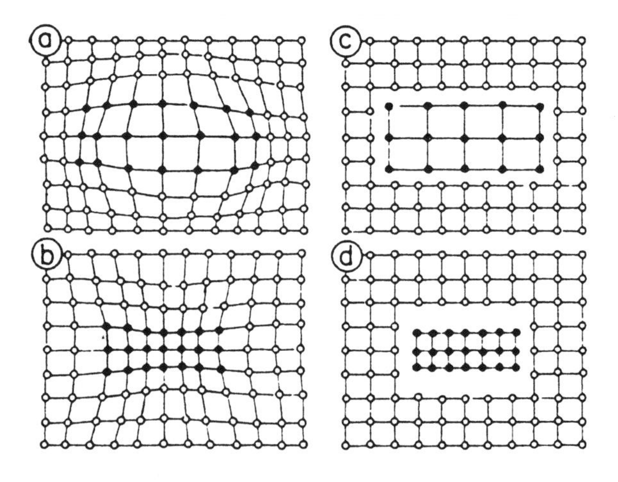

Fig. 6-7 Coherent (a and b) and noncoherent (c and d) precipitation

ened by solution treating and aging. Some of these alloys are artificially aged, and some are naturally aged. General requirements for typical heat treating processes are described in the paragraphs that follow.

Annealing of wrought, cold-worked alloys of the nonaging type calls for heating to a temperature at which recrystallization occurs. The cooling rate is not critical.

Precipitation-hardenable alloys have a somewhat higher annealing temperature, but are below the solution treating temperature. Alloys in this class must be held at the annealing temperature for the time needed to permit precipitation or full loss of coherency, as well as subsequent growth of the precipitate to a size where it has little effect on increasing hardness. Because some solute phase is in solid solution at all elevated temperatures, slow cooling (30 °C, or 50 °F, per hour) down to 260 °C (500 °F) frequently is recommended.

Solution heat treatment. In this instance, alloys are heated to a temperature at which all hardening constituents are in solid solution. A rapid quench, usually in water, follows. Wrought aluminum alloys in this condition are designated by a T4 temper, which indicates that natural aging follows solution treating. Artificially aged alloys are in the T6 temper condition.

Age hardening treatment. Natural aging takes place at room temperature over a period of days or weeks. In artificial aging, alloys are heated to a moderately elevated temperature of 160 to 240 °C (320 to 460 °F) for 1 to 50 hours, depending on alloy type. Naturally aged alloys are indicated by the T4 temper. Artificially aged alloys are in the T6 temper.

General Heat Treating Practice

Aluminum alloys are heat treated in molten salt baths or in hot air furnaces. Salt baths contain sodium and potassium nitrates. The ambient atmosphere in the heat treating shop is used in open furnaces if certain requirements are met: The shop atmosphere must contain sufficient carbon dioxide to serve as an atmosphere and must be free of sulfur compounds and excess moisture. In addition, furnaces should have good circulation to ensure uniform heating of the metal being treated. Furnaces may be heated by oil, gas, or electricity.

In both solution treatment and aging, accurate temperature control is required. For example, an excessively high solution treating temperature can cause eutectic melting, resulting in a grain-boundary condition (brittleness) that weakens alloys. Low solution temperatures may be detrimental to the aging response. Automatic temperature controllers are preferred because a record is kept; this can be helpful in determining whether the material has been treated at the proper temperature for the proper length of time.

Material should be quenched as quickly as possible after it is removed from the furnace on completion of the solution treatment cycle. Optimum corrosion resistance and mechanical properties are obtained by quenching in cold water. To prevent warping of parts being treated, however, a less drastic quench—in hot water, oil, or air—may be desirable.

Considerable caution in the use of nitrate baths is recommended. Oil, wood, charcoal, coke, and other carbonaceous materials react violently with molten nitrate baths. Parts must be cleaned before they are placed in the bath. Water can also cause violent reactions in salt baths. As a precautionary measure, a supply of dry sand should be kept on hand for control of fire.

Finally, it should be noted that the mechanical properties of some common precipitation-hardenable aluminum alloys are stronger in terms of yield strength than many low-carbon steels and even some low-alloy steels.

Heat Treating of Beryllium-Copper Alloys

Dispersion strengthening is introduced here with the aid of a partial phase diagram for beryllium-copper (Fig. 6-8). Key features in the phase diagram are:

- Temperature is in the vertical axis. Beryllium content is in the horizontal axis.
- Beta (β) identifies beryllium in the bcc phase.
- Gamma (γ) identifies beryllium in the fcc phase.
- Reading from left to right, note that the copper phase contains about 1.55% beryllium at a temperature of 604 °C (1121 °F) and about 2.7% beryllium at 865 °C (1590 °F).

The phase diagram shows that the maximum solid solubility of beryllium in copper is about 2.7% at a temperature of 865 °C (1590 °F). At 3.31% beryllium content, the beryllium-copper alloy contains more than the maximum amount of beryllium that can be dissolved in the copper. Hardness is increased by the presence of the undissolved gamma phase (see Cu + γ in the middle of the bottom part of the phase diagram). The resulting gain in hardness is called *dispersion hardening*. The explanation is that the undissolved γ is harder than the copper matrix surrounding it. Final hardness of the Cu + γ mixture depends on how much undissolved γ is present and on the size, shape, and distribution of the phase.

Figure 6-9 shows aging curves for simple binary beryllium-copper alloys containing from 0.77 to 4.0% beryllium. Solution treatment was at 800 °C (1470 °F), with aging at 350 °C (660 °F) up to 36 hours. The aging curves demonstrate that as the percentage of beryllium increases, the aging time required to reach maximum hardness is shortened and maximum hardness is increased.

Hardening does not take place at 0.77% beryllium with aging at 350 °C (660 °F) mainly because copper can hold only about 0.5% beryllium in solid solution at this temperature. By comparison, when beryllium content is increased to 1.32% and aging time is 16 hours, hardness increases significantly. At 1.82% beryllium a great deal of hardening occurs after aging for 4 hours, and hardness remains at essentially the same level for the duration of the aging period.

Peak hardness is reached at a beryllium content of 2.39%. At the 4% level, a higher hardness is obtained, but the gain is temporary. The alloy is overaged and, as Fig. 6-9 shows, hardness drops sharply. Figure 6-9 reveals that the solubility of beryllium in the copper phase drops with decreasing temperature—the first requirement for precipitation hardening. Dispersion is a type of precipitation.

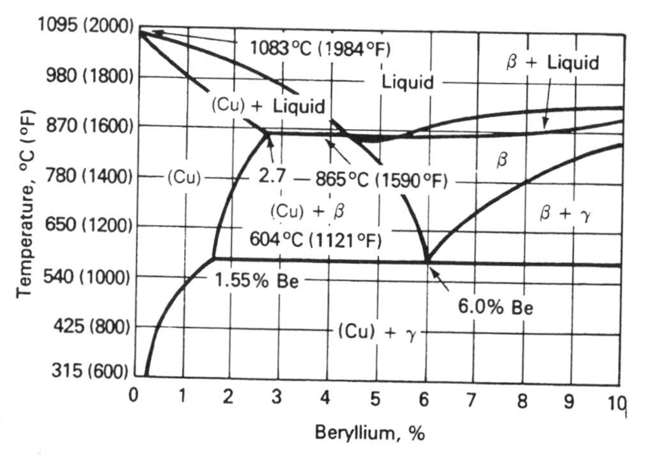

Fig. 6-8 Modified beryllium-copper phase diagram

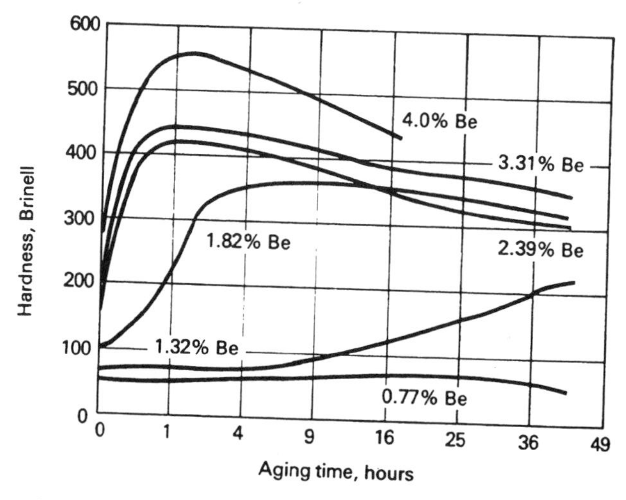

Fig. 6-9 Aging curves for beryllium-copper binary alloys

Effects of Aging on Tensile Properties

Changes in tensile properties of a 2.39% beryllium-copper alloy with aging are shown in Fig. 6-10. Maximum tensile strength reached on aging at 300 °C (570 °F) is about 1310 MPa (190 ksi). Aging time is about 100 hours. Yield strength approaches a similar value after 100 hours of aging. These are high levels of strength. For comparison, consider that plate made of aluminum alloy 7050, which is widely used in aircraft structures because of its very high strength, has a tensile strength of 510 MPa (74 ksi) and a yield strength of 455 MPa (66 ksi). A megapascal (MPa) is the metric equivalent of nearly 7000 pounds per square inch (1000 psi = ksi).

Figure 6-10 also shows that increases in strength are accompanied by decreases in ductility, as indicated by declining values for elongation with increasing aging time. In the solution-treated condition, elongation of the beryllium-copper alloy was about 40%; by the time the alloy reached maximum strength, elongation was in the 2 to 3% range.

The most widely used high-strength beryllium-bronze alloys contain about 2% beryllium (about the maximum on the practical level). A third element (nickel or cobalt) can be used to extend the time during which an alloy is at maximum hardness. Cobalt also ensures that precipitation occurs more uniformly throughout the structure, rather than concentrating primarily in grain boundaries. Beryllium content above 2% would make the alloy very difficult to fabricate and very expensive.

In Fig. 6-11, hardness versus aging time is plotted for two beryllium-copper alloys. Curve 1 is a conventional copper plus 2% beryllium alloy; curve 2 is the same alloy with a 0.38% addition of cobalt. Curve 1 shows that after 1 hour, hardness drops rapidly. After 2 hours, the alloy is softer than it was before the aging process was started. Curve 2 shows that with the cobalt addition, aging time is not so critical, and hardness continues to increase for up to about 3 hours. This behavior is attributed to the formation of a cobalt-beryllide precipitate that resists aging.

More on Mechanical Properties

Beryllium-copper alloys are unique among engineering alloys. Copper is used extensively by itself throughout the electrical industry because it conducts electric current better than other metals, with the exceptions of gold and silver. By itself, copper does not have much strength, and it is too soft to be used in most structural applications. However, copper is strengthened by alloying elements such as beryllium.

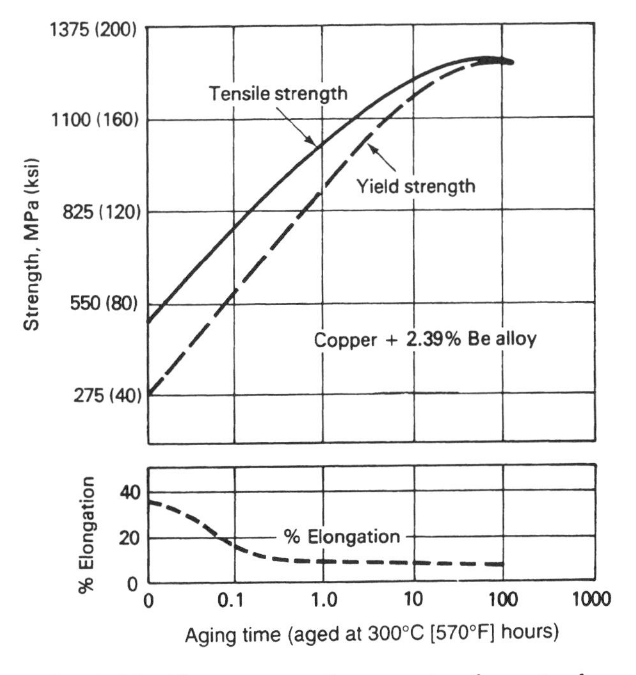

Fig. 6-10 Changes in tensile properties after aging for a 2.39% beryllium-copper alloy

Beryllium-copper alloys are among the strongest known copper alloys. High strength combined with good electrical conductivity makes them particularly suited for electrical contacts and relays, which must open and close many times each minute and may be stressed a billion times or more during years of continuous service. High fatigue strength is also important in this application.

Tools are another significant application for these alloys. When struck together they do not spark, meaning they do not present a safety hazard in environments where fumes from fuels or flammable solvents may be present. The alloys are also nonmagnetic, have good resistance to corrosion, and retain their strength at higher temperatures better than aluminum alloys.

Heat Treating of Nickel-Base Superalloys

The three types of superalloys are based on nickel, iron-nickel, and cobalt-nickel. All are characterized by their retention of strength at elevated temperatures and resistance to surface degradation, making them suitable for jet engine applications, for example. They are identified only by generic names or tradenames. The cobalt-base alloys are classified as solid-solution alloys; the iron-nickel-base and nickel-base alloys are classified as either solid-solution or precipitation-hardening alloys.

As a general rule, superalloys require higher solution temperatures and higher aging temperatures than aluminum or copper alloys. For example, aluminum alloy 2024 is solution treated at about 510 to 540 °C (950 to 1000 °F), while most wrought nickel-base alloys are solution treated at about 955 to 1150 °C (1750 to 2100 °F).

In addition, nickel-base superalloys are often air cooled to room temperature because of the high temperatures in solution treating, the sluggishness of precipitation, and the danger of cracking in some instances. These nickel-base superalloys are reheated to about 700 to 815 °C (1300 to 1500 °F) for aging, which provides excellent strength in many critical high-temperature applications.

Many of the superalloys are hardened by the precipitation of coherent ordered phases, Ni_3Al or Ni_3Ti. Precipitation hardening differs from that of aluminum alloys and other age-hardenable systems. In aluminum alloys, overheating causes overaging, with an accompanying decrease in strength. To recover these properties, it is necessary to reheat to the solution treating temperature, cool rapidly, then age again. By comparison, when nickel-base alloys are overaged the precipitate redissolves. How-

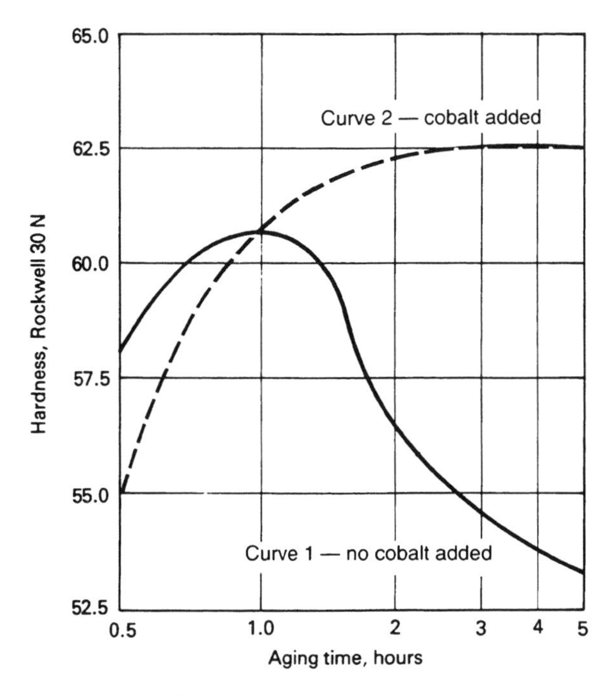

Fig. 6-11 Effect of cobalt addition on beryllium-copper alloys

ever, it is not necessary to quench rapidly to a low temperature to reprecipitate the hardening elements, because precipitation occurs again and the strength of the alloy is restored with cooling to a somewhat lower range of temperatures. This phenomenon prevents excessive creep in structures that may be overheated in service.

Example: Hastelloy B

Hastelloy B is a nickel-base solid-solution alloy containing chromium, nickel, cobalt, molybdenum, iron, carbon, and vanadium. The as-cast microstructure of this alloy is shown in Fig. 6-12(a) and the structure of the casting after annealing in Fig. 6-12(b).

Recommended heat treating practice is as follows.

Solution treating is at 1175 °C (2145 °F) for 30 minutes. Cooling rapidly to below 540 °C (1000 °F) is necessary to prevent precipitation in the intermediate temperature range. For most sheet metal parts, rapid air cooling is sufficient. Oil or water quenching media are often required for heavier sections that are not subject to cracking. The reheating requirement in aging is met when the part is subjected to elevated temperatures in service.

Full annealing (complete recrystallization and maximum softness) is required in stress relieving because intermediate temperatures cause aging.

Annealing is at 1175 °C (2145 °F); holding time is 1 hour per inch of section.

Example: Astroloy

Astroloy is a nickel-base precipitation-hardening alloy that contains chromium, nickel, cobalt, molybdenum, titanium, aluminum, iron, carbon, boron, and zirconium. Figure 6-13(a) shows the microstructure of a forging that was solution annealed for 1 hour at 1150 °C (2100 °F) and air cooled; it consists of grain boundaries and fine carbides in a gamma-phase (fcc) matrix. The mi-

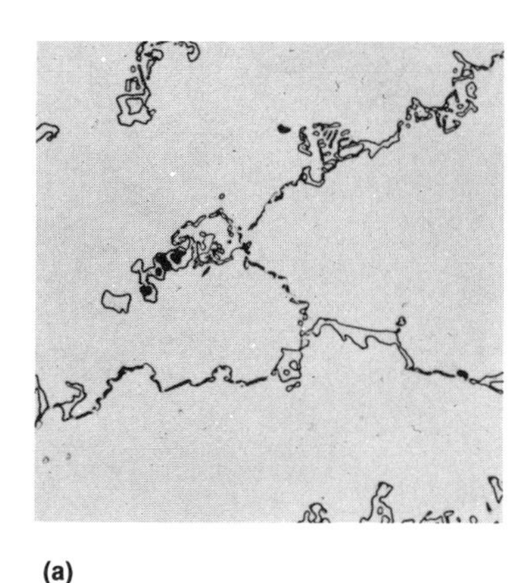

(a)

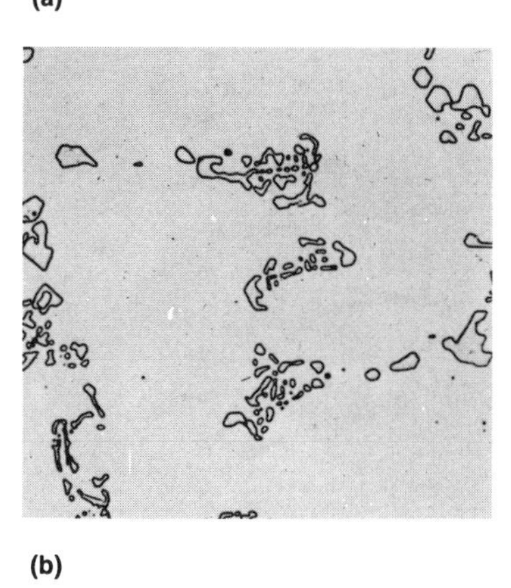

(b)

Fig. 6-12 (a) As-cast structure of Hastelloy B. (b) Structure after annealing

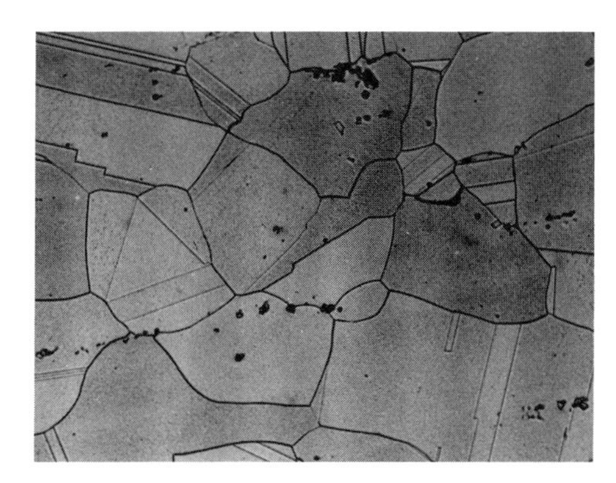

(a)

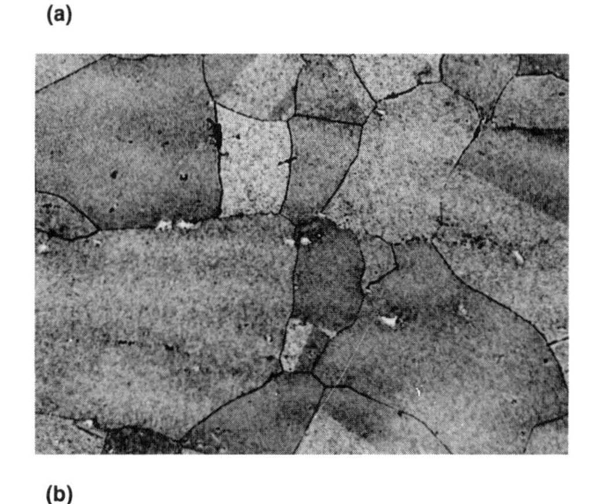

(b)

Fig. 6-13 Microstructure of solution-annealed Astroloy forging after different annealing times. See text for details. 100×

crostructure in Fig. 6-13(b) was obtained by this process:

- Solution annealed 4 hours at 1150 °C (2100 °F), air cooled
- Aged 4 hours at 1080 °C (1975 °F), oil quenched
- Aged 4 hours at 845 °C (1555 °F)
- Aged 16 hours at 760 °C (1400 °F), air cooled

Recommended heat treating practice is as follows:

Solution Treating and Aging. Alternative treatments are available:

- Solution heat treat at 1175 °C (2145 °F) for 4 hours, air cool; or, solution heat treat at 1080 °C (1975 °F) for 4 hours, air cool
- Age at 845 °C (1555 °F) for 24 hours, air cool; or age at 760 °C (1400 °F) for 16 hours, air cool

Stress Relieving. Full annealing is recommended because intermediate temperatures cause aging.

Annealing is at 1135 °C (2075 °F); holding time is 4 hours per inch of section.

Heat Treating of Copper-Zinc Alloys

In the solution treating and aging of nonferrous alloys, the solvent is the major phase and the solute precipitate is the minor phase, in terms of percentage of content. Not so for some of the two-phase titanium-base alloys (discussed in the next section) and copper-zinc alloys high in zinc. In this instance, both phases are similar in quantity. Muntz metal, an alloy containing 60% copper and 40% zinc, is an example. The structure and amount of each phase are determined by the treatment temperature and the cooling rate from that temperature.

The area of interest in this instance is identified in the simplified version of the copper-zinc phase diagram in Fig. 6-14. Note that from a temperature of 903 °C (1655 °F) down to about 450 °C (840 °F), the alloy is in the alpha plus beta (α β) phase. The alpha phase is fcc,

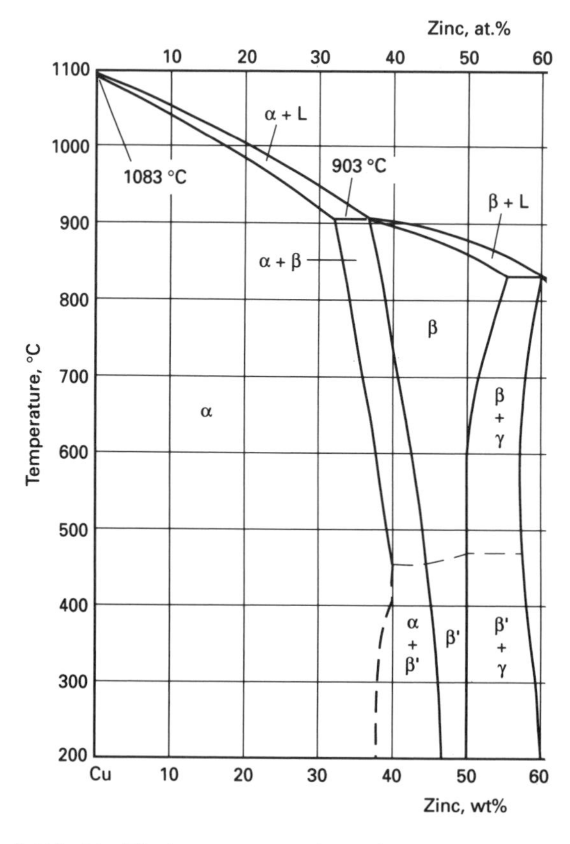

Fig. 6-14 Modified copper-zinc phase diagram

whereas the beta phase is bcc. At this time, the copper and zinc atoms are located at random in the lattice sites.

The copper-zinc alloys on cooling below about 450 °C (840 °F) in the area below the horizontal dashed line in Fig. 6-14 take up new positions, forming an ordered structure called a *superlattice*. Note that this area is identified as alpha plus beta prime (α + β') and that zinc content ranges from about 38% to about 48%. The area to the immediate right of this is designated beta prime (β'). It extends from about 48 to 50% zinc.

Alloys completely in beta prime form are not suitable for commercial use, because the structure is brittle. By comparison, the alpha-beta prime (α-β') combination is important commercially. Muntz metal is an example.

The typical microstructure of annealed Muntz metal is shown in Fig. 6-15. White regions are beta prime (β'); dark and gray regions are alpha (α). Applications range from architectural panel sheets and tubing for heat exchangers, to hot forgings and brazing rod for copper alloys and cast iron. The general corrosion behavior of Muntz metal is typically good. Its forgeability rating is 90% of the standard forging brass. Annealing temperatures range from 425 to 600 °C (800 to 1110 °F), and hot working temperatures range from 625 to 800 °C (1155 to 1470 °F).

Heat Treating of Titanium-Base Alloys

Based on microstructural phases retained at room temperature as a result of alloying, titanium alloys fall into three categories: alpha (α), alpha-beta (α-β), and beta (β). In many respects, phase changes parallel those in iron. Titanium has two allotropic forms: At room temperature the stable form is the alpha phase, which is hcp; above 885 °C (1625 °F), in the beta phase, the structure is bcc.

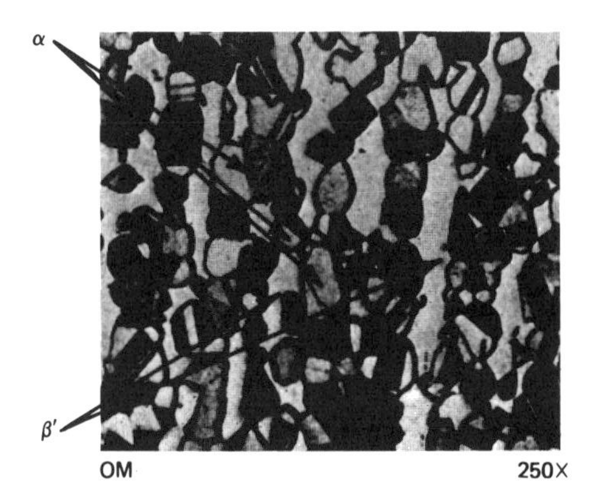

Fig. 6-15 Typical microstructure of Muntz metal

Alpha-type alloys cannot be hardened significantly by heat treatment. However, as with copper, nickel, and aluminum alloys, they can be strengthened with cold work and recrystallization annealing. The alpha-beta and beta titanium alloys can be strengthened by heat treatment.

Example: Ti-3Al-2.5V

Ti-3Al-2.5V is an example of a near-alpha alpha-beta alloy that is generally used in the cold-worked and stress-relieved condition. This alloy can be heat treated to high strength but has limited hardenability. Cold forming properties are rated excellent. Yield strengths obtained are 30 to 50% higher than those of unalloyed titanium. Moderately high strength and good ductility are obtained with cold working reductions of 75 to 85%.

The alloy was developed originally as seamless tubing for aircraft hydraulic and fuel systems and has performed well in high-technology military aircraft, spacecraft, and commercial aircraft. Other tubing applications include golf club shafts, tennis racquets, bicycle frames, medical and dental implants, and barrels for expensive ballpoint pens.

Recommended Heat Treating Practice. Ti-3Al-2.5V can be used in the annealed or in the cold-worked plus stress-relieved conditions. Typical heat treating procedures include:

- *Minimum stress relief* is at 315 °C (600 °F) for 30 seconds, followed by air cooling.
- *Typical stress relief* is at 370 to 650 °C (700 to 1200 °F) for 30 seconds to 2 or 3 minutes, followed by air cooling. Heating above 540 °C (1000 °F) substantially reduces strength and hardness.
- Annealing is at 650 to 760 °C (1200 to 1400 °F) for 30 seconds to 2 minutes, followed by cooling of the same duration. Heating to 705 °C (1300 °F) for 2 hours develops a fully annealed condition. There is no advantage in annealing above 800 °C (1470 °F).

Example: Ti-6Al-4V

A widely used alpha-beta alloy, Ti-6Al-4V (Fig. 6-16) is solution treated and aged following standard procedures. Solution treating usually is below the *beta transus* (the temperature above which the alloy is in the 100% beta phase) in the alpha-beta field, because at higher temperatures grain growth is excessive.

Recommended Heat Treating Practice. This alloy is used in many fracture toughness appli-

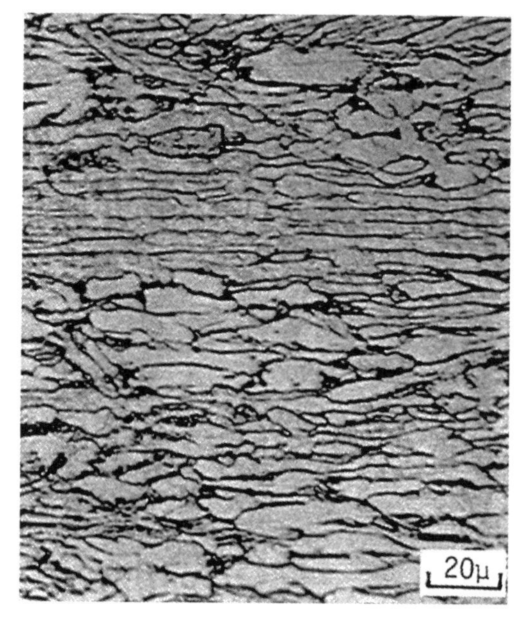

(a)

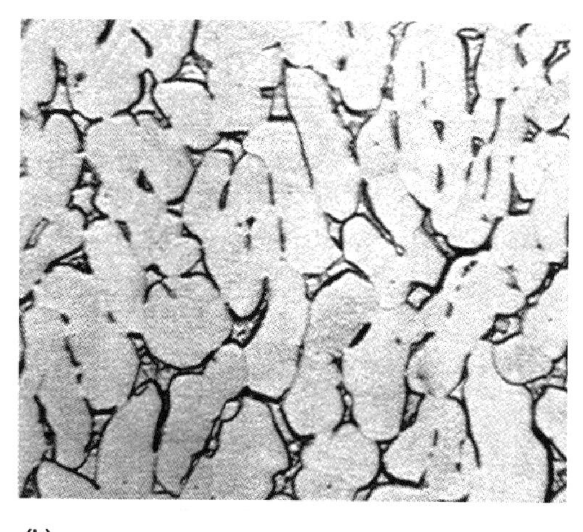

(b)

Fig. 6-16 Typical microstructures of Ti-6Al-4V. (a) Mill-annealed textured plate. (b) Conventionally processed plate after recrystallization annealing at 980 °C (1795 °F) for 30 minutes. 500×

cations. For optimum fracture toughness properties, one of two different heat treating cycles is used:

- *Recrystallization annealing.* Parts are heated high in the alpha-beta field, then air cooled to ambient temperature. Parts are then reheated in the range of 730 to 790 °C (1350 to 1450 °F) for a short time, followed by air cooling. This treatment produces a recrystallized alpha-beta structure with high fracture toughness.
- *Beta annealing is the alternative treatment. Parts are heated about 30 °C (50 °F) above the beta transus temperature for a short time, then air cooled. Annealing is at 730 to 790 °C (1350 to 1450 °F).*

Caveats: Titanium has a strong affinity for oxygen, nitrogen, and hydrogen gases. Oxygen and nitrogen in the air create surface contamination at temperatures above 540 °C (1000 °F). The contaminated surface is brittle and must be removed by mechanical or chemical means, or both, prior to further processing or fabrication. Hydrogen can be absorbed during thermal treatments. However, control of furnace atmospheres and use of protective coatings reduces hydrogen pickup. The alternative is to degas at 650 to 815 °C (1200 to 1500 °F) in a vacuum environment.

Chapter 7

Hot Working and Cold Working of Ferrous and Nonferrous Metals

Both hot working and cold working are involved in the conversion of metals and alloys to semifinished (semifabricated) wrought products such as sheet, plate, and bar. These products are subsequently used in the manufacture of articles ranging from pop and beer cans to panels destined for car and pickup truck bodies.

By comparison, the semifinished stage is skipped in making parts in the "other than wrought" category—namely castings and powder metallurgy parts. These are ready for use "as is" or "almost as is"; further processing, such as machining, may not be required. Castings and P/M parts are known as near-net shapes, a topic outside the scope of this chapter.

Hot Working Technology

Cast ingots mark the first step in converting liquid metal to solid wrought metal, in which two casting processes are used: static casting and continuous casting.

Static Casting Process

In static casting, the typical practice is to pour, or teem, molten metal into a vertical, open-ended mold (Fig. 7-1). The mold is made of metal, usually cast iron, and is tapered along its length so that solidified metal can be easily withdrawn.

A number of undesirable events may take place during the solidification of a cast ingot. The most common unwanted occurrence is the formation of a conical cavity, called *pipe,* at the open end of an ingot. Pipe may involve up to 20% of the total length of an ingot; before further processing can proceed, the entire section must be removed, or cropped. Cropped metal can be recycled, but it represents considerable waste.

Static casting has an inherent shortcoming: It is a batch process, making it less than ideal for volume production. This niche has been filled by the development of the continuous casting process.

Continuous Casting Process

Refer to Fig. 7-2(a), starting at the top and moving downward, reading each label along the way. Molten metal—indicated by the white stream with black dots, known as the *melt*—is poured in from the top. The temperature of the metal is reduced in the water-cooled mold. Note that from the middle of the mold, solidification has started, beginning at the outer surface of the mold and moving toward the middle of the embryonic ingot. From this point, the liquid-solid

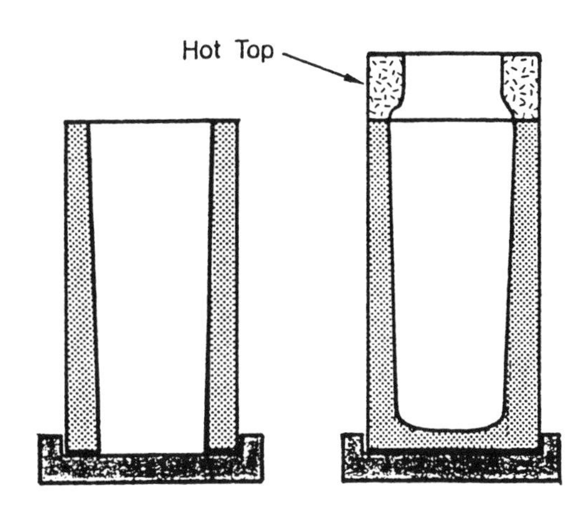

Fig. 7-1 Statically cast ingots

metal assumes a wedge shape as it passes through the solidification zone and through the first part of the water spray zone. All the metal is solidified from this point on. The shape of the ingot is indicated by solid black.

The solidified ingot may be handled in several ways. For instance, the ingot, which is growing in length, can be directed into a holding pit. At the appropriate time, the continuous caster is shut down; a section of the ingot is cropped and removed from the holding pit, and then the process is restarted. This is known as *semicontinuous casting,* which was pioneered by the aluminum industry.

In another process, the ingot is bent (Fig. 7-2a) so that it emerges horizontally from the casting machine. With this setup, casting time is limited only by the amount of liquid metal that can be supplied to the caster. The ingot can be cut to any desired length, or the growing ingot can be fed directly into other processing equipment. Aluminum, steel, and copper are continuously cast.

Horizontal casting, a third approach, is shown in Fig. 7-2(b). This design is limited to metals with lower melting points, such as copper, aluminum, and cast iron.

The term *continuous* is a misnomer. The process is continuous only as long as liquid metal is being fed into the system. An infinite source of liquid metal probably is out of the question. Practice in the steel industry, for example, is to pour liquid steel from steelmaking furnaces into holding vessels, which are transported to the site of the casting operation. At this point, the steel is refined (i.e., some impurities are removed) before it is poured into the caster. In fact, the quality of continuously cast metal is better than that which is static cast. For one thing, pipe is virtually eliminated. Significant production advantages also may be realized by reducing or simplifying subsequent operations traditionally involved in converting to semifinished products. The shape and length of an ingot, for example, can have a desirable impact on the nature and number of rolling operations traditionally needed in producing plate and sheet.

Next Step: Hot Working Ingots

Of course, not all ferrous and nonferrous metals are continuously cast. Many ingots continue to be statically cast.

Ingots are hot worked, or plastically deformed, chiefly by rolling, forging, and extrusion. Ingots of aluminum and copper are allowed to cool prior to hot working. They are reheated for hot working.

In the steel industry, on the other hand, operating temperatures are high and large quantities of materials must be handled. To maintain the hot ingots at the uniform temperature required in subsequent processing, they are placed in heated refractory-lined chambers known as soaking pits; in this instance, heat is the soaking medium. The practice saves energy. However, massive equipment still is required in the first stages of hot working. Large, heavy sections of metal are handled, and the objective is to obtain large reductions. Properly heated metal is in its most plastic condition.

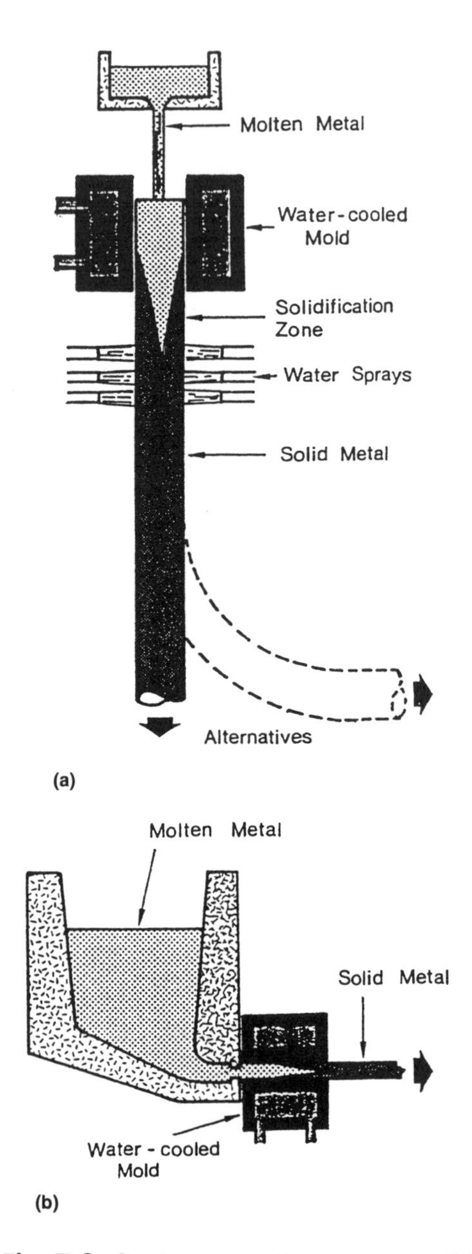

Fig. 7-2 Continuous casting processes. (a) Semicontinuous casting. In the alternative method (dashed lines), the ingot is bent. (b) Horizontal casting

Hot Rolling Technology

Hot ingot produced by static casting or by continuous casting is squeezed in rolling by forcing it through a gap between two rollers that are turning in opposite directions. The gap between the rollers is smaller than the thickness of the ingot, which is reduced in thickness and increased in length. Changes in width are minor. To get the desired result, the process is repeated as needed. With each pass through the rollers, gap width is reduced. The operation is stopped if the material has cooled too much to be worked further. At this point, the material may be reheated and annealed to restore ductility. The amount of reduction experienced by the material is expressed as a percentage: final thickness versus original thickness.

The equipment used in the first stage of breaking down an ingot is called a primary rolling mill, which is massive in both size and investment. Smaller secondary rolling mills take over if further work is necessary. Continuously cast ingots may be fed directly into secondary mills. Flat products such as sheet, plate, or strip are first rolled in primary mills, then reduction is continued using cylindrical rolls. Work may be done by a single rolling mill or by a number of mills set up in series. In the latter procedure, roll gaps are progressively smaller than those of preceding stands. The resulting long strip can be coiled and handled as a continuous strip in the next stage of fabrication, or it can be cut into sheets of desired length.

Continuous strip mills are commonly used in the steel industry in the production of thin, flat products. Virtually all steel sheets used in car bodies and food containers are rolled in an intermediate reduction stage in mills of this type.

Sections other than simple plate or sheet are rolled in secondary mills through rolls in which pairs of grooves have been cut. The rolls shown in Fig. 7-3 are used in a three-high, nonreversing mill designed to produce railroad rails. A square billet is passed backward and forward through grooves 1 to 4 in succession.

Hot Extrusion Technology

Ingots are round or prerolled billet, an intermediate form, heated to specified temperatures and placed in an extrusion press, which is an open-ended cylinder. A horizontal press that produces aluminum alloy bars and sections is shown in Fig. 7-4. The ram projects through one end of the cylinder, and the other end is closed by a plate in which a central hole has been machined. As the ram advances, metal is extruded by the die, like toothpaste emerging from a tube. The die determines the shape of the extrusion. Complicated sections, including tubes, are within the capability of the equipment.

The process has two main limitations: (1) the massive forces that are required to force the hot metal through the die, necessitating massive presses, and (2) the availability of extrudable metals. As expected, lower-melting-point alloys, such as lead, aluminum, and copper, are relatively easy to extrude. By comparison, steel is difficult to extrude. To some extent, extrusion of steel is facilitated by coating extrusion billets

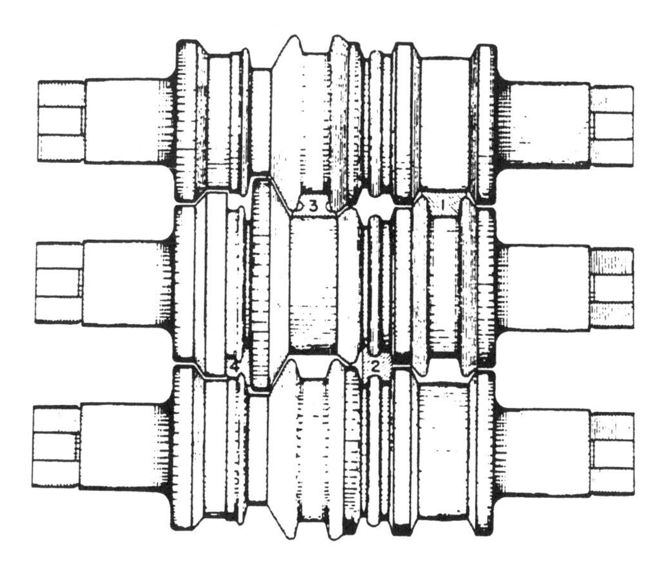

Fig. 7-3 Grooved rolls, used in secondary mills to produce complex shapes

with glass, which partially protects the die by serving as a lubricant.

Partmaking Technology

Thus far, the discussion has centered on the production of semifinished ferrous and nonferrous metal shapes, such as plate, sheet, strip, and bar. All are one or more steps away from evolving into end products.

Forging and sheet metal forming are among the important partmaking processes. Forged products include crankshafts for car engines (Fig. 7-5), adjustable wrenches, hammers, blades for steam turbines, jet engines, and landing gear components for aircraft. Sheet metal typically is formed into end products that require contoured shapes, ranging from saucepans to car bodies.

Stretch forming is used in making products ranging from brackets to shaping skins for airplane wings from sheet too large to be handled by conventional press equipment. Five methods of stretch forming are illustrated in Fig. 7-6. Sheet, bar, and rolled or extruded sections are formed over a form block of the required shape, while the workpiece is held in tension. Metal is stretched just beyond its yield point to retain the contour of the form block.

Open-Die Forging

In forging ingots, metal is squeezed between two anvils, one of which is actuated by a large press, such as the steam-driven power hammer shown in Fig. 7-7. Faces of the anvils may be flat or recessed with V-shape or concave depressions. Reductions are obtained by deforming lo-

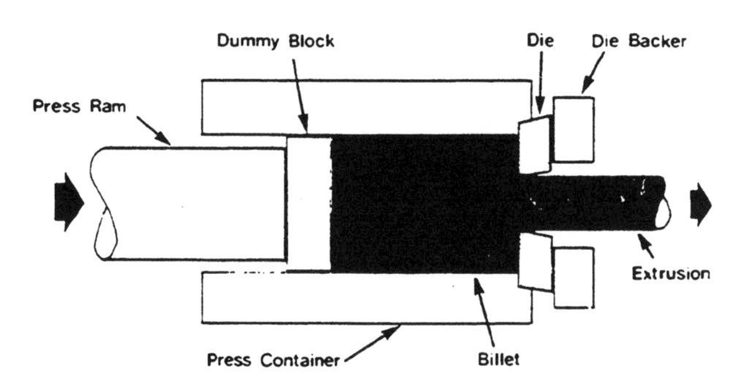

Fig. 7-4 Horizontal extrusion press

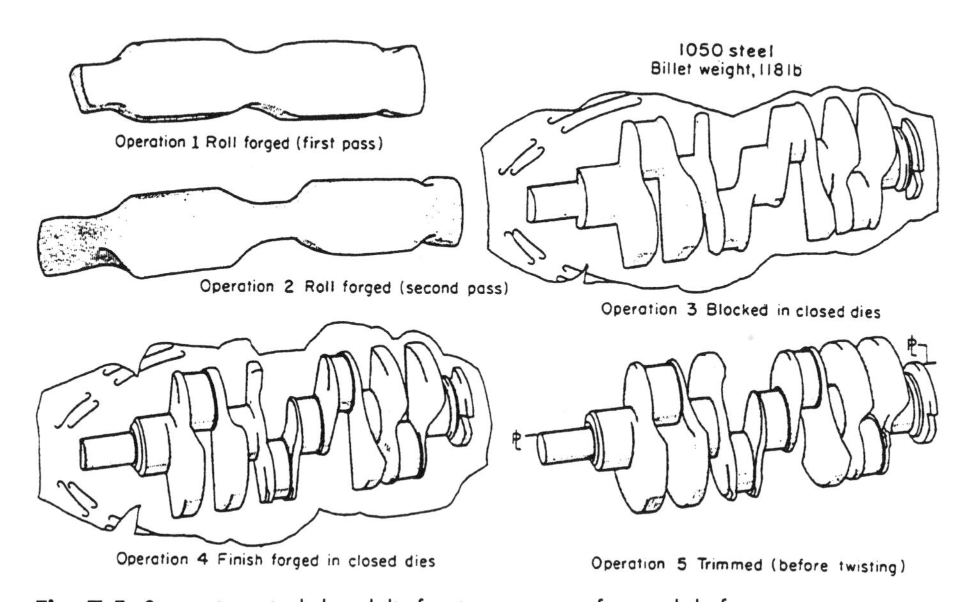

Fig. 7-5 Stages in typical closed-die forging sequence for crankshafts

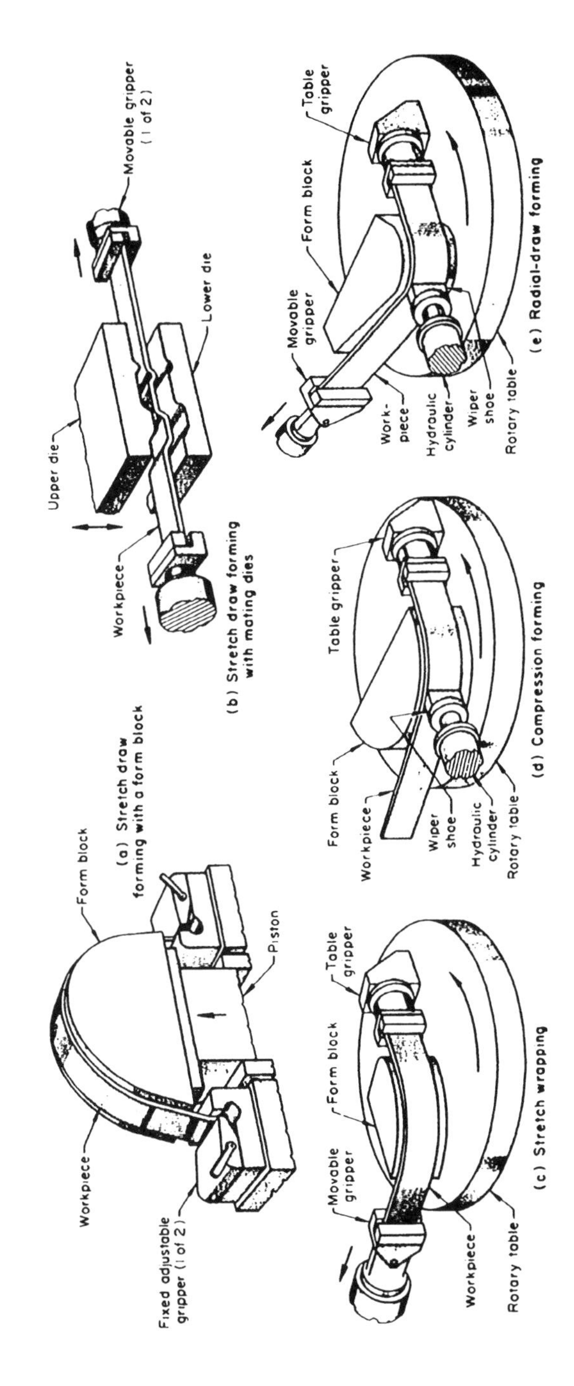

Fig. 7-6 Five methods of stretch forming

calized areas of the workpiece.

This is an operator-critical operation. The operator has complete control over the final configuration of the forging. Shapes are comparatively simple, and control of dimensions tends to be modest. Holes may be punched and forgings may be bent. Applications for such forgings include very large components such as rotors for power-generating turbines and propeller shafts for ships. A large ingot is being forged in the open-die press shown in Fig. 7-8. Figure 7-9 illustrates typical dies and punches for open-die forging.

Closed-Die Forging

In this operation, hot metal is shaped within the walls or cavities of two dies that come together to enclose the workpiece on all sides. Forging stock, which is usually round or square bar, is cut to provide the amount of metal needed to fill the die cavities, plus an allowance for *flash*—the metal that escapes in the space between the mating surfaces of dies, and which is typically removed in a subsequent operation. Flash also acts as a brake to slow the outward flow of metal and permit complete filling of thin sections. A hydraulic press used in closed-die forging is shown in Fig. 7-10. Figure 7-11 illustrates the closed-die forging process.

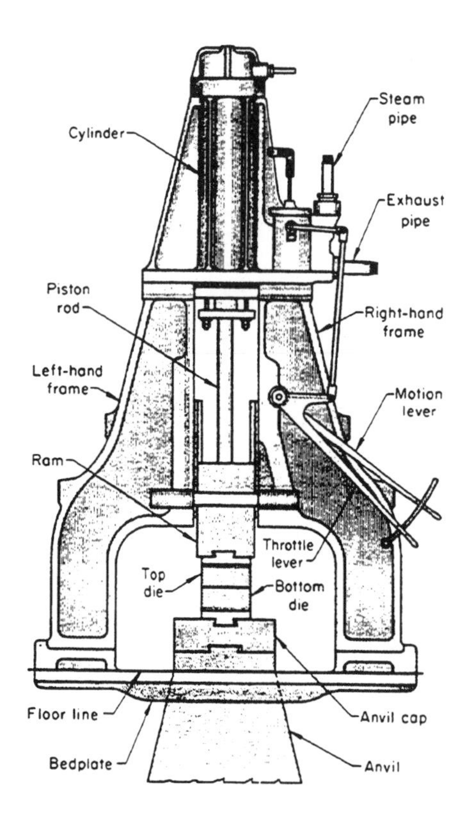

Fig. 7-7 Steam-driven power hammer for open-die forging

With closed dies, complex shapes and heavy reductions can be made in hot metal within closer dimensional tolerances than are usually feasible with open dies. Open dies are used primarily for forging simple shapes or for making forgings too large to be contained in closed dies. Closed-die forgings usually are designed to minimize follow-up machining.

Closed-die forging is suitable for either low-volume or high-volume production. In addition to producing net or near-net shapes, this process allows control of grain flow direction, which can improve mechanical properties in the longitudinal direction of the workpiece. An etched section of a forged steel crane hook is shown in Fig. 7-12. Note that the flow lines are parallel to the surface of the hook, improving strength in tension, the direction in which the hook needs to be strong in service. Forgeability varies considerably among the different ferrous and nonferrous metal alloys, as indicated in Table 7-1.

Closed dies generally are classified as blocker, conventional, and close-tolerance types. These are discussed in the paragraphs that follow.

Blocker forgings are produced in relatively inexpensive dies, but in comparison with conventional closed-die forgings they weigh more and are somewhat larger in dimension. Blocker-type forgings are fairly close to near net-shape. Sometimes they are specified when production volume is low and the cost of machining to net shape is not excessive.

Conventional closed-die forgings are produced to comply with commercial tolerances within the broad range of general forging practice. They are closer to net shape than blocker-type forgings, are lighter, and have more detail.

Table 1 Classification of alloys for forgeability

Alloy group	Approximate forging temperature range °C	Approximate forging temperature range °F
Aluminum alloys (least difficult)	400–550	750–1020
Magnesium alloys	250–350	480–660
Copper alloys	600–900	1110–1650
Carbon and low-alloy steels	850–1150	1560–2100
Martensitic stainless steels	1110–1250	2010–2280
Maraging steels	1100–1250	2010–2280
Austenitic stainless steels	1100–1250	2010–2280
Nickel alloys	1000–1150	1830–2100
Semiaustenitic PH stainless steels	1100–1250	2010–2280
Titanium alloys	700–950	1290–1740
Iron-base superalloys	1050–1180	1920–2160
Cobalt-base superalloys	1180–1250	2160–2280
Niobium alloys	950–1150	1740–2100
Tantalum alloys	1050–1350	1920–2460
Molybdenum alloys	1150–1350	2100–2460
Nickel-base superalloys	1050–1200	1920–2190
Tungsten alloys (most difficult)	1200–1300	2190–2370

Close-tolerance forgings usually are held to closer dimensional tolerances than conventional forgings. They are in the near-net or net shape. Close-tolerance forgings cost more than conventional types, but lower finishing costs often are a trade-off.

Hot Upset Forging

Hot upset forging is essentially a process for enlarging and reshaping part of the cross-sectional area of a bar or tube. Heated forging stock is held between grooved dies, and pressure is applied to the end of the stock in the direction of its axis. Work is done by a heading tool that spreads the end of the metal by displacement.

Cold Forming Technology

Plastic deforming via cold working is one of the basic methods of strengthening metals. Hardness, tensile strength, yield strength, and electrical resistivity are increased. On the deficit side, ductility is reduced. The process goes by such names as *cold work, work hardening,* and *strain hardening*—technically, the last is the most descriptive of the process. Plastic deformation not accompanied by work hardening is known as *hot working*. Work hardening is discussed in the section on hot working technology in this chapter.

Cold and hot working can be compared in Fig. 7-13. The intention is to depict hot working by rolling. However, part of the figure also illustrates what happens when strain is applied to a normal grain structure in cold working. In this instance, steel is being rolled. Steel, aluminum, and brass are among the alloys that are processed by both hot and cold rolling. In Fig. 7-13, stages of the process run left to right, starting with "original grains." Note the size, shape, and orientation of the grains. Also note that as the steel passes through the rollers, the grains are deformed and change in size, shape, and orientation due to the applied strain. In cold rolling, it is at this point that strengthening is initiated. However, the scenario is different in hot rolling from this point of the process and until the process is complete. Successive changes (labeled in Fig. 7-13) occur until finally the structure is composed of entirely new grains—a result similar to that obtained in recrystallization annealing.

Cold work and annealing theory are closely related. For example, go back to the left side Fig. 7-13, where the original grain structure is plastically deformed by the rolls. Grains in this configuration impede the flow of electricity—hence, an increase in electrical resistivity. Conversely, electrical conductivity is restored by annealing when the condition of the original grain is restored.

To explain further: Most of the energy used in cold working increases the temperature of the metal. Try bending coat hanger wire back and forth rapidly until it breaks. Note that the metal in the break areas is hot to the touch. Part of the heat generated in cold working is stored in strained or deformed metal in the form of lattice distortion, which increases strength. Annealing can be used to restore ductility by supplying enough energy to expedite atom movement,

Fig. 7-8 Huge ingot being forged in an open-die press

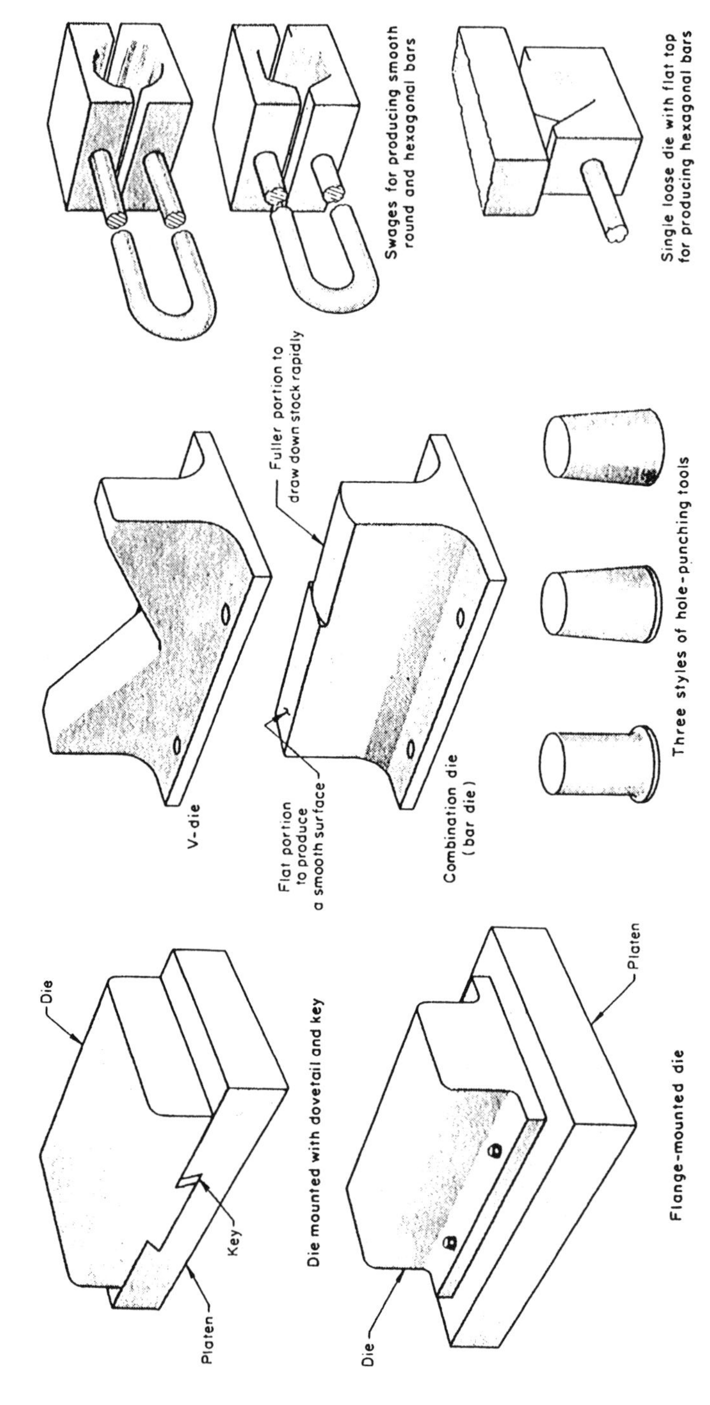

Fig. 7-9 Dies and punches used in open-die forging

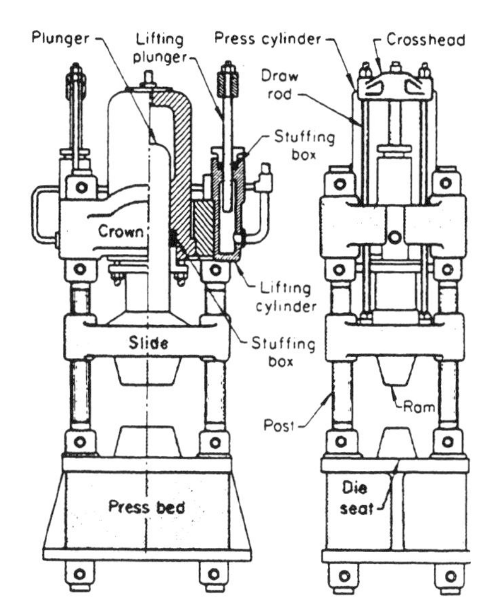

Fig. 7-10 Hydraulic press used in closed-die forging

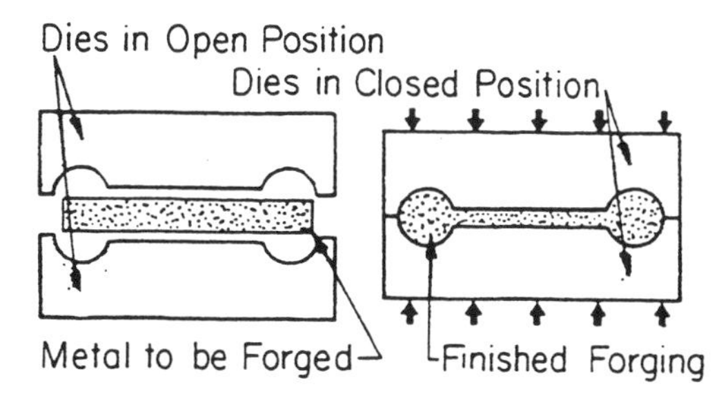

Fig. 7-11 Closed-die forging of metal

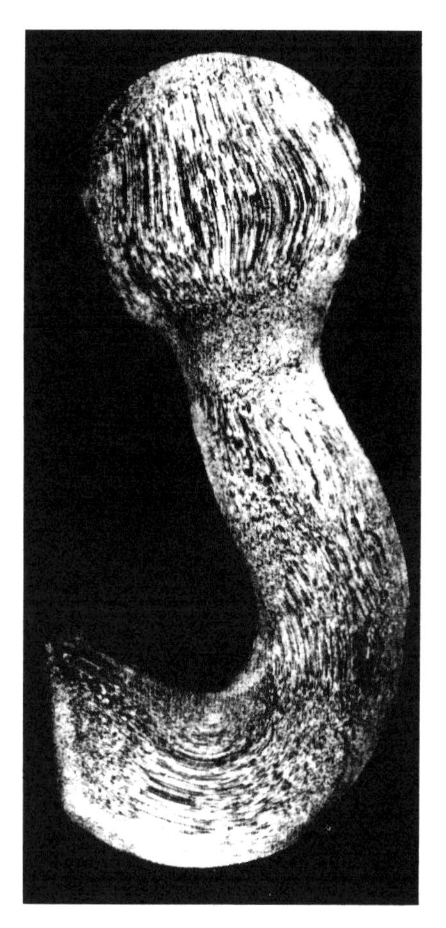

Fig. 7-12 Forged crane hook. Note direction of flow lines.

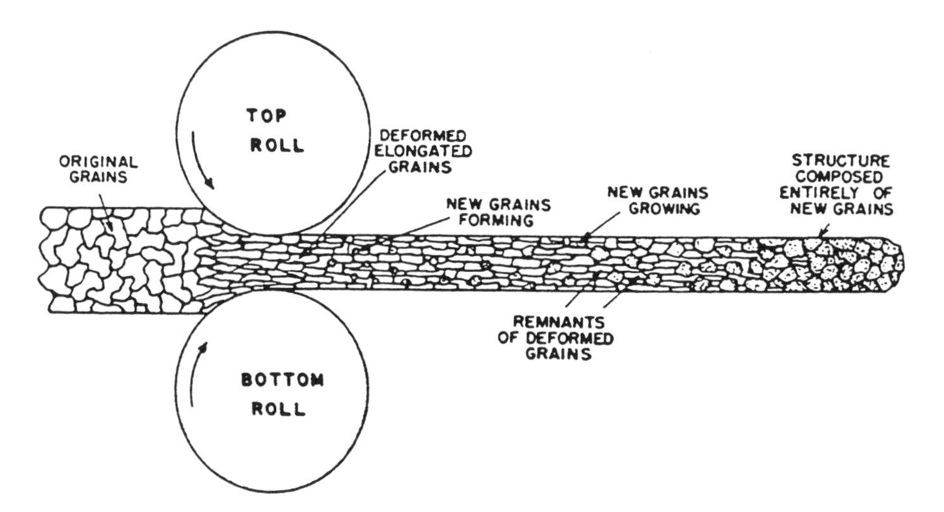

Fig. 7-13 What happens in hot (and cold) rolling, showing the combined effects of the mechanisms of hot rolling and recrystallization on the grain structure of steel. Grain size is exaggerated for clarity.

which eventually removes strains from deformed metals.

The essential function served by annealing in conjunction with cold working is explained by another example. Plastic deformation cannot be continued to infinity. At some point, the metal fractures. Annealing restores metal to a structural condition similar to that present prior to deformation, making additional cold work possible. In the process, hardness increases slightly, but the values of other properties return to where they were prior to cold work. This process is known as *recovery annealing*.

With longer times at temperature, or at higher temperatures, resulting changes in lattice structure are more pronounced. Crystals begin to nucleate (form embryos) in some areas. With time, grains grow to the point where they touch one another, and all evidence of cold work has disappeared. Formation of new grains is called *recrystallization annealing*. Completion of this phase marks the beginning of growth in grain size (refer to the right side of Fig. 7-13). With further annealing, grains continue to get larger.

The microstructure of the copper-zinc alloy shown in Fig. 7-14 was cold rolled to 60% reduction in thickness; annealing was at 400 °C (750 °F) for different times. Figures 7-14 and 7-15 are related. The purpose of the latter is to show the effect of annealing time on the same alloy at a fixed temperature. All other data are identical.

Note that recovery, recrystallization, and grain growth stages are identified at the top of Fig. 7-15; times are at the bottom. Points 3 and 5 are located in the flat part of the curve. Also note the growth in crystal size from points 3 to 5.

The effect of annealing temperature on the hardness and electrical resistivity of nickel is shown in Fig. 7-16. The metal was cold worked at 25 °C (75 °F) almost to fracture. Annealing time was 1 hour. Note the relationship of hardness (top curve) and electrical resistivity (bottom curve).

Cold Working

Deformation processes include those based on compression, tension, and bending.

Compression Processes. All compression operations reduce metal thickness by squeezing—either slowly as in press operations, or faster as in hammer forging. Rolling, another process of this type, is used to make bar, sheet, strip, plate, and other shapes known generically as semifinished or semifabricated products. Further work must be done to make articles of commerce.

Tension-type processes are characterized by pulling or stretching. Examples include stretch forming and the straightening of parts that were warped out of dimension due to prior processing. Processes in which metal is subjected only to tensile forces are few in number; usually, compressive and tensile forces coexist. Exceptions are drawing, wire drawing, tube drawing, deep drawing, and spinning.

Bending is involved in a number of metal-forming operations. Deep drawing and spinning are examples.

New Topic: Tempers

Typically, the term *temper* indicates that a metal has undergone a degree of cold deformation, has been subjected to a particular type of

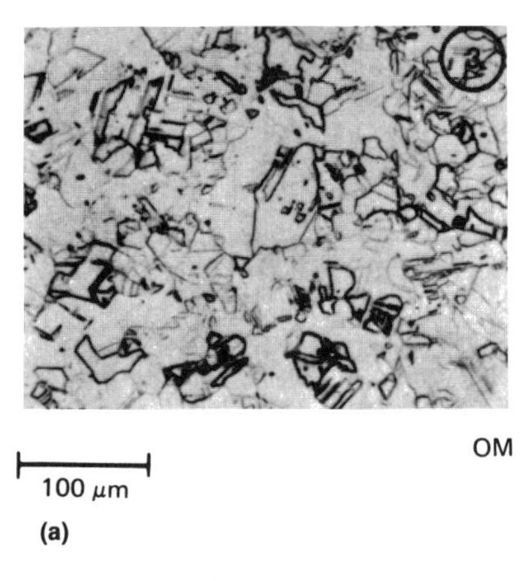

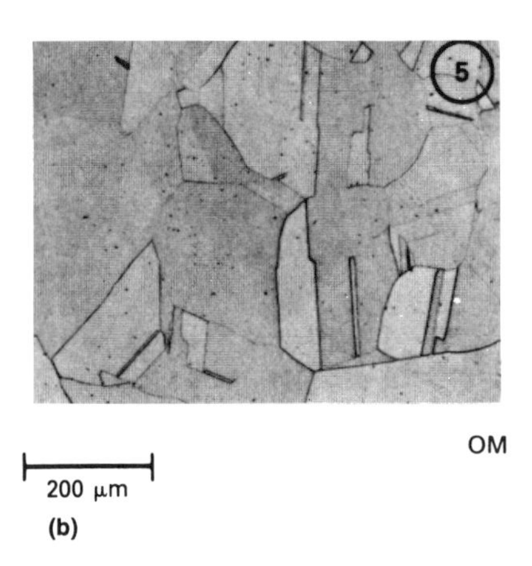

Fig. 7-14 Copper-zinc alloy where recrystallization is essentially complete and grain growth is starting. (a) Grain growth after 20% reduction. (b) Grain growth after 60% reduction

working process, or has been put through a particular heat treatment. In other words, *temper* describes how properties have been modified. For example, intermediate increases in strength are obtained by adjusting the amount of cold work, in percentage of reduction, put into a metal. The material is then described as being in a particular temper.

The T designation of temper in aluminum is an example:

- *T2* tells us that the metal has been cooled from an elevated temperature (as in hot work), then cold worked, then naturally aged. Results of any cold work are reported.
- *T3* metal has been solution heat treated, cold worked, then naturally aged.
- *T8* metal has been solution heat treated, cold worked, then artificially aged.

Cold Rolling

Production of thin strip starts with the output from a hot strip mill. Before cold rolling, the metal usually must be pickled to remove surface oxide or scale produced during hot rolling. In pickling, scale is removed by chemical or electrochemical means.

Cold rolling with cylindrical rolls is similar to the principle used in hot rolling. Roll surfaces must have a good finish and be flooded with lubricant to make product with a fine finish.

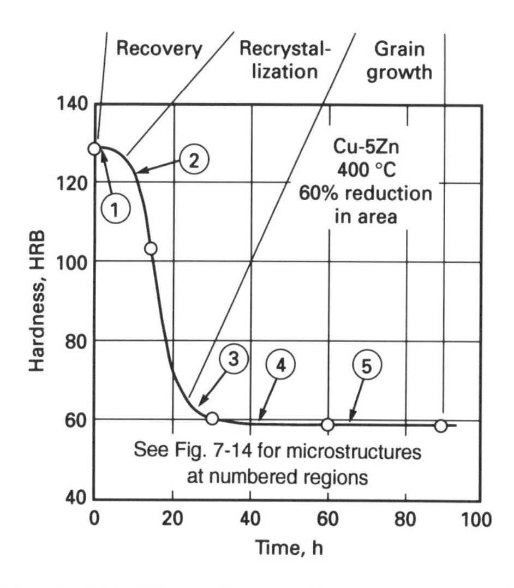

Fig. 7-15 Effect of annealing time on hardness of copper-zinc alloy at 400 °C (750 °F). Reduction is 60%. Note: The corresponding microstructures for points 3 and 5 are shown in Fig. 7-14.

Sheet may be reduced on a single-stand rolling mill, but the trend is toward four-high mills to get maximum uniformity in thickness across a strip.

Sheet and strip may be annealed at intermediate stages of reduction, or after the final reduction. Typically, this process takes place in a controlled atmosphere to keep surfaces free of oxidation and scaling. When a metal bar is rolled, any increase in width is minimal; however, the metal becomes thinner and longer. For all practical purposes, the volume of metal does not change.

Defects in rolling almost always are associated with excessive deformation, which causes the metal to fracture. The most common defect is *edge cracking,* or transverse cracks at the edge of rolled strip. Preventive measures include sufficient annealing between rolling operations and use of clean ingot surfaces.

The most important rolling operations are those in which cast ingots are converted into semifinished and finished products such as sheet, strip, rod, and tube. After cold work, metal is stronger and much more *homogeneous*—that is, uniform in grain structure—than it was at the ingot stage.

Contour Roll Forming

In this process, also known as *cold roll forming,* sheet or strip is formed into shapes nearly uniform in cross section. Rolling stock is fed longitudinally through a series of roll stations with contoured rolls (sometimes called roller dies), with two or more rolls per station. For an example of rolls for making shapes, see Fig. 7-3.

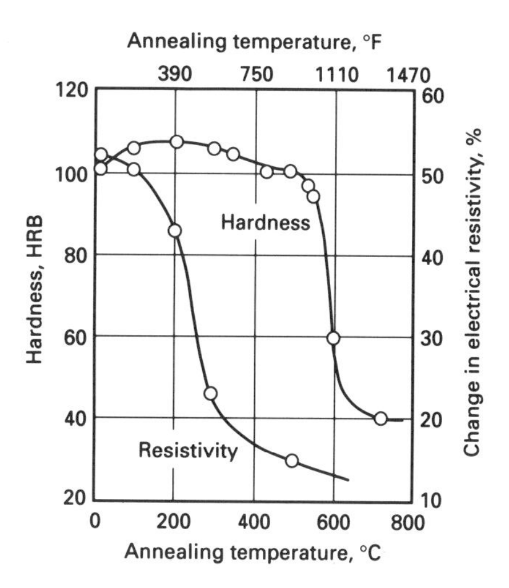

Fig. 7-16 Effect of annealing temperature on hardness and electrical resistivity of nickel

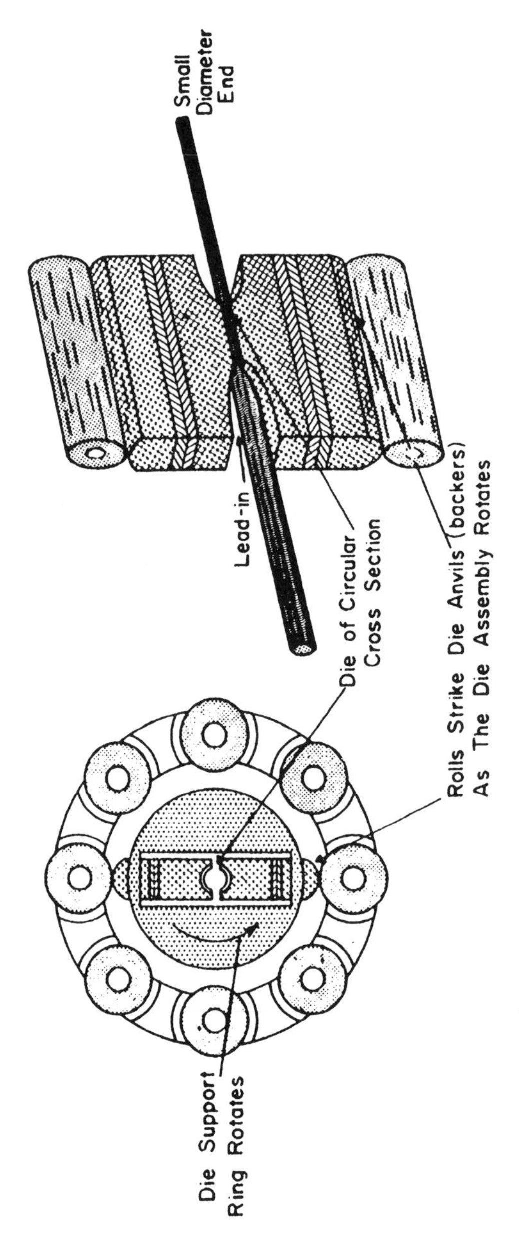

Fig. 7-17 Die used in swaging process

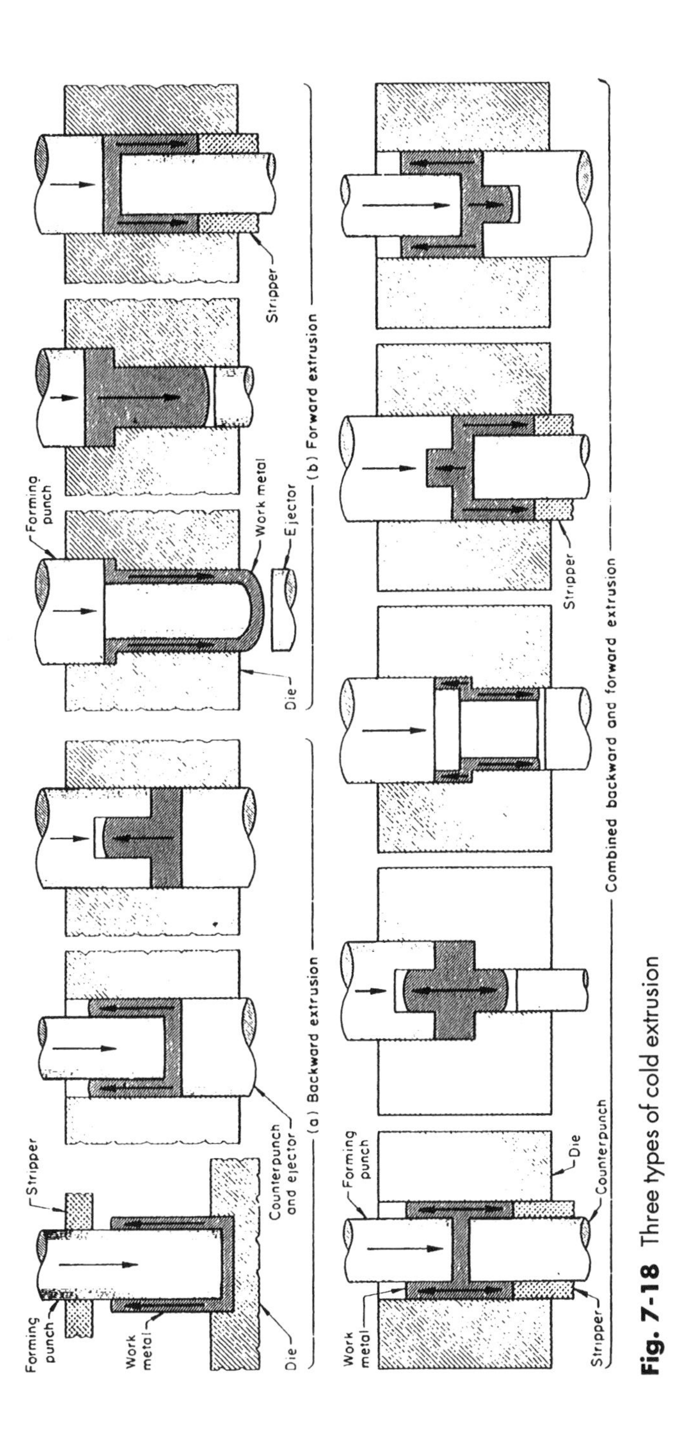

Fig. 7-18 Three types of cold extrusion

All metals suitable for common cold forming processes can be contour roll formed. The list includes carbon steel, stainless steel, aluminum alloys, and copper alloys. In the "less frequent" category are titanium and nickel-base superalloys.

Cold Forging

Cold forging has advantages over machining as a partmaking process: Waste material is not generated (machining produces chips); the tensile strength of forging stock is improved; and grain

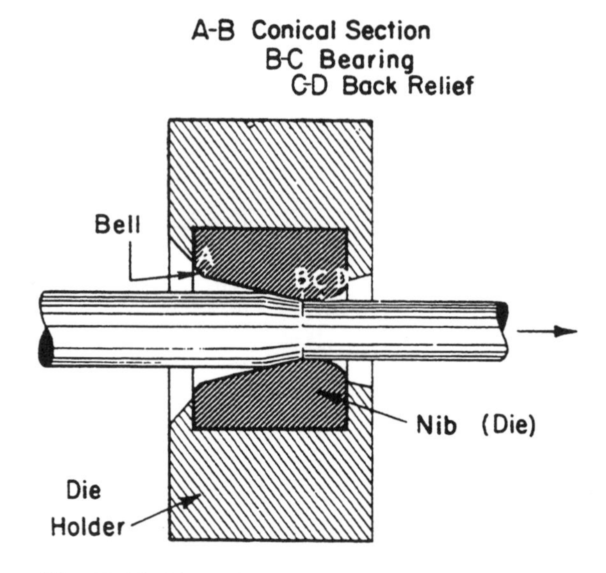

Fig. 7-19 Wire drawing process. Orville Wright took advantage of the high strength of drawn steel wire in support members of the Wrights' first airplane

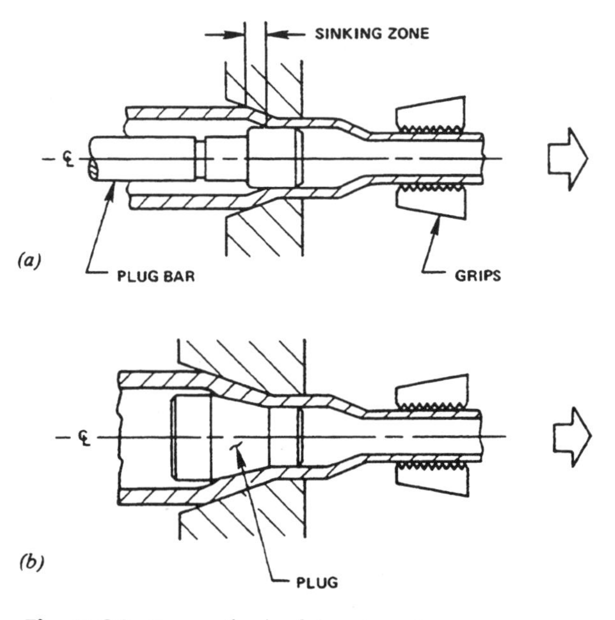

Fig. 7-20 Two methods of drawing tubes

flow can be controlled, as in conventional forging.

Cold heading is a cold forging process used to make a variety of small and medium-size hardware items, such as bolts and rivets. Heads are formed on the end of wire or rod forging stock by one or more strokes of a heading tool. Forming is by upsetting or displacement of metal. Heads thus formed are larger in diameter than the forging stock. In other applications, forging stock is displaced at a single point or several points along the length of a workpiece.

Low-carbon steel of a specified hardness is the major cold heading material. Copper, aluminum, stainless steel, and some nickel alloys also are cold headed. Titanium, beryllium, magnesium, and refractory metals, such as tungsten and molybdenum, are difficult to cold head at room temperature and are susceptible to cracking. They sometimes are warm formed.

Wire for cold heading generally is available in five levels based on surface quality. In order of increasing quality and cost, they are:

- Industrial quality
- Cold heading quality
- Recessed head or scrapless nut quality
- Special head quality
- Cold turned, ground, or shaved wire (seamless)

Swaging

In swaging, as in hammer forging, metal is deformed in compression in a series of rapid blows. Parts are circular in cross section and relatively small in diameter. The process is shown in Fig. 7-17. The die support ring rotates, and centrifugal force throws the die halves outward. As the die backers or anvils strike the hardened steel rollers, the dies are closed on the

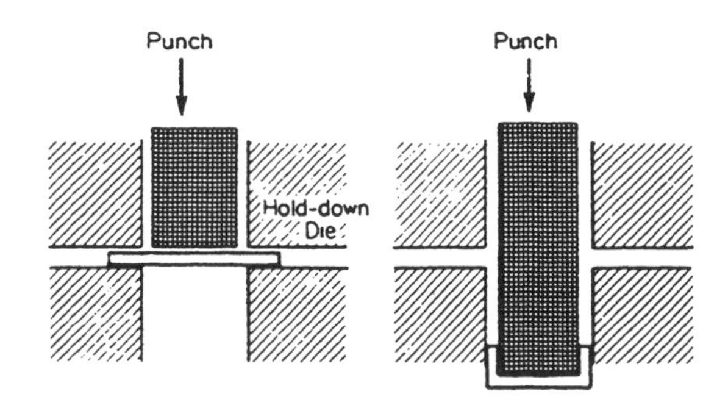

Fig. 7-21 Deep drawing process

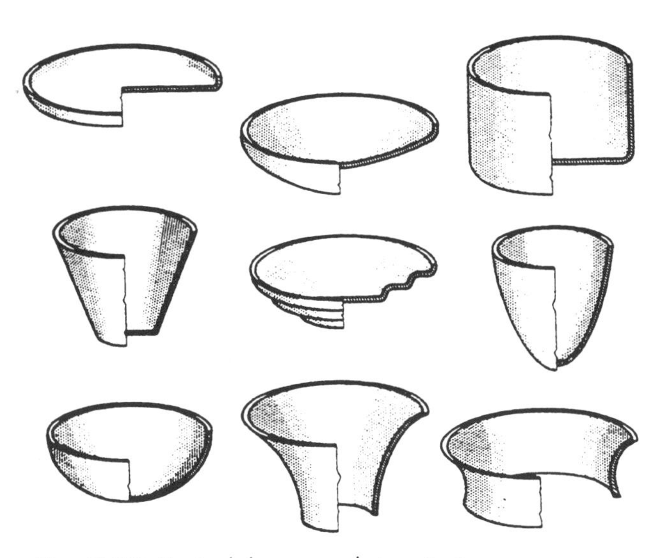

Fig. 7-22 Typical shapes made in spinning

work (right side in Fig. 7-17) under great pressure, reducing the diameter of the bar.

Swaging machines operate at several hundred strokes per minute. As the die assembly rotates, dies are forced together by striking the rolls several times in revolution. In the process, a square or hexagonal bar is reduced uniformly in diameter to a round bar.

Swaging may be done hot or cold. The objective is to reduce diameter to the point where a workpiece can be threaded through the dies of a wire drawing machine.

Coining

Coining is a compression-type operation in which a design or form is pressed or hammered into the surface of a workpiece. Coining is actually a closed-die squeezing operation in which all surfaces of the work are confined or restrained. It is also a striking operation, used to sharpen or change a radius or profile. In coining, the surface detail in dies is copied with a dimensional accuracy seldom equaled by any other process. Final products include coins and decorated items such as patterned tableware.

In minting coins, a small disk is placed in a die set. One half has the "heads," the other half the "tails." Dies are brought together and metal flows to fill the die cavities. Coining is one of the most severe of all metalworking processes.

When decorative articles with a design and a polished surface are required, coining is the only practical method available. It is also well suited to the manufacture of extremely small items, such as locking-fastener elements.

Dimensional accuracy in coining is equal to that in the very best machining practice. Many automotive elements are sized by coining. Semifinished products are often sized to meet specified dimensions. Powder metallurgy parts also are sized by coining.

Cold Extrusion

In this process, metal is squeezed or squirted through a hole in a die. Low-melting-point and relatively ductile metals such as aluminum, tin, and lead are easily extruded. With the help of special glass lubricants, some steels can be extruded.

Cold extrusion gets its name from the nature of the process. A slug or preform enters the die at room temperature or at a temperature appreciably below the recrystallization temperature. Any other rise in temperature is due to heat generated in plastic deformation, or by friction between die and workpiece.

There are three types of extrusion (Fig. 7-18): backward, forward, and combined backward-forward. Metal is displaced by plastic flow under steady though not uniform pressure, as follows:

- In *backward extrusion,* metal is displaced in the direction opposite that of punch travel (Fig. 7-18a). Workpieces are often cup shaped and have a wall thickness equal to the clearance between the punch and the die.
- In *forward extrusion,* metal is forced in the direction of punch travel (Fig. 7-18b).
- Sometimes both methods are combined so that some metal flows backward and some flows forward, as also shown in Fig. 7-18.

The most commonly cold-extruded metals are aluminum and aluminum alloys, copper alloys, low-carbon and medium-carbon steels, modified carbon steels, low-alloy steels, and stainless steels. The listing is in the order of decreasing extrudability.

Drawing

Drawing is one of the oldest metalforming operations. Long products are produced with excellent surface finishes, closely controlled dimensions, and constant cross sections. In drawing, a previously rolled, extruded, or fabricated product is pulled through a die at relatively high speed. In drawing steel or aluminum wire, for instance, exit speeds greater than 1000 meters per minute (more than 3000 feet per minute) are commonplace.

Wire Drawing

Industry uses cold working to strengthen semifinished products such as wire. Wire used in steel ropes and cables is made by cold drawing steel of controlled composition and microstructure, and is produced in a range of strengths up to 1380 MPa (200 ksi). Steel at this strength level is used in aircraft landing gear. Piano wire used in musical instruments is two and a half times stronger—the highest strength level that can be obtained with metals.

In wire drawing, a pointed rod or wire is pulled through a circular hole or die, and the diameter of the wire is reduced—as is its ductility. After a few reductions, the wire must be annealed to allow further reduction. Wire drawing is illustrated in Fig. 7-19. The conical section of the die lies at an angle of about 17°. The exit section (C-D in Fig. 7-19) is shown as back relief, and acts to strengthen the circular edge of the hole and prevent breakage.

Wire drawing dies typically are made of tool steel or sintered carbides, which are extremely hard alloys. Very small dies may be made from diamond. Surfaces must be clean and free of imperfections before wire or rod is drawn.

Lubricants reduce friction between the wire and die surfaces.

Tube Drawing

Wire and tubes are drawn in a similar manner. General practice is to place a mandrel (also called a plug) or a solid bar with a blunt end inside the tube. As the tube is drawn, metal flows past the mandrel, which controls the inside diameter of the tube and supports it during deformation.

Drawing with a fixed plug is illustrated in Fig. 7-20(a). This process is widely used in drawing large- to medium-diameter straight tubes. In drawing long and small-diameter tubes, the plug bar that holds the fixed plug in place may stretch or even break. In such cases, a floating plug is used (Fig. 7-20b). Tubing of any length may be coiled as it is drawn.

Deep Drawing

Deep drawing, also known as *cupping,* is similar in concept to tube drawing. The difference: Instead of being pushed through the die from the outside, the metal is passed through by a punch from the inside (Fig. 7-21). The punch takes the place of the mandrel.

As shown in Fig. 7-21, a metal sheet is held between two hold-down dies while a formed punch stretches the metal into an open space or die. As the sheet stretches, it is drawn out from the hold-down die and into the die cavity. The cup is elongated by drawing it in successively smaller dies, and its wall thickness is reduced simultaneously. The size and depth of the cup that can be drawn depend on the yield strength and thickness of the sheet. Applications include metal toys, aircraft sheet parts, and car fenders.

Spinning

Sheet metal or tubing is converted into seamless hollow cylinders, cones, hemispheres, or other circular shapes by a combination of rotation and force. Two methods are used: manual spinning and power spinning. Typical shapes produced are shown in Fig. 7-22.

Manual spinning usually takes place in a lathe. A tool is pressed against a circular metal blank that is rotated by the headstock. The blank is normally forced over a mandrel of preestablished shape, but simple shapes can be spun without a mandrel.

Manual spinning is used to make flanges, rolled rims, cups, cones, and double-curved surfaces such as bells. Applications include light reflectors, tank ends, covers, housings, shields, parts for musical instruments, and aircraft and aerospace components.

Power spinning is used for virtually all ductile metals. Products range from small hardware items made in great quantities, such as metal tumblers, to large components for aerospace service. Blanks as large as 6 m (240 in.) in diameter and plate up to 25 mm (1 in.) thick have been power spun without the assistance of heat. With heat, blanks as thick as 140 mm (5.5 in.) thick have been power spun.

Chapter 8

Fabricability of Materials: A Key Factor in Selection

Fabricability is one of several important considerations in the selection of materials for a given application. The subject is treated separately in this chapter as background for the next chapter.

In choosing a metal or alloy for a production part, the metallurgist and design engineer must first determine the ability of a material to be worked or shaped into the part in question. Forming and joining are the generic topics.

Forming methods include casting, drawing, extruding, forging, and machining. All have been discussed in prior chapters. *Joining methods* include welding, brazing, and soldering. They will be discussed in the concluding sections of this chapter.

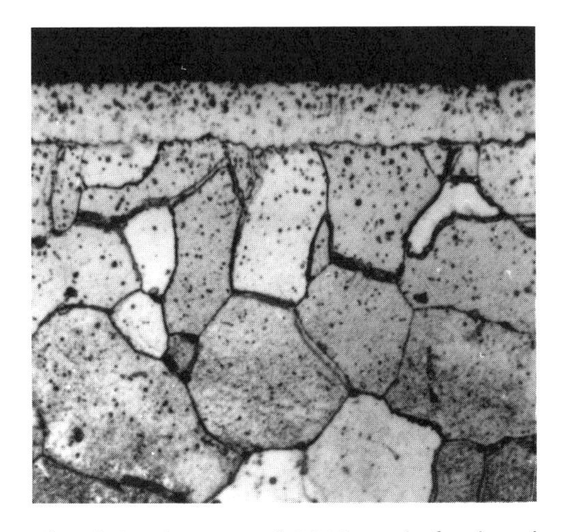

Fig. 8-1 Structure of 1010 steel after liquid nitriding in salt bath for an hour at 570 °C (1060 °F). Dark area at top is nitrided case. Lower area is made up of blocky ferrite and grain-boundary carbides.

Fabricability generally involves these questions:

- What is the relative ease or difficulty to be expected in casting, machining, or welding a material?
- What is the range of mechanical or physical properties of a material, such as ductility, fluidity, and hardenability, keeping in mind that results may vary with the fabrication method used?
- What are the trade-offs, if any, in choosing material A over material B or C?
- What is the importance of adjusting elements up or down in the composition of a material?

Fabrication Properties of Ferrous Alloys

The first four materials discussed in this section are nonresulfurized grades in the 1000 series of low-carbon steel. Carbon content ranges from 0.08 to 0.20%; typical sulfur content is 0.050 maximum. Sulfur content determines machinability. Nonresulfurized grades do not contain additional sulfur for this purpose. Grades with added sulfur, known as the 1100 series, typically contain 0.08 to 0.13% sulfur.

Example: 1010 Steel

Composition: Range of carbon content is 0.8 to 0.13%, sulfur content is 0.050% maximum, and manganese content is 0.30 to 0.60%.

Fabrication Properties

Cold Formability: Excellent, because the alloy is extremely ductile in the annealed or normalized condition. There is a trade-off: As carbon content increases, cold formability decreases.

Joinability: Welding and brazing, excellent. Low carbon content favors weldability.

Machinability: Poor, due to low sulfur content.

Heat Treatability: Can be case hardened. Methods include carbonitriding, liquid salt bath nitriding, liquid carburizing, and flame hardening. In normalizing, 1010 steel is heated to 925 °C (1695 °F) and cooled in still air. In annealing, it is heated to about 910 °C (1670 °F) and slow cooled in a cooler section of a continuous furnace.

Example: 1012 Steel (Versus 1010 Steel)

Composition: Carbon content in 1012 is higher (0.10 to 0.15% versus 0.8 to 0.13%). Sulfur content is the same (0.050% maximum), as is manganese content (0.30 to 0.60%).

Fabrication Properties

Cold Formability: Excellent for both grades. One difference: 1012 is better suited for production of stamped parts when drawing operations are minimal.

Joinability: Welding and brazing, excellent.

Machinability: Poor, due to low sulfur content.

Heat Treatability: Can be case hardened using the same methods as for 1010.

Example: 1016 Steel (Versus 1012 Steel)

Composition: Carbon content in 1016 is higher (0.13 to 0.18% versus 0.10 to 0.15%). Sulfur content is the same (0.050% maximum), but manganese content is higher (0.60 to 0.90%).

Fabrication Properties

Cold Formability: Reasonably good.

Forgeability: Excellent.

Joinability: Welding and brazing, excellent.

Machinability: Relatively poor, compared to 1100 to 1200 grades of steel, due to low sulfur content.

Heat Treatability: Can be case hardened using the same processes as for 1010.

Example: 1018 Steel (Versus 1016, 1012, and 1010 Steels)

Composition: The carbon content of each grade is progressively higher: 0.08 to 0.13% (1010 steel); 0.10 to 0.15% (1012 steel); 0.13 to 0.18% (1016 steel); 0.15 to 0.20% (1018 steel). Sulfur content is the same for all four grades of steel (0.050% maximum). The manganese content of 1010 and 1012 is the same (0.30 to 0.60%), and the manganese content of both 1016 and 1018 is 0.60 to 0.90%.

Fabrication Properties

Cold Formability: Reasonably good, but not as good as for 1010 and 1012 steels due to increased carbon content.

Forgeability: Excellent:

Joinability: Welding, excellent.

Heat Treatability: Case hardening methods include liquid and gas carburizing. (By comparison, 1010, 1012, and 1016 grades also can be carbonitrided, nitrided in liquid salt baths, and flame hardened.) Note: Higher manganese content explains the slight increase in strength in the normalized or annealed condition, as well as the mild increase in hardenability (indicating depth of hardening). This grade is commonly used when deep case depths are wanted in carburizing.

Resulfurized Carbon Steels (1100 Series)

Sulfur is added to these grades to improve machinability. Typical sulfur content is 0.08 to 0.13% (compared to a typical content of 0.050% maximum in 1000 series carbon steels).

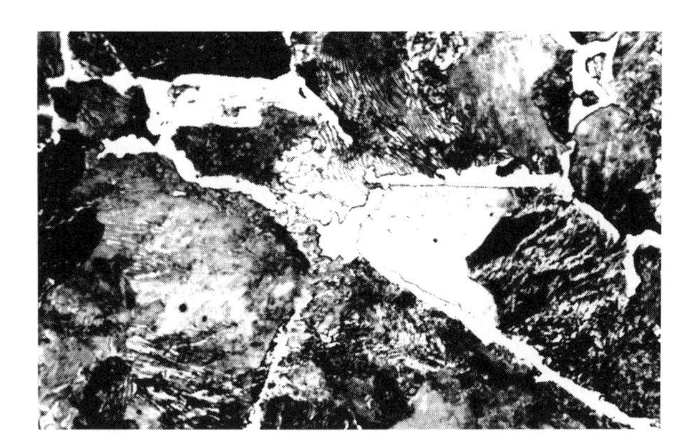

Fig. 8-2 Structure after carburizing for 8 hours. Surface carbon ranges from 0.60 to 0.70%. Light areas indicate ferrite. Dark areas are pearlite.

Example: 1110 Resulfurized Grade

Composition: Carbon content is 0.08 to 0.13%, sulfur content is 0.08 to 0.3%, and manganese content is 0.30 to 0.60% (same as that of 1016 and 1018 nonresulfurized grades because of the addition of sulfur to improve machinability).

Fabrication Properties

Higher Material Costs: One of several trade-offs.

Sulfur: Detrimental to some mechanical properties, such as impact strength and ductility in the transverse (diagonal) direction.

Forgeability, Cold Formability, and Weldability: Losses are incurred in these areas.

Applications: Typically limited to those where machining is the main fabrication method.

Heat Treatability: Suitable case hardening processes are carbonitriding, salt bath nitriding, and flame hardening.

Example: 1117 Resulfurized Grade

Composition: Carbon content is higher than that of 1110 (0.14 to 0.20% versus 0.08 to 0.13%), sulfur content is the same (0.08 to 0.13%), and manganese content is higher (1.00 to 1.30% versus 0.30 to 0.60%).

Economics: Grade 1117 costs more than 1020 or 1021.

Fabrication Properties

Plus Side: Slight improvement in hardenability, or depth of case. In addition, manganese combines with sulfur to form manganese sulfide stringers. These stringers promote chip breaking, which benefits machining.

Machinability: Excellent.

Trade-offs: Losses in mechanical properties, forgeability, and cold formability.

Example: 1137 Resulfurized Grade

Composition: Carbon content is higher than other grades: 0.32 to 0.39% versus 0.08 to 0.13% for 1110 and 0.14 to 0.20% for 1117. Manganese content is also higher: 1.35 to 1.65% versus 0.30 to 0.60% for 1110 and 1.00 to 1.30% for 1117. Sulfur content is the same in all cases (0.08 to 0.13%).

Fabrication Properties

Plus side: See grade 1117.

Trade-offs: Carbon content is too high for good weldability, and sulfur content can cause *hot shortness* (weakness in grain boundaries when a part is stressed or deformed at temperatures near the melting point). Because of its high carbon content, this alloy is not suitable for welding.

Example: 4340 Alloy Steel

Composition: High carbon content (0.38 to 0.43%); also contains silicon, nickel, chromium, and molybdenum.

Fabrication Process

Hardenability: Highest among all standard AISI grades due to the carbon content.

Joinability: Welding by conventional processes is not recommended because of the high carbon content, but electron beam welding is possible.

Machinability: Relatively poor.

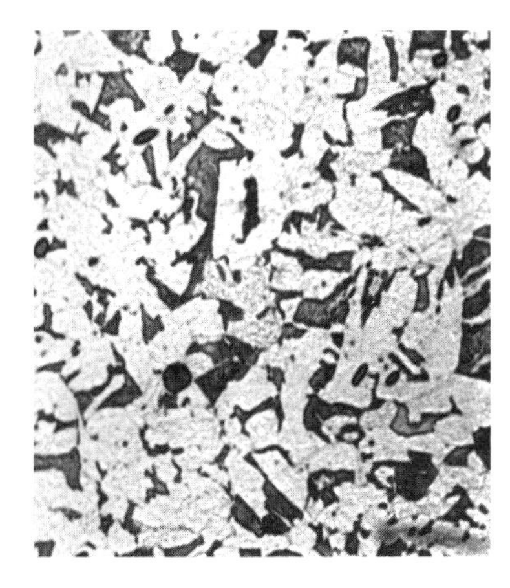

Fig. 8-3 Structure after heat treatment. Blocky ferrite is white; pearlite is dark; round particles are manganese sulfide.

Fig. 8-4 Structure is tempered martensite. Alloy was quenched in oil from a metal temperature of 845 °C (1555 °F) and tempered at 315 °C (600 °F).

Formability: Forging is not a problem, but forging equipment must be powerful due to alloy strength. Forging is at 1230 °C (2250 °F) and is not recommended when the temperature of the metal drops below 900 °C (1650 °F).

Fabrication Properties of Nonferrous Metals and Alloys

In the following examples, machinability and forgeability are rated as a percentage of standard alloys: free-cutting brass for machinability, and forging brass for forgeability. Free-cutting brass (alloy 36000) is the subject of the last example. All compositions listed are nominal.

Example: C10100 (Pure Copper)

Common Name: Oxygen-free copper.

Composition: 99.9% copper (minimum), plus fractional amounts of other elements.

Fabrication Properties

Machinability: 20% that of C36000 (free-cutting brass).

Forgeability: 65% that of C37700 (forging brass).

Formability: Readily formed by both hot and cold working. Easily stamped, bent, coined, sheared, spun, upset, swaged, roll threaded, and knurled.

Joinability: Readily soldered, brazed, gas tungsten arc welded, and upset welded. Welding processes such as oxyfuel gas, shielded metal arc, and most resistance methods are not recommended.

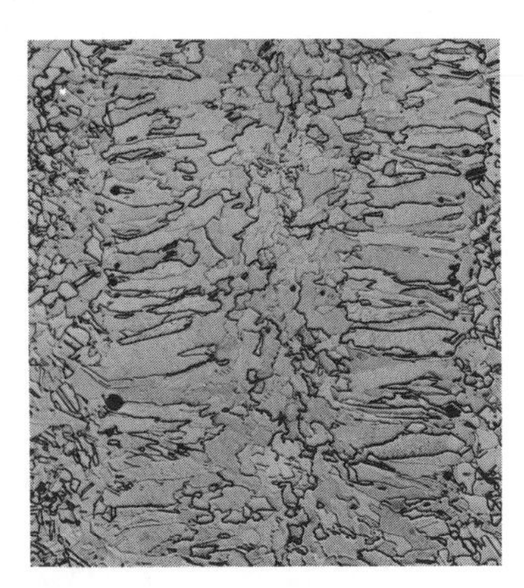

Fig. 8-5 EB welded copper bar. In mid-area, columnar grains are in fusion zone. Equiaxed grains (all similar in shape) are base metal. Gas porosity is visible as scattered black dots at edge of fusion zone.

Example: C16200 Copper Alloy

Commercial Name: Cadmium copper.

Composition: 99% copper, 1% cadmium, and a small amount of iron.

Typical Uses: Trolley wire, heating pad and electric blanket elements, high-strength transmission lines, cable wrap, and waveguide cavities.

Fabrication Properties

Machinability: 20% that of C36000.

Formability: Cold working, excellent; hot forming, good.

Joinability: Soldering and brazing, excellent. Welding processes such as oxyfuel gas, gas shielded arc, and resistance butt welding methods are rated good. Not recommended: shielded metal arc, resistance spot, and resistance seam welding.

Example: C35000 Brass Alloy

Common Name: Medium-leaded brass.

Composition: 62% copper, 36.4% zinc, 1.1% lead.

Typical Uses: Bearing cages, book dies, clock and engraving plates, gears, hinges, keys, lock parts, sink strainers, nuts, and washers.

Fabrication Properties

Machinability: 70% that of C36000.

Forgeability: 50% that of C37700.

Formability: Cold and hot working, fair.

Joinability: Soldering, excellent; brazing, good; resistance butt welding, fair. Other welding processes are not recommended.

Example: C36000 Brass Alloy

Common Name: Free-cutting brass.

Composition: 61.5% copper, 35.5% zinc, 3% lead.

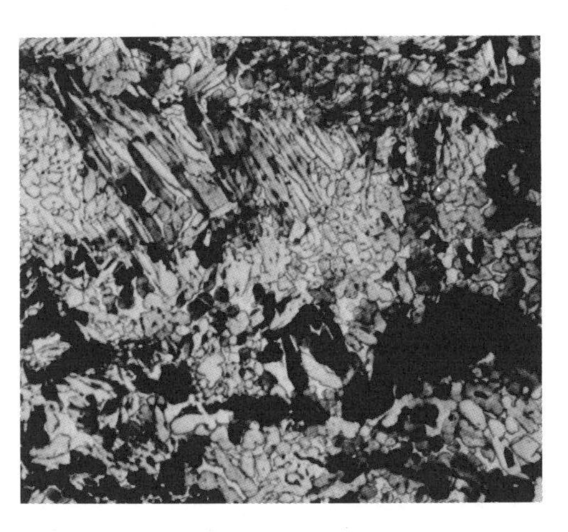

Fig. 8-6 Structure of as-cast free-cutting brass

Typical Uses: Gears and pinions; industrial parts made with automatic, high-speed screw machining process.

Fabrication Properties

Machinability: 100% (the standard for all copper alloys).

Formability: Cold working, poor; hot working, fair.

Joinability: Soldering, excellent; brazing, good; resistance butt welding, fair. All other processes are not recommended.

Trade-offs: Several, for sake of machinability.

Joining Processes: Welding, Brazing, and Soldering

Three of the five families of conventional joining processes are discussed in this section. Excluded are mechanical joining (as with screws or nuts and bolts) and adhesive bonding, which are outside the scope of metallurgy per se. Do-it-yourself details are avoided.

Of the 25 or so available welding processes, only six are reviewed here: three electric arc welding processes (shielded metal arc, gas metal arc, and gas tungsten arc); one process (oxyfuel gas welding) in which heat is provided by gas flames; and two resistance welding processes (spot welding and seam welding). The focus in the discussion of brazing is on the joining of five metals and alloys (steel, cast iron, stainless steel, aluminum alloys, and copper and its alloys). The section on soldering emphasizes the unique features of the process.

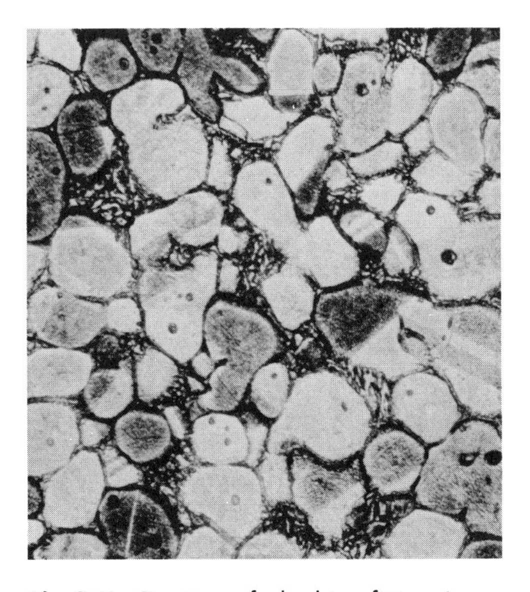

Fig 8-7 Structure of plumbing fitting. Large particles (white and light gray) were present before alloy was cast. Rapid solidification produced matrix structure.

Arc Welding

Arc welding is defined as a group of processes that fuse or join metals together by heating them with an electric arc, with or without the application of pressure, and with or without the use of *filler metal* (metal added in making a welded, brazed, or soldered joint).

Shielded metal arc welding (SMAW) is a manual process (Fig. 8-8) in which heat for welding is generated by an arc established between a flux-covered consumable electrode and the workpiece. *Flux* is a material that either prevents the formation of oxides and other undesirable elements, or dissolves oxides and facilitates their removal. An *electrode* is a current-carrying rod that supports the arc between the rod and the workpiece.

The electrode tip, molten weld pool, arc, and adjacent areas of the workpiece are protected from atmospheric contamination by a gaseous shield obtained from the combustion and decomposition of the electrode covering. Electrodes must be discarded when they reach 50 mm (2 in.) in length.

Additional shielding is provided for the molten metal in the weld pool by a covering of molten flux or *slag*—a nonmetallic product resulting from the mutual dissolution of flux and nonmetallic impurities. Filler metal is supplied by the core of the consumable electrode (Fig. 8-8) and from metal powder mixed with flux coverings on certain types of electrodes. Electrodes have many different compositions of core wire and a wide variety of types and weight of flux covering.

Process Capabilities. Because of its versatility, portability, simplicity, and low cost, SMAW is the most widely used process for joining metal parts. Welding can be done indoors or outdoors. A joint in any position that can be reached with an electrode can be welded.

To start the welding process, an arc is struck by briefly touching the workpiece with the tip of the electrode. The welder guides the electrode by hand and controls its position, direction, travel speed, and arc length. In welding low-carbon steels, acceptable results are obtained with either alternating or direct current.

Gas metal arc welding (GMAW) (Fig. 8-9), also known as metal inert gas (MIG) welding, differs from the SMAW process in that its electrode is a bare solid wire that is continuously fed to the weld area and becomes the filler metal as it is consumed. By comparison, in SMAW the electrode must be discarded when it reaches a minimum length.

In GMAW, the electrode, weld pool, arc, and adjacent areas of the base metal workpiece are protected from atmospheric contamination by a gaseous shield provided by a stream of gas, or mixture of gases, fed through the welding gun. The gas shield must provide full protection, because even a small amount of entrained air can contaminate the weld deposit.

Process Capability. Gas metal arc welding is widely used in semiautomatic, machine, and automated modes. The semiautomatic process is the most popular. The operator guides the gun along the joint and adjusts welding conditions. The wire feeder continuously feeds the filler wire electrode, and arc length is maintained by the power source. In the automated mode, the machinery controls welding parameters, arc length, joint guidance, and wire feed.

Applications include all ferrous metals and most nonferrous metals. Workpiece thicknesses vary widely.

Equipment includes a power supply that provides the voltage needed to push the electrical current across the gap that makes the arc, sufficient current to melt the electrode to make the weld deposit, a wire feeder that continuously advances the electrode as it melts, a smooth flow of shielding gas, and a welding gun that carries the current, electrode wire, shielding gas, and (depending on gun design) cooling water (Fig. 8-9).

Many types of direct-current power sources can be used. Welding guns transmit welding current to the electrode. Because wire is fed continuously, a sliding electrical contact is used. Welding current is passed to the electrode through a copper alloy contact tube. Contact tubes have various hole sizes, corresponding to the diameter of the electrode wire. Guns also have a gas supply connection and a nozzle to direct the shielding gas around the arc and weld pool. To prevent overheating, welding guns must be cooled. Shielding gas or water, or both, are used for cooling; some guns are also air cooled. Hand-held semiautomatic guns usually have a curved neck, making it possible to weld in all positions (Fig. 8-10).

Electrodes usually are quite similar or identical in composition to those used in welding with

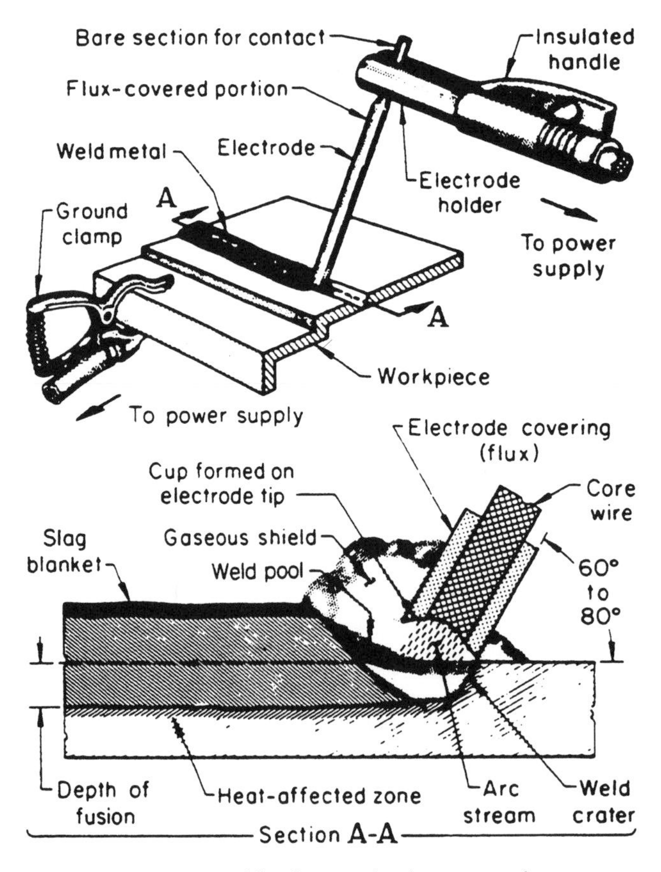

Fig. 8-8 Setup and fundamentals of operation for SMAW

most other bare electrode methods. Wires may be chosen to match the chemical composition of the base metal or workpiece as closely as possible. Different chemical compositions may be used to obtain maximum mechanical properties or improved weldability.

Shielding gases protect molten weld metal and the heat-affected zone from oxidation and other types of contamination. The *heat-affected zone* is that part of the base metal not melted in welding but whose microstructure and mechanical properties are changed by the heat.

Originally, only inert gases such as argon and helium were used for shielding. Today, carbon dioxide is also used, and may be mixed with the inert gases. Welding-grade argon is 99.99% pure. In addition to being inert, it is insoluble in molten metal. Helium, used primarily in welding aluminum, magnesium, and copper alloys, is obtained by separation from natural gas and generally is used as compressed gas in cylinders. Used alone or in combination with inert gases, carbon dioxide improves arc action and metal transfer, in addition to serving as a shielding gas.

Gas Tungsten Arc Welding (GTAW). In this process, also known at tungsten inert gas (TIG) welding, heat is produced between a nonconsumable electrode and the workpiece metal. Because the electrode is nonconsumable, a weld can be made by *fusion* (where the metal interfaces are melted) of the workpiece metal with-

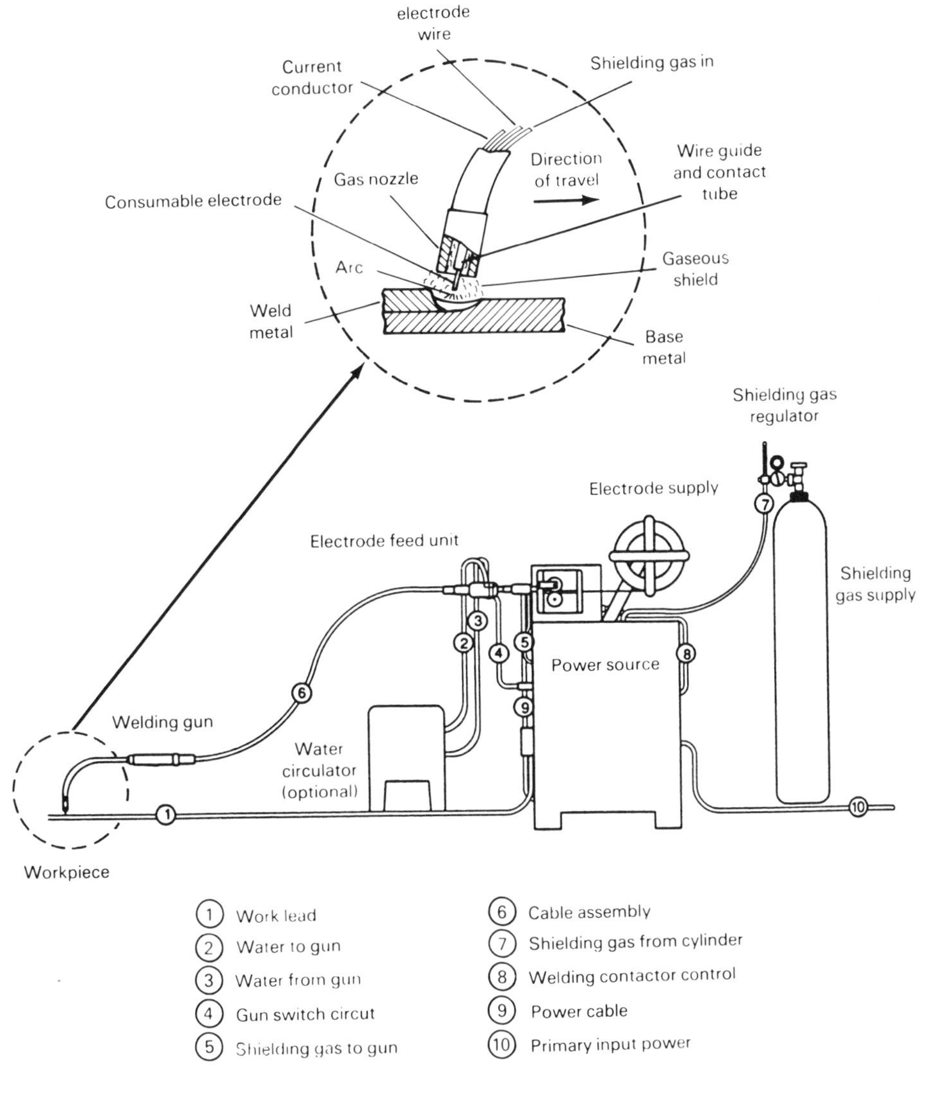

Fig 8-9 Schematic of GMAW process

out the addition of a filler metal. The electrode, weld pool, arc, and heat-affected zone are protected from atmospheric contamination by a gaseous shield provided by a stream of gas that is usually inert, or by a mixture of gases. Full protection must be provided; even a minute amount of entrained air can contaminate the weld.

Manual, semiautomatic, machine, and automated methods are available. A hand-held torch is used in the manual method (Fig. 8-11).

Applications. Most metals and alloys can be welded with the GTAW process, including carbon and alloy steels, stainless steels, heat-resistant alloys (also known as superalloys), refractory metals, aluminum alloys, copper alloys, magnesium alloys, nickel alloys, titanium alloys, and zirconium alloys.

The process is particularly suited to the welding of thin metals—often as thin as 0.125 mm (0.005 in.). Maximum base metal thickness is 6.4 mm (¼ in.). Other welding processes are used for thicker metals.

Process Fundamentals. An electric arc is produced by the passage of current through the inert shielding gas. Electrode and filler metal positions in the commonly used manual process are shown in Fig. 8-12. Once the arc is started, the torch is held so that the electrode is positioned about 75° to the surface of the workpiece and points in the direction of welding. Control of welding current is important; this variable influences depth of penetration, welding speed, deposition rate, and weld quality.

Torches for the manual process should be compact, light in weight, and fully insulated. Other requirements are a handle for holding the torch, a means of conveying shielding gas to the arc area, and a collet, chuck, or other means of securing the tungsten electrode and conducting welding current to it (Fig. 8-13).

Electrodes. Use of a nonconsumable electrode is the major difference between the GTAW process and other metal-arc processes. Tungsten, which has a melting point of 3410 °C (6170 °F), is the best material for the electrode. In addition to its very high melting point, tungsten is a strong emitter of electrons, which contributes to the maintenance of a stable arc. Tungsten electrodes containing about 2% thoria (a rare earth metal) are superior to pure tungsten types.

Shielding gases do not directly add heat to the weld, but they do affect heat input. Gases ordi-

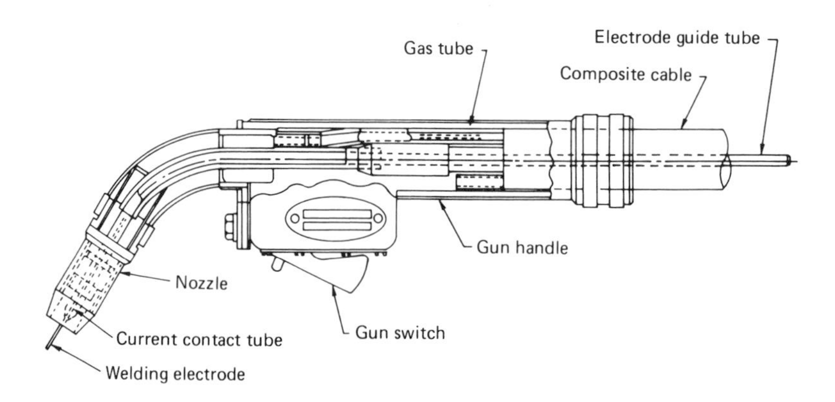

Fig. 8-10 Cross section of an air-cooled GMAW gun

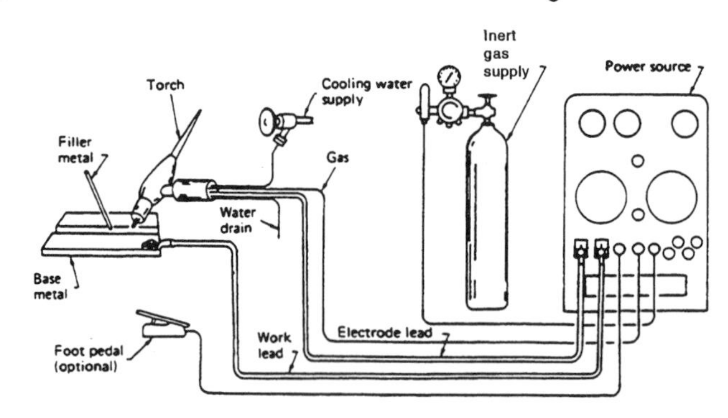

Fig 8-11 Manual GTAW equipment

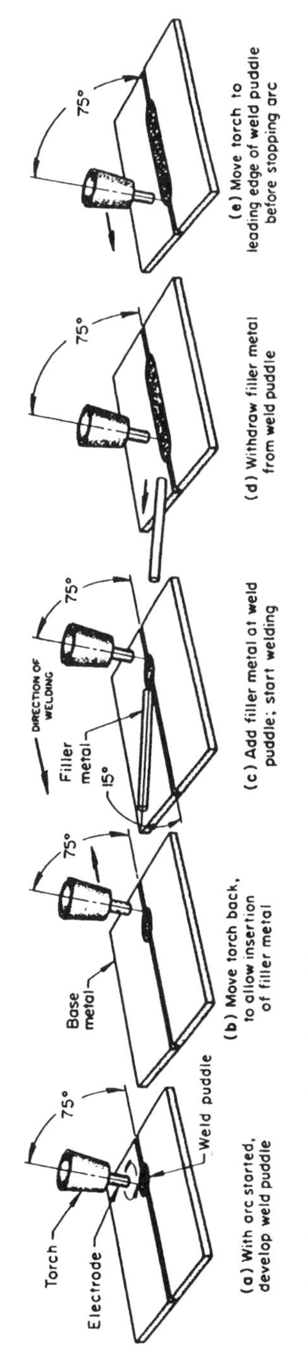

Fig. 8-12 Positions of torch and filler metal in manual GTAW

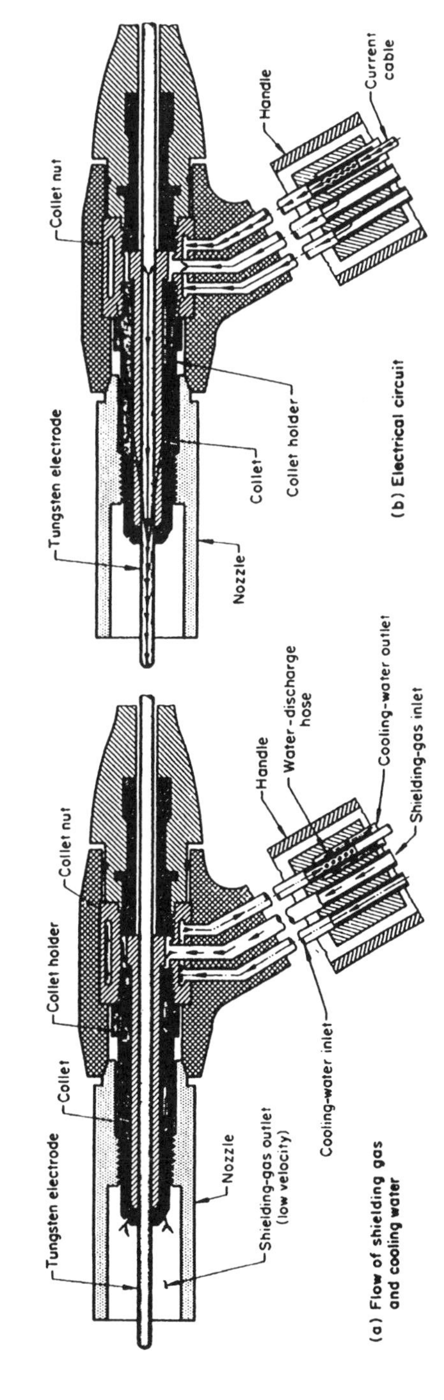

Fig. 8-13 Sectional views of a typical torch assembly for GTAW

Table 8-1 Recommended tungsten electrodes and shielding gases for welding different metals

Type of metal	Thickness	Type of current(a)	Electrode	Shielding gas
Aluminum	All	Alternating current	Pure or zirconiated	Argon or argon-helium
	Thick only	DCEN	Thoriated	Argon-helium or argon
	Thin only	DCEP	Thoriated or zirconiated	Argon
Copper, copper alloys	All	DCEN	Thoriated	Argon or argon-helium
	Thin only	Alternating current	Pure or zirconiated	Argon
Magnesium alloys	All	Alternating current	Pure or zirconiated	Argon
	Thin only	DCEP	Zirconiated or thoriated	Argon
Nickel, nickel alloys	All	DCEN	Thoriated	Argon
Plain carbon, low-alloy steels	All	DCEN	Thoriated	Argon or argon-helium
	Thin only	Alternating current	Pure or zirconiated	Argon
Stainless steel	All	DCEN	Thoriated	Argon or argon-helium
	Thin only	Alternating current	Pure or zirconiated	Argon
Titanium	All	DCEN	Thoriated	Argon

(a)DCEN, direct current electrode negative; DCEP, direct current electrode positive

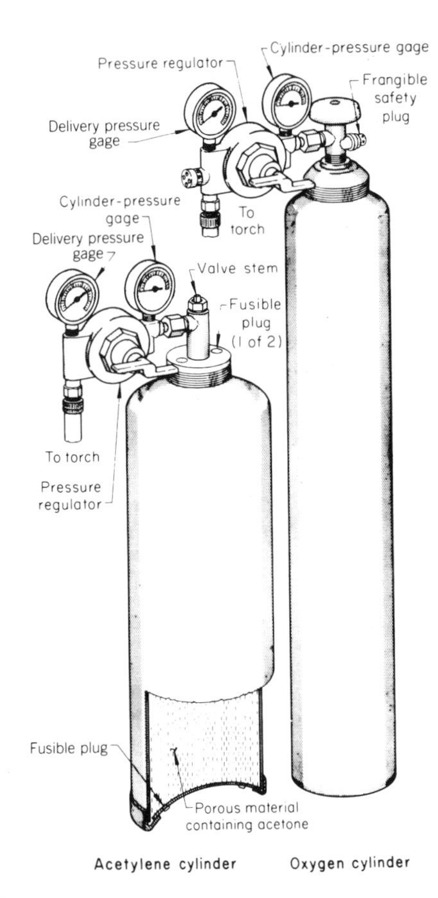

Fig. 8-14 Gas cylinders and regulators used in oxyfuel gas welding

narily used are argon, helium, argon-helium, and argon-hydrogen. The shielding gas used can significantly affect weld quality and welding speed. Argon, helium, and argon-helium mixtures do not react with tungsten or tungsten alloy electrodes and have no adverse effect on the quality of the weld metal (Table 8-1).

Filler Metals usually are matched in composition with the base metal as closely as possible. Control of their composition, purity, and quality is tighter than for the base metal, and their choice depends on the application. Important properties include tensile strength, impact toughness, electrical and thermal conductivity, corrosion resistance, and weld appearance.

Oxyfuel Gas Welding

Oxyfuel gas welding is a manual process in which the metal surfaces to be joined are melted progressively by heat from a gas flame, with or without filler metal. The molten metal subsequently solidifies, making a welded joint without the application of pressure.

Applications. The process is used to join thin carbon steel sheet and carbon steel tubing and pipe. Pipe may be up to 75 mm (3 in.) in diameter.

Gases Used. Oxygen and acetylene are commonly used. Oxygen supports the combustion of fuel gases. Acetylene provides both heat intensity and the atmosphere needed in welding steel. Hydrogen, natural gas, and proprietary gases are used to a limited extent. If gas consumption requirements are low, oxygen and acetylene are supplied and stored in standard steel containers (Fig. 8-14).

Oxygen must have a purity of 99.5% or higher; minute amounts of contaminants have a detrimental effect on combustion efficiency. Acetylene is a hydrocarbon gas. It is unstable under pressure, and a slight shock may cause it to explode. Safety rules for handling the gas and supporting equipment are extremely important.

Torches control the operating characteristics of the flame and enable it to be manipulated

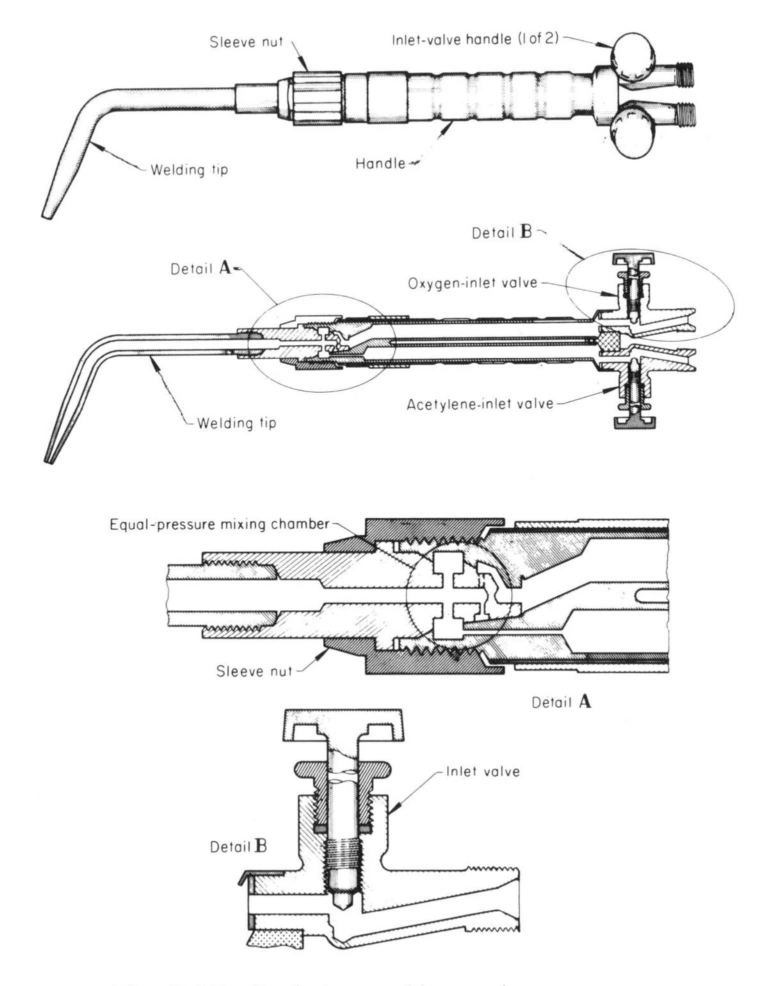

Fig. 8-15 Oxyfuel gas welding torch

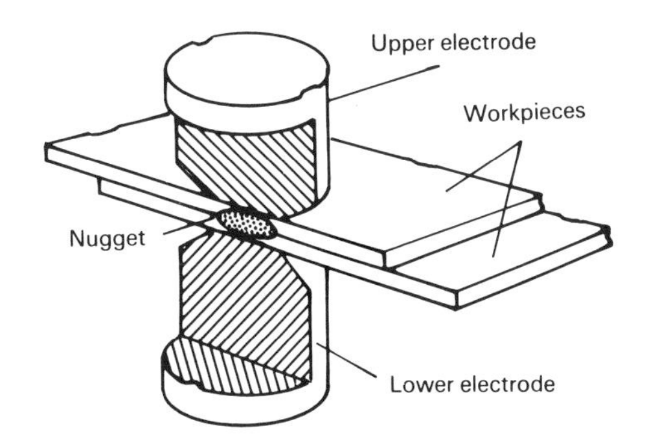

Fig. 8-16 Sectional view showing shape of nugget and position of nugget relative to inner and outer surfaces of workpieces in resistance spot welding

during welding. Choice of torch size and style depends on the application. The general features of a torch are shown in Fig. 8-15.

Low-carbon steel filler metals are available in the form of cold-drawn rods measuring 0.9 m (36 in.) long by 1.6 to 6.4 mm ($1/16$ to $1/4$ in.) in diameter.

Resistance Welding

The four welding processes discussed thus far are *fusion* processes: The edges of the base metals being welded and the filler metal, if used, are in the molten state when they are joined. In resistance welding, heating is followed by very brief melting and the simultaneous application of pressure. Two such processes are discussed here: spot welding and seam welding.

Resistance Spot Welding. Closely fitted surfaces are joined in one or more spots by heat generated by resistance to the flow of electric current through the workpieces held together under force by electrodes. Contacting surfaces

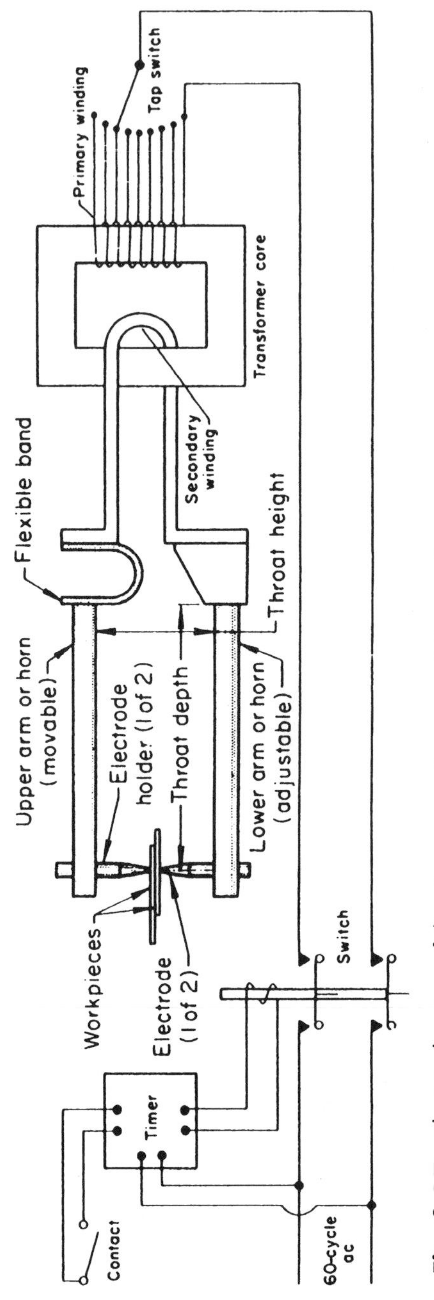

Fig. 8-17 Electrical system of direct-energy resistance welding machine

are heated by a short-time pulse of low-voltage, high-amperage current to form a fused nugget of weld metal. When the flow of current is stopped, electrode pressure is maintained while the weld metal cools rapidly and solidifies. Electrodes are retracted after each weld is completed—a matter of seconds.

A setup for spot welding is shown in Fig. 8-16. Note the shape of the nugget and its position relative to the inner and outer surfaces of the two workpieces. Electrodes are located above and below the two workpieces.

Applications. Sheet steel up to about 3.2 mm (1/8 in.) thick is the typical application. On occa-

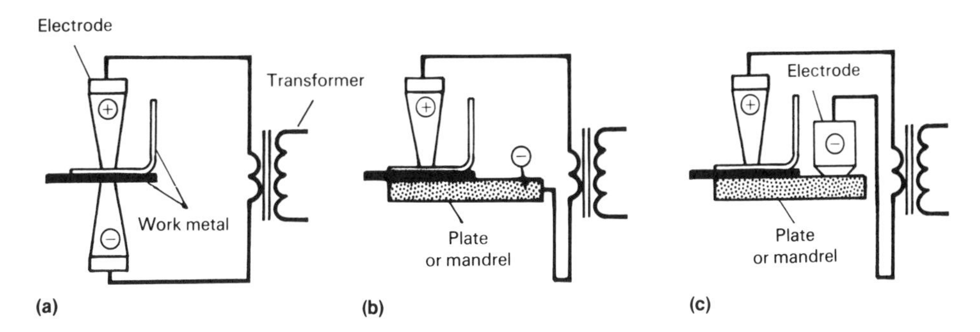

Fig. 8-18 Setup of work metal and electrodes for making single spot welds

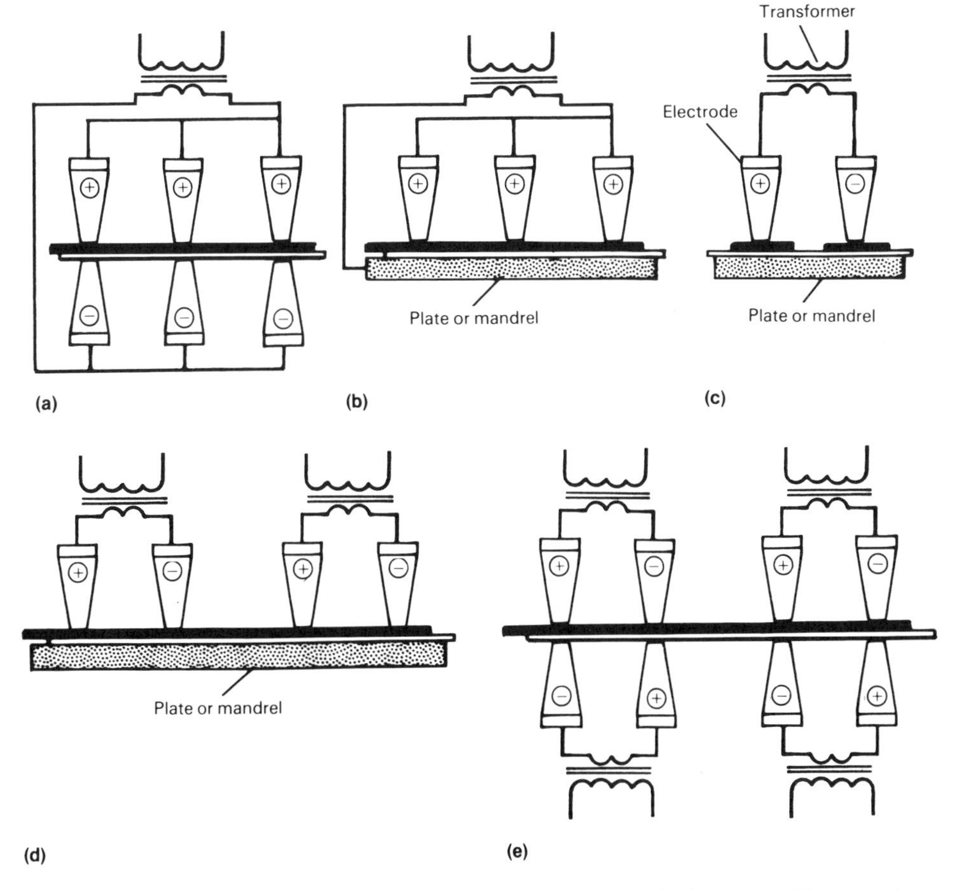

Fig. 8-19 Setup of work metal and electrodes for making multiple spot welds using direct and series welding

sion, thickness ranges up to 6.4 mm (¼ in.). With special equipment, thicknesses of 25 mm (1 in.) or more have been joined.

Spot Welding Equipment. A system called *direct-energy welding* (Fig. 8-17) is commonly used to join low-carbon steel. Single- or three-phase 60-cycle current, ordinarily drawn from 220 to 440 volt in-plant lines and stepped down to about 2 to 20 volts, is fed directly to the electrodes. Equipment may be simple and inexpensive, or complex and expensive.

The principal components of a direct-energy system are:

- *Electrical circuit:* Consists of welding transformer, tap switch, and secondary circuit, which includes electrodes that conduct current to workpieces
- *Control equipment:* Initiates and times duration of current flow and may be used instead of, or in addition to, a transformer tap switch to regulate welding current
- *Mechanical system:* Includes frame, fixtures, and other devices and may be used instead of, or in addition to, a transformer tap switch to regulate welding current

Single and Multiple Weld Setups. Three different setups for single spot welds are shown in Fig. 8-18. The simplest and most common arrangement of two workpieces sandwiched between opposing upper and lower electrodes is shown in Fig. 8-18(a). The setup in Fig. 8-18(b), where a conductive plate is used as the lower workpiece and conducts heat away from the weld rapidly, may be necessitated by the shape of the workpiece. In Fig. 8-18(c) a conductible plate that also acts as a heat sink is used in combination with a second upper electrode. Setups for making multiple spot welds are shown in Fig. 8-19.

Resistance Seam Welding. A seam consists of a series of overlapping spot welds, and the resulting structure normally is gastight or liquidtight. Seams are produced by rotating electrode wheels that transmit current to the work metal (Fig. 8-20). A series of welds is made without retracting the electrode wheels or releasing the electrode force between spots, but the wheels may advance either intermittently or continuously.

Advantages of seam welds over spot welds include:

- Gastight and liquidtight joints can be produced.
- Overlap may be less than in spot welds, and seam width can be less than the diameter of spot weld nuggets.

Applications. Metals suitable for seam welding include low-carbon steels, high-carbon steels, high-strength low-alloy steels, stainless steels, and many coated steels. However, carbon levels above 0.15% can tend to form areas of hard martensite upon cooling, which could cause brittleness. Galvanized and tin-coated steels have been seam welded, as have aluminum and magnesium alloys.

Equipment. Spot welders and seam welders are similar in construction except for differences in electrode geometry. Generally, seam welding is done in press-type machines.

Brazing

Brazing is broadly defined as a group of joining processes that produce bonding by the melting and subsequent resolidification of a filler metal in the space between surfaces to be joined. The filler metal must have a lower melt-

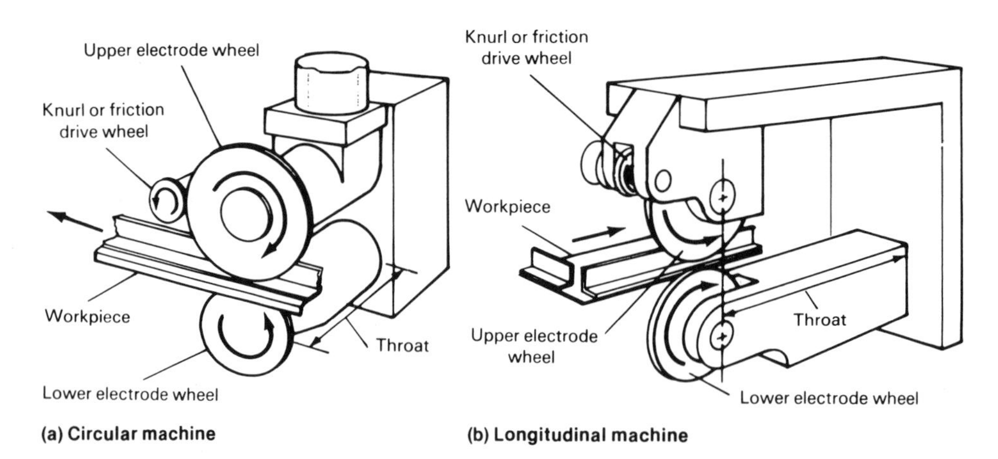

(a) Circular machine

(b) Longitudinal machine

Fig. 8-20 Position of electrode wheels on resistance seam welding machines

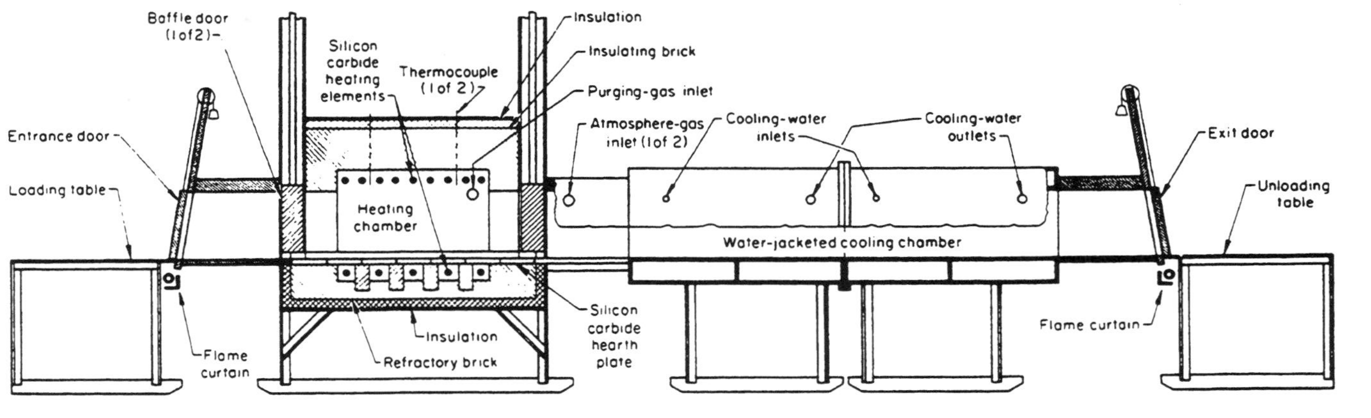

Fig. 8-21 Box-type batch brazing furnace

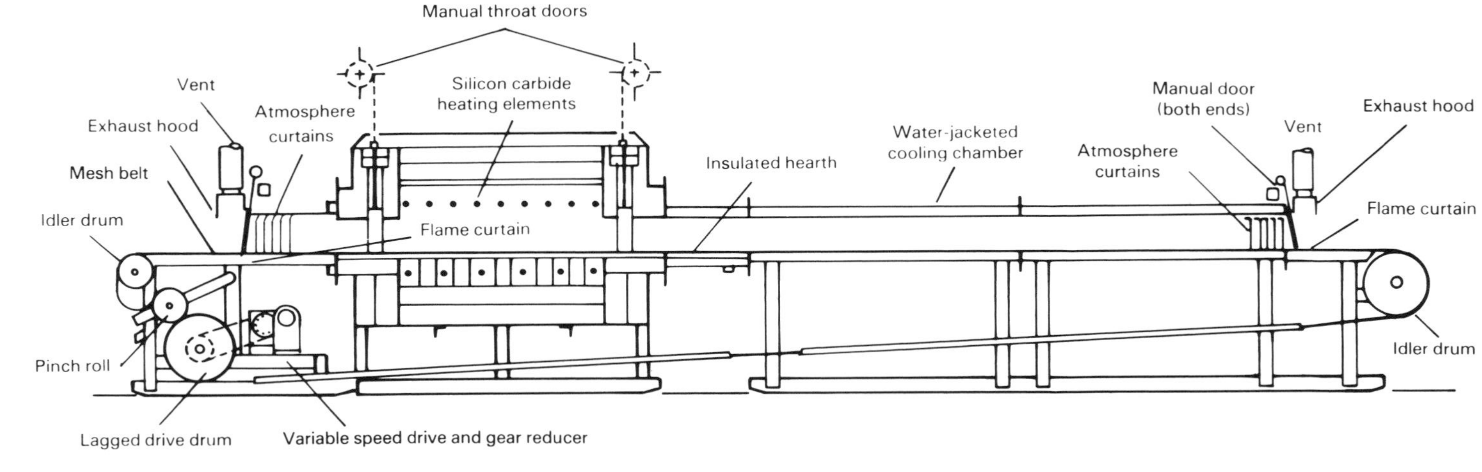

Fig. 8-22 Mesh-belt conveyor brazing furnace utilizing a water-jacketed cooling chamber

ing point than the material being joined, because in brazing the base metal does not melt. In conventional brazing, molten filler metal is distributed between closely fitted surfaces or the base metal of the joint by *capillary action,* in which the combined forces of adhesion and cohesion cause molten metals to flow between very closely spaced surfaces—even against gravity.

Most of the common metals can be brazed, but not necessarily with the same equipment, method, or procedure. The discussion that follows is limited to several commercial methods of brazing steels, cast iron, stainless steel, aluminum alloys, and copper alloys.

Brazing of Steel. *Furnace brazing* is a mass production method for producing small steel assemblies, using a nonferrous filler metal (i.e., copper) as the bonding material and a furnace as the heat source. Furnace brazing is only practical if the filler metal can be placed on the joint before brazing and retained in position during brazing.

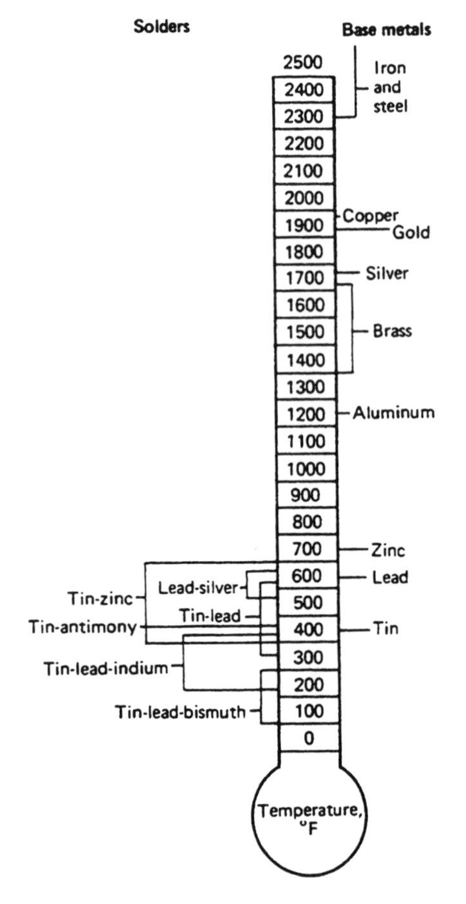

Fig 8-23 Soldering temperature ranges compared with base metal melting points

Generally, small assemblies that weigh less than 2.3 kg (5 lb) are the leading candidates for brazing, due to a combination of process efficiency and favorable economics. Larger assemblies require special furnaces.

Cleaning generally is limited to the removal of oils used in machining operations. Preferred cleaning methods are alkaline cleaning, solvent cleaning, and vapor degreasing. It is important that all alkaline cleaning compounds be removed from the workpiece before it enters the brazing furnace. Pigmented drawing compounds containing lead generally are removed using mechanical cleaning methods.

Assembling and Fixturing. Components are generally designed for assembly by press fitting, expanding, swaging, or other means that eliminate the need for fixturing.

Brazing. Assemblies are moved into the brazing chamber, where they are heated under a protective atmosphere. When the assembly reaches a temperature higher than the melting point of the filler metal, the filler metal wets and flows over the steel surfaces and is drawn into joints by capillary action. In making the bond, the filler metal forms a solid solution with, but does not melt, the steel. Heating time usually ranges from 10 to 15 minutes.

Brazing temperatures are considerably higher than those required in heat treating, which imposes limitations on furnace design and operation. Brazing temperatures range from 1095 to 1150 °C (2000 to 2100 °F).

Batch furnaces heat each workload separately and normally include an external reinforced steel shell, a heating system for the chamber, and one or more access doors to the heated chamber. A box-type batch furnace and its features are shown in Fig. 8-21.

Typically, a furnace of this type accommodates four trays at a time: one in the heating chamber and three in the cooling chamber. As soon as the tray in the heating chamber reaches brazing temperature, the operator pulls the end tray out of the cooling chamber and pulls the other two trays closer to the end. The operator then pushes the hot tray of brazed assemblies into the empty space in the cooling chamber and pushes a new tray of unbrazed assemblies into the heating zone.

Continuous furnaces receive a steady flow of incoming assemblies. The two common furnace types for copper brazing of steels are mesh-belt and roller-hearth conveyor furnaces. The mesh-belt type (Fig. 8-22) has the advantage of continuous operation at high capacity and accurate automatic cycle timing in both heating and cooling chambers.

Atmospheres. Gas atmospheres primarily protect steel assemblies from oxidation or scaling and assist the flow of filler metal by promoting the wetting of steel surfaces. Technically, *wetting* involves the interfacial surface tension between a liquid and a solid. A *wetting agent* is a surface-active agent that decreases cohesion within the liquid metal. Atmospheres may also maintain the carbon content of the steel by preventing carburization or decarburization at elevated temperatures.

Filler Metals. Because of its low cost and its ability to produce high-strength joints, copper is the preferred filler metal in furnace brazing of carbon and low-alloy steel assemblies without flux in protective atmospheres. Significant amounts of two trace elements, arsenic and phosphorus, should be avoided because they form brittle compounds in brazed joints. Copper should be essentially arsenic-free, and if it is deoxidized with phosphorus, residual phosphorus content should be low.

Cooling. After brazing, assemblies are moved to the cooling chamber of the furnace, where they are cooled under a protective atmosphere (usually the same as that used in the brazing chamber). Assemblies remain in the cooling chamber until they have cooled enough to prevent discoloring when exposed to air, usually to about 150 °C (300 °F).

Brazing of Cast Irons. Brazing of gray, ductile, and malleable irons differs from brazing of steels in two important respects: (1) Special precleaning methods are needed to remove graphite from the surfaces of iron, and (2) the brazing temperature should be kept as low as possible to avoid reducing the hardness and strength of the iron.

Furnace brazing is among the processes used in joining cast irons. As with other metals, selection of the brazing process depends largely on the size and shape of the assembly, quantity to be brazed, and equipment available.

Irons relatively high in silicon content with sand inclusions on as-cast surfaces have some adverse effects on brazeability. However, these disadvantages are less significant than the adverse effects of graphite, which is present in all cast irons. Generally, malleable iron is the most brazeable, while gray iron is the least brazeable.

Surface Preparation. Most methods are only partly successful. In the preparation of ductile and malleable iron surfaces, abrasive cleaning with steel shot or grit is reasonably effective. Results with gray iron surfaces are marginal. Electrolytic treatment in molten salt baths is the most effective method of treating all graphitic cast irons. Alkaline-base salts operating at 370

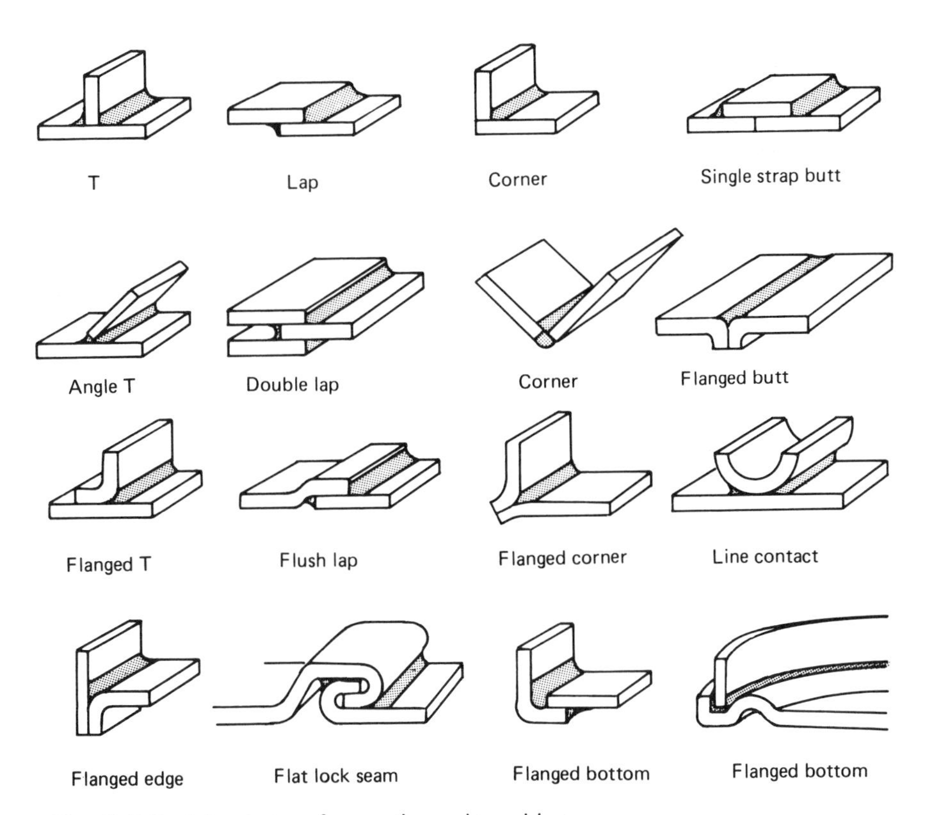

Fig. 8-24 Joint designs frequently used in soldering

to 480 °C (700 to 900 °F) are used to remove surface oxides and sand.

Brazing of Stainless Steels. These alloys are brazed by all the conventional processes, including furnace, torch, induction, resistance, and salt bath brazing. Furnace brazing is the most widely used method, because most applications require atmospheres.

Filler metals include silver alloys, nickel alloys, copper, and gold alloys. In most applications, selection is based on mechanical properties, corrosion resistance, and service temperature of the brazed assembly.

Silver alloys are the most widely used filler metal, but service temperatures are limited to a maximum of 370 °C (700 °F). Nickel alloys rank second, producing joints with excellent corrosion resistance and high-temperature strength. However, nickel alloys in combination with stainless steel can result in brittle structures.

High brazing temperatures and protective atmosphere requirements dictate the use of furnace brazing when copper filler metals are used. Copper is not recommended for applications that involve exposure to certain corrosive substances, such as sulfur in jet fuel. These filler metals have poor resistance to oxidation at service temperatures above 425 °C (800 °F).

Gold is used at times in the fabrication of aerospace equipment, producing joints with good ductility. Its high cost, however, restricts its use.

Brazing of aluminum alloys was made possible by two developments: (1) the invention of fluxes that disrupt the oxide film on aluminum without harming the underlying metal, and (2) aluminum alloy filler metals with suitable melting ranges and other desirable properties.

Aluminum-base filler metals have liquidus temperatures much closer to the solidus temperature of the base metals than those encountered in the brazing of most other metals. *Liquidus* is the temperature at which metals begin to freeze on cooling, or finish melting on heating. *Solidus* is the temperature at which metals finish freezing on cooling, or begin to melt during heating.

The general requirement is to control brazing temperatures to about 40 °C (70 °F) below the solidus temperature of the base metal. With accurate controls and short brazing cycles, temperatures can be held within 5 °C (10 °F) of the solidus. Generally, brazing is limited to parts thicker than 0.38 mm (0.015 in.).

Applications. The most successfully brazed non-heat-treatable wrought alloys are those in the 1000 and 3000 series and the low-magnesium members of the 5000 series. Alloys higher in magnesium are more difficult to braze by the usual flux methods because of poor wetting and excessive penetration by the filler metal.

Filler Metals and Fluxes. Commercial filler metals are aluminum-silicon alloys containing 7 to 12% silicon. Filler metals for vacuum brazing usually contain magnesium.

Conventional brazing in air or other oxygen-containing atmospheres requires the use of a chemical flux. Fluxes become active before the brazing temperature is reached and are molten over the entire brazing range. They penetrate the thin oxide film on aluminum, exclude air, and promote wetting of the base metal by the filler metal.

Brazing of Copper and Copper Alloys. This family of metals is suited to brazing by all common methods, including furnace brazing. Their brazeability ratings range from good to excellent.

Filler metals include copper-zinc, copper-phosphorus, and copper-silver-phosphorus alloys; silver alloys; and gold alloys. Copper-silver-phosphorus and silver alloys are the most widely used.

Fluxes used in brazing coppers and copper alloys are proprietary and do not have standard composition ranges.

Soldering

In *soldering,* metals are joined by heating them to a suitable temperature below the solidus of the base metals and by applying a filler metal having a liquidus temperature that does not exceed 450 °C (840 °F). Molten metal is distributed between closely fitted surfaces of the joint by capillary action.

The major soldering alloys are combinations of tin and lead or alloys of the same. Tin reacts with the base metal, forming a metallurgical bond, as opposed to a mechanical bond. Variations in tin and lead content and the addition of various alloying elements result in different melting ranges and joining characteristics. Solders may contain antimony, silver, zinc, indium, and bismuth. Soldering temperatures and base metal melting points are compared in Fig. 8-23.

Principles of Soldering. When solder and a flux are melted, the liquid materials flow over the surfaces to be joined. The cleanliness and chemical composition of these surfaces are critical to the process. One function of the flux is to ensure that the base metal is sufficiently clean to provide adequate spread and flow of the soldering alloy in order to promote joint formation.

At the soldering temperature, liquid metal displaces flux from joint surfaces and is in intimate contact with the base metal, making possible a metallurgical reaction between the liquid solder and base metal. The intermetallic compound formed—a mixture, as opposed to an alloy—continues to grow at a substantial rate until solidification of the joint takes place. When the joint is completely solidified, diffusion between the base metal and soldered joint continues until the assembly is cooled to room temperature. As a result, mechanical properties of soldered joints are generally related to, but not equivalent to, those of the soldering alloy.

Most soldering operations are carried out in air, with flux acting as a barrier to surface oxidation and interaction with the atmosphere.

Tin-Lead Solders. Care is required in selecting solders because each alloy is unique in terms of composition and properties. Solders in the tin-lead system are the most widely used. Tin content is customarily listed first; for example, 40-60 solder refers to an alloy of 40% tin and 60% lead. These soldering alloys are available with melting temperatures as low as 180 °C (360 °F) or as high as 315 °C (600 °F). With some exceptions, all soldering alloys melt within a temperature range that varies according to the composition of the alloy.

Applications. Soldering alloys containing less than 5% tin are used to join tin-plated containers and in car radiator manufacture. General-purpose soldering alloys contain 40 to 50% tin; applications include plumbing repairs, making electrical connections, and general soldering of domestic items. The 60-40 and 63-37 alloys are used most extensively by the electronics industries.

Fluxes. This technology is complex. Corrosive, general-purpose fluxes are effective on low-carbon steel, copper, brass, and bronze. Applications include the production of car radiators, air-conditioning and refrigeration equipment, and sheet metal assembly.

Noncorrosive fluxes are resin based. In all electronic and critical soldering applications, water-white resin dissolved in an organic solvent is the safest known flux. These fluxes are effective on clean copper, brass, bronze, tinplate, terneplate or galvanized product, electrodeposited tin, cadmium, nickel, and silver.

Precleaning and Surface Preparation. Oil, film, grease, tarnish, paint pencil markings, cutting lubricants, and general atmospheric dirt interfere with soldering. A clean surface is a must to ensure a sound, uniform joint. Fluxing alone cannot substitute for proper precleaning. The most common cleaning methods are degreasing, acid cleaning, mechanical cleaning with abrasives, and chemical etching.

Joint Design. Soldering alloys generally are lower in strength than the materials to which they are joined. Joints must be capable of carrying service loads for the expected life of the product. Stress rupture and creep properties are critical. A variety of joint designs are illustrated in Fig. 8-24. The lap joint, for example, is considered a safe, conservative design.

Chapter 9

The Material Selection Process

Material selection is a complex process with a multiplicity of parts, each of which calls for separate consideration. One of the complicating factors in material selection is that virtually all material properties, including fabricability (the topic of Chapter 8) are interrelated. Substituting one material for another, or changing some aspect of processing to effect a change in one particular property, generally affects other properties simultaneously.

Similar interrelations that are more difficult to characterize exist among the various mechanical and physical properties and variables associated with manufacturing processes. For example, cold drawing a wire to increase its strength also increases its resistance to electricity. Steels high in carbon and alloying elements for high hardenability and strength generally are difficult to machine and weld. Additions of alloying elements such as lead to enhance machinability generally lower long-life fatigue strength and make welding and cold forming difficult. The list of interrelationships is nearly limitless.

Further light is cast by Fig. 9-1, which links relationships among material properties, design geometry, and manufacturing processes. Another perspective is added by Fig. 9-2, which summarizes the process of selecting materials for the main landing gear used in a helicopter.

Not surprisingly, compromise is common in material selection. Generally, the material most

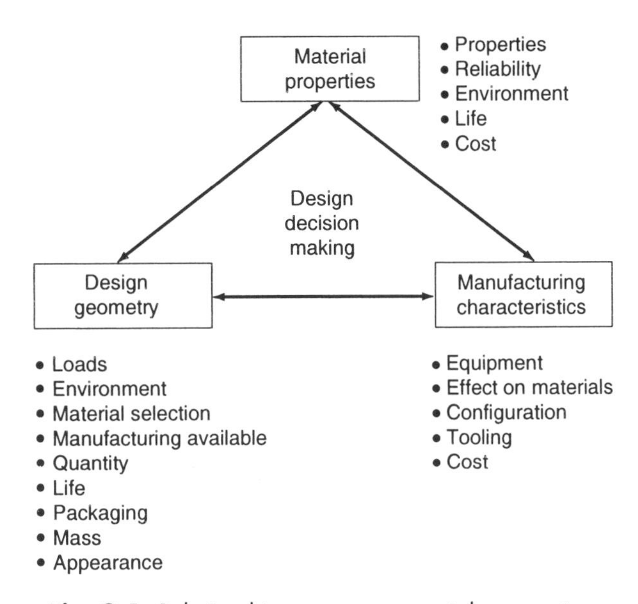

Fig. 9-1 Relationships among material properties, design geometry, and manufacturing characteristics

suitable for a given use will be the material that most nearly supplies the necessary properties and durability with a satisfactory appearance at the lowest cost.

Selection considerations in this chapter include mechanical properties, durability, producibility economics, availability, energy requirements, design configuration, appearance, and conformance to specifications and standards. Considerations in later chapters are:

- *Chapter 10:* causes of failures in service
- *Chapter 11:* corrosion and its prevention
- *Chapter 12:* quality control practices and technology

The Materials Battle

"Materials" is an all-purpose term with different meanings in different contexts. For the sake of convenience it is applied to both metals or nonmetals. Nonmetals, which include structural plastics, structural ceramics, and possibly metal-nonmetal composites, are beyond the scope of a book dedicated to metals and metallurgy. However, it must be acknowledged here that nonmetals are in stiff, daily competition with metals in traditional applications ranging from airplanes and cars to various and sundry items such as flatware and ribs for umbrellas. Finally, the contested materials turf is under the siege of highly innovative materials professionals. The next four figures provide a peek at several aspects of the ongoing materials war:

- *Figure 9-3* compares the relative cost of structural plastics and various metals.
- *Figure 9-4* compares the stiffness (modulus) of different metal and nonmetal materials of construction used in military applications at different temperatures.
- *Figure 9-5* indicates variations in the strength of metals and nonmetals used in military applications at different temperatures.
- *Figure 9-6* compares the cost of short-fiber composite parts with those of their magnesium counterparts for a radar antenna.

Selection Factors

Mechanical Properties

Does the material have the strength needed to withstand stresses imposed in service loading?

Although strength is an important consideration, of possible equal importance are factors such as toughness, corrosion resistance, electrical conductivity, magnetic characteristics, thermal conductivity, strength-to-weight ratio, or some other property. In general, a combination of properties, rather than just one, must be justi-

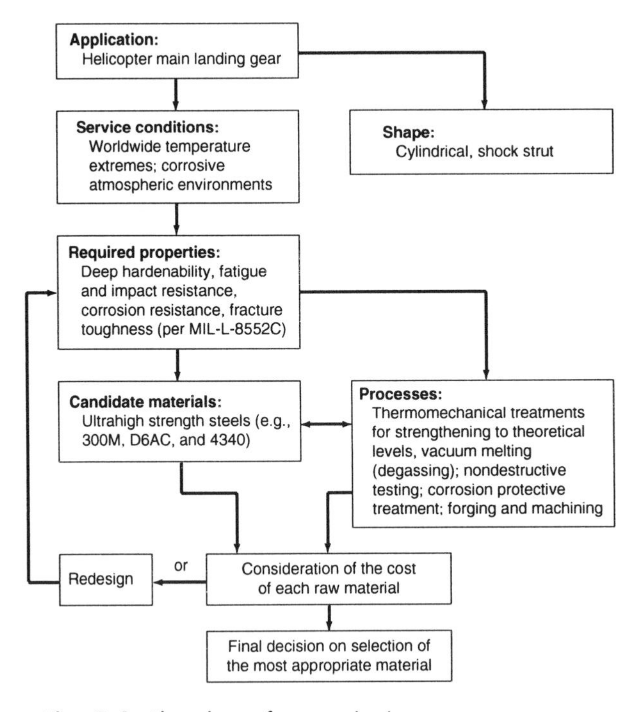

Fig. 9-2 Flowchart of material selection process

fied. Additional factors such as manufacturing characteristics and cost are other essential parts of the selection equation.

Mechanical properties of particular interest include ultimate tensile strength, yield strength, hardness, and ductility.

Ultimate Tensile Strength. Does this property indicate the ability of a material to withstand loads? The correct answer requires qualification.

First, the direct application of tensile strength data to design problems is extremely difficult. *Ultimate tensile strength* is defined as the maximum stress (based on initial cross-sectional area) that a specimen can withstand before it fails. The condition occurs at the onset of plastic instability—a point in time that is inherently difficult to pinpoint. Further, any component loaded to its ultimate strength is likely to fracture immediately.

Second, if a design is based on a fraction of ultimate strength, the question becomes: what fraction provides adequate strength and safety? Finally, there seems to be only a rough correla-

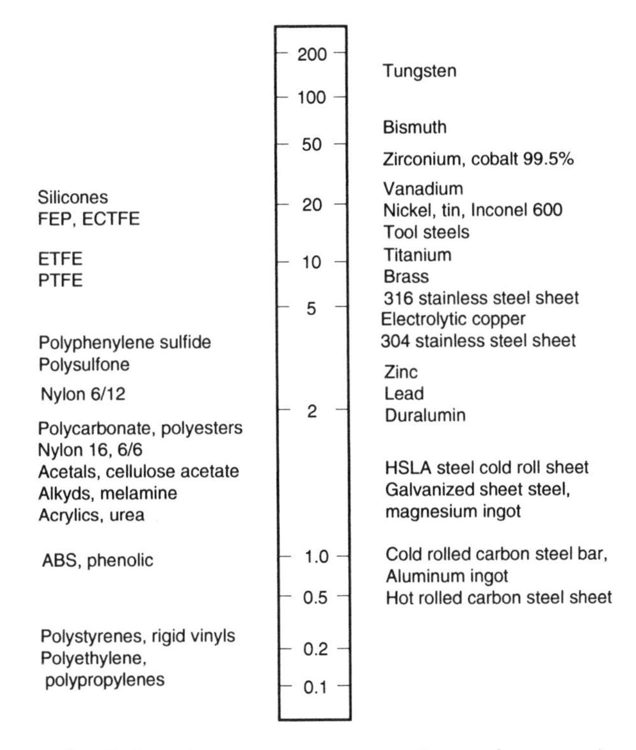

Fig. 9-3 Relative cost per unit volume of structural plastics (left) and metal (right) relative to cost of hot-rolled carbon steel. Basis: prices at end of 1972

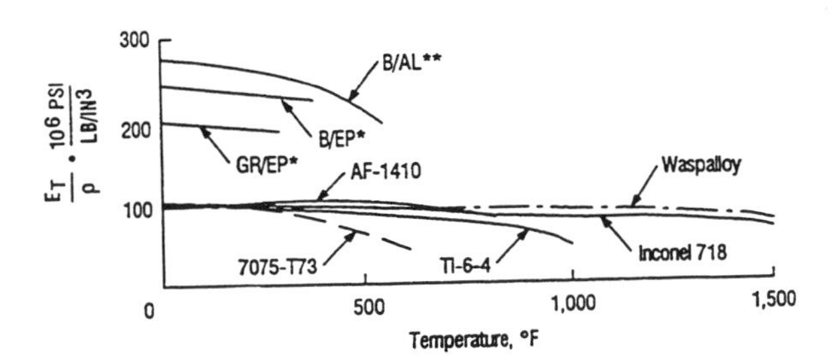

Fig. 9-4 Modulus (rigidity) of different military materials of construction at different temperatures. B/EP, boron-epoxy; B/Al, boron-aluminum; GR/EP, graphite epoxy

tion between tensile strength and material properties such as hardness and fatigue strength at a specified number of cycles. In addition, there is no correlation between tensile strength and properties such as resistance to crack propagation, impact resistance, or proportional limit (the maximum stress at which strain remains directly proportional to stress).

Yield Strength. The information needed here concerns the lowest stress at which measurable permanent deformation occurs. This knowledge is necessary in estimating the forces required in forming operations. Most structures must be designed so that a foreseeable overload does not exceed yield strength.

Hardness is useful in estimating the wear resistance of materials and the approximate strength of steels. Greatest usage of this value is in quality control in heat treating. For example, hardness testing is used to determine consistency in maintaining a specified hardness or range of hardness. However, hardness does have limitations as an indicator of other properties. Only a rough correlation can be made between hardness and other mechanical properties or between hardness and the behavior of materials in service.

Ductility. Percent elongation in area or elongation that occurs in a tensile test is the usual definition of this property, which often is considered an important factor in material selection. One assumption is that if a metal has a certain minimum elongation, it will not fail in service through brittle fracture. Another assumption is that if a little ductility is good, a lot is better. Neither assumption is confirmed by experience. In fact, several key questions await answers:

- How much ductility is actually usable under service conditions?
- How can this value be measured?
- What is the relationship of ductility to fabricability and formability?

Word of Caution. Engineers responsible for the selection of materials should keep in mind that all engineering materials show a certain variability in mechanical properties. Also, data can be misleading for many reasons; for example, strength data often are obtained in uniaxial tests, whereas stresses in service generally are complex. In addition, it may not be possible to estimate applied service loads. Stresses must be approximated, making it necessary to provide a margin of safety and to protect against failure from unpredictable causes. Allowable stresses must be lower than those causing failure (called the *factor of safety*). Careful consideration should also be given to the consequences of failure.

Durability

Durability is a generic term for the performance of a material in service. To put it another way, a material has durability if it continues to function properly during the design life of the structure of which it is a part.

The four major causes of failure are wear, corrosion, rupture, and exposure to a range of

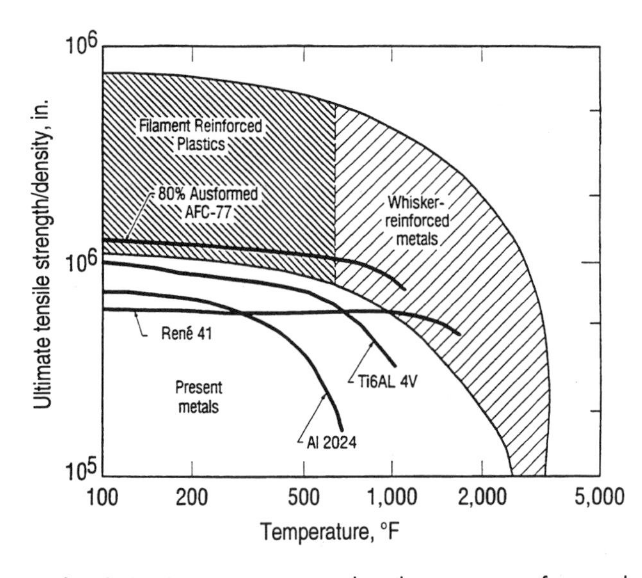

Fig. 9-5 Variations in strength with temperature for metal and nonmetal military materials of construction

service temperatures. These factors are examined in some detail in subsequent chapters. Relationships between modes of failure and material properties are shown in Fig. 9-7.

Wear and corrosion probably account for more withdrawals from service than fracture. Environmental factors such as humidity or chemicals can cause deterioration and subsequent failure. High or low temperatures can downgrade service performance. To be durable, a material must resist all forms of destruction to such a degree that the system or device will not be made unsafe or inefficient at any time during its prescribed life.

Designers can add service life to a product by taking advantage of available metal treating technology. For example, the cam follower portion (wear surface) of a steel valve lifter could be shot peened, flame hardened, or case hardened, depending on the service life desired or the severity of the service environment. In *shot peening,* metal shot cold-works material surfaces (Fig. 9-8). In *flame hardening,* the desired microstructure, such as martensite, is obtained by using an intense flame to heat the surface layers of a part. In *case hardening,* wear properties are improved by applying a surface layer of carbon or nitrogen, or a mixture of the two, by diffusion.

The designer may also consider alternative materials for a part to improve its service life potential. Say the designer chooses phosphor bronze over steel for a spring application. The steel has superior fatigue life, but is more susceptible to corrosion. This will seriously reduce fatigue life, since in a corrosive environment the steel will deteriorate and eventually fail. On the other hand, bronze resists corrosion and retains

Anatomy of a radar antenna

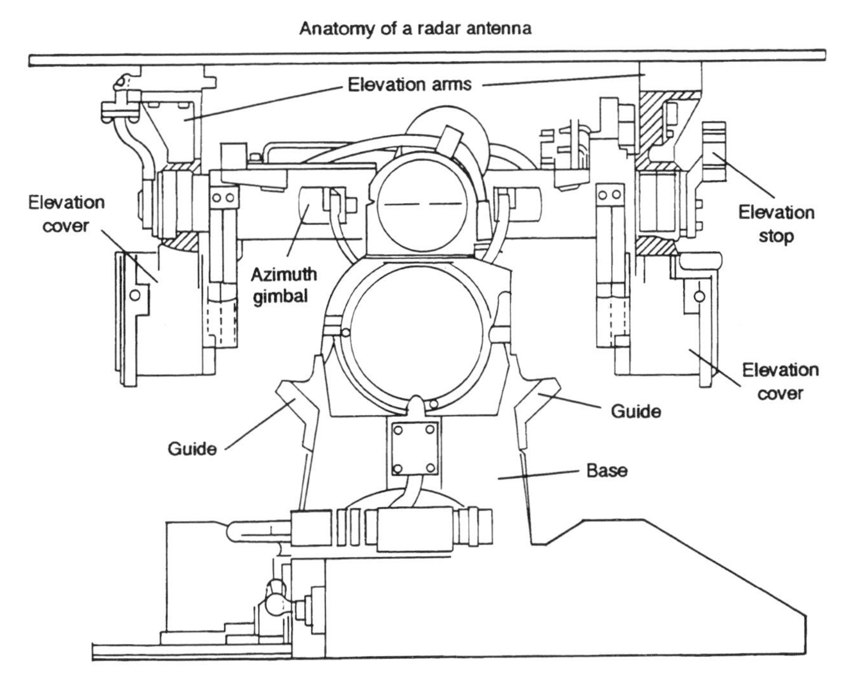

Part	Cost ratio (metal:composite)
Base	1:2
Elevation arm	1:4
Azimuth gimbal	1:4
Elevation stop	2:1
Elevation cover	5:1
Guides	1.5:1

Includes material, manufacturing, and tooling costs

Fig. 9-6 Materials in parts identified in drawing are short-fiber composites. The ratio of their costs vs those of magnesium counterparts are compared in the numbers that follow. Composite costs are listed second.

	Material Property														
Failure Mode	Ultimate tensile strength	Yield strength	Compressive yield strength	Shear yield strength	Fatigue properties	Ductility	Impact energy	Transition temperature	Modulus of elasticity	Creep rate	Fracture toughness K_{IC}	Stress corrosion fracture toughness K_{ISCC}	Electrochemical potential	Hardness	Coefficient of thermal expansion
Gross yielding															
Buckling															
Creep															
Brittle fracture															
Fatigue, low-cycle															
Fatigue, high-cycle															
Contact fatigue															
Fretting															
Corrosion															
Stress corrosion cracking															
Galvanic corrosion															
Hydrogen embrittlement															
Wear															
Thermal fatigue															
Corrosion fatigue															

Fig. 9-7 Relationships between modes of failure and material properties

its original properties. Table 9-1 lists some corrosion-resistant materials and their characteristics.

Producibility

Does the technology exist to produce the desired quantity of product at an economical cost? Does the factory have the capability to produce the desired number of parts? A third question concerns the utilization of existing equipment and personnel.

Obviously, this factor affects engineering design at several levels. In some instances, answers will descend from upper levels of management. However, it usually is desirable for the designers and engineers to analyze the various possibilities and present management with alternatives, together with possible consequences, advantages, and disadvantages of each.

Economics

All things considered, the material selected for a given application should be the most economical one that satisfies the requirements for functionality, durability, and appearance. Designing a product for minimum cost consistent with fulfilling the functional requirements of the product is an accepted principle in engineering design.

Material costs are based on such factors as quality, availability, and workability. Low initial cost is seldom the conclusive factor. For example, a high-cost material can be easier to manufacture and process than the bargain material, wiping out any anticipated savings in material cost. Prices for structural materials can be obtained by referring to purchasing manuals or by contacting the manufacturer.

Cost savings usually can be realized in material selection by changing the fabrication procedures, by changing the configuration of the product, or by the simple method of using an alternative material without substantially changing the design or processing procedure. If greater strength is the objective, the cost of the alternative will probably be higher. The magnitude of change in cost depends on the material, as indicated in Table 9-2. Of course, the ultimate challenge is getting a combination of reasonable cost and reliability in service.

Availability of Materials

The material selected must be readily available in the required form and quantity and be deliverable at the required time and agreed-on price. Such conditions usually do not pose a problem. However, there are exceptions: wartime for one. Material supplies shrink and prices soar. The plus side is that shortages and skyrocketing prices tend to promote innovation. A typical remedy is to locate substitute materials from the list of those still available that will provide the desired properties despite shortages of key elements.

EX steels, for example, were developed in the first decade of the 1970s during a period of worldwide shortages of practically all materials used in the manufacturing industry. These steels (Table 9-3) are alternatives to certain standard SAE (Society of Automotive Engineers) grades. EX steels not only reduced the high cost of standard grades due to the shortages of expensive alloying elements such as nickel and molybdenum, but also provided the desired hardness, strength, ductility, and metallurgical response to heat treatment. Jominy curves in Fig. 9-9 show that EX steels have the same hardenability characteristics as SAE grades 8620, 8640, and 52100. EX steels, which are still available, are higher in manganese content to compensate for lower percentages of nickel, chromium, and molybdenum.

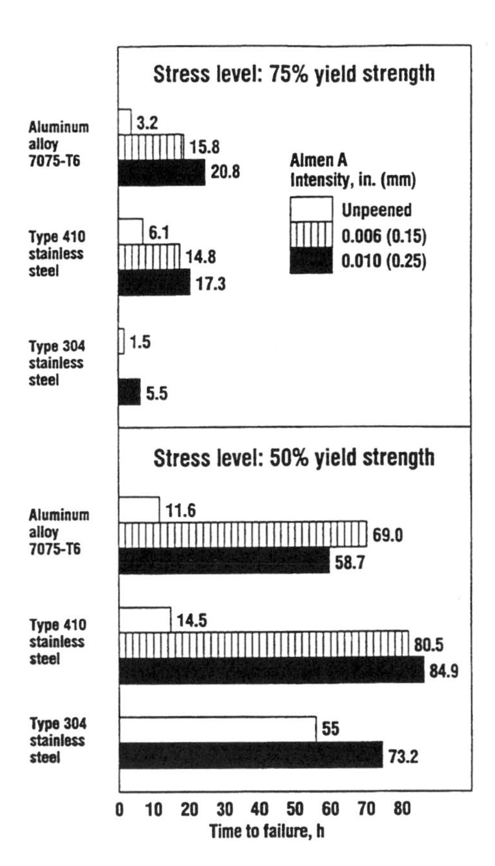

Fig. 9-8 Effect of shot peening intensity rates on failure rates of materials exposed to stress-corrosive environments

Energy Requirements

Justification is compelling for the conservation of scarce resources, especially those subject to manipulation for economic or political gain. All production operations generate scrap—such as metal chips in machining operations, and gates and risers (waste by-products) in the casting of parts. Any change in process, component design, or form of material used has the potential to reduce scrap, which in turn conserves finite resources and reduces costs. For

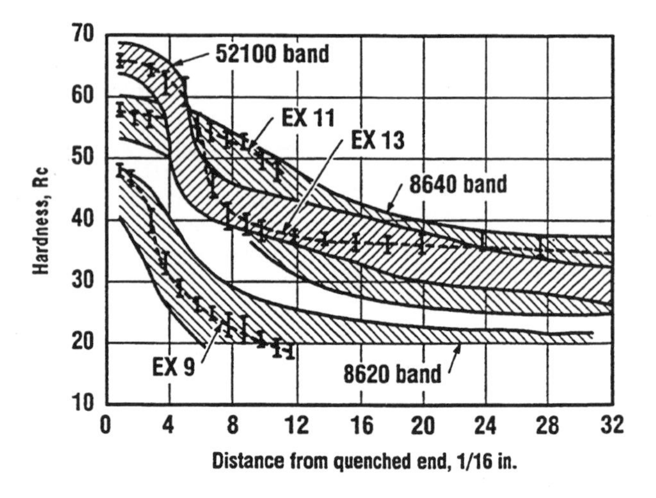

Fig. 9-9 Jominy curves for EX 9, 11, and 13 compared with those for SAE 8620, 8640, and 52100, respectively

Table 9-1 Corrosion-resistant materials

Material	Characteristics
Coated steel	Ultimate tensile strength, 550 to 860 MPa (80 to 125 ksi). Low-carbon, medium-carbon, and low-alloy steels can be made resistant to atmospheric corrosion by coating them. Examples: A325, A490, SAE J429 materials
Stainless steel	
Austenitic	Ultimate tensile strength, 515 to 825 MPa (75 to 120 ksi). Most common of the stainless steels, and more corrosion resistant than the three listed below. Nonmagnetic. Cannot be heat treated, but can be cold worked. Good high- and low-temperature properties. Type 321 can be used up to 816 °C (1500 °F), for example. Examples: A193 B8 series, A320 B8 series, any of the 300 or 18-8 series materials, such as 303, 304, 316, 321, 347
Ferritic	Ultimate tensile strength, 480 MPa (70 ksi). Cannot be heat treated or cold worked. Magnetic. Examples: 430 and 430F
Martensitic	Ultimate tensile strength, 480 to 1240 MPa (70 to 80 ksi). Heat treatable, magnetic. Can experience stress-corrosion if not properly treated. Examples: 410, 416, 431
Precipitation hardening	Ultimate tensile strength, 930 MPa (135 ksi). Heat treatable. More ductile than martensitic stainless steels. Examples: 630, 17-4PH, Custom 455, PH 13-8 Mo, ASTM A453-B17B, AISI 660
Nickel-base alloys	
Nickel-copper	Ultimate tensile strength, 480 to 550 MPa (70 to 80 ksi). Can be cold worked, but not heat treated. Example: Monel
Nickel-copper-aluminum	Ultimate tensile strength, 895 MPa (130 ksi). Can be both cold worked and heat treated. Good low-temperature material. Example: K-Monel
Titanium	Ultimate tensile strength, 930 to 1380 MPa (135 to 200 ksi). Good corrosion resistance. Low coefficient of expansion. Has a tendency to gall more readily than some other corrosion-resistant materials. Expensive. Example: Ti-6Al-4V
Superalloys	Ultimate tensile strength, 1000 to 1965 MPa (145 to 285 ksi). High-strength materials with excellent properties at high and/or low temperatures. Primarily used in aerospace applications. Expensive. Some, such as MP35N, are virtually immune to marine environments and to stress-corrosion cracking. Examples: H-11, Inconel, MP35N, A286, Nimonic 80A. MP35N, Inconel 718, and A286 are especially recommended for cryogenic applications.
Nonferrous materials	Many nonferrous fastener materials can provide outstanding corrosion resistance in applications that would rapidly destroy more common bolt materials. The main drawback to these materials is a general lack of strength, but this can sometimes be offset by using fasteners of larger diameter or by using more fasteners. Examples: silicon bronze, ultimate tensile strength, 480 to 550 MPa (70 to 80 ksi); aluminum, ultimate tensile strength, 90 to 380 MPa (13 to 55 ksi); nylon, ultimate tensile strength, 76 MPa (1 ksi)

example, a part normally produced by machining an aluminum bar can be made from an impact extrusion, resulting in a gain in material utilization.

In addition, consideration must be given to the disposal of materials after a part or machine has outlived its usefulness. Recycling is an obvious answer. Many communities and industrial firms now have facilities for recycling materials that can be resurrected in new applications.

Design Configuration

This is an important factor in material selection. For example, design configuration can contribute significantly to service demands placed on aerospace structures. The probability of structural failure due to wear or corrosion, for instance, is very dependent on the shape (configuration) of a part. For this reason, design engineers specify configurations that include

Table 9-2 Strength and cost relationships of various high-strength low-alloy (HSLA) steels

Specification	Type	Minimum yield point		Strength increase over A36,	Cost, increase over A36,	Other features
		MPa	ksi	%	%	
ASTM A36	Plain carbon	248	36	...	......	Lowest cost per pound
ASTM A572, grade 50	Nb/V microalloy	345	50	38.9	9.4	Good weldability
ASTM A440	Mn-Cu	345	50	38.9	13.4	Not for welded structures. Corrosion resistance twice that of A36
ASTM A441	Mn-V-Cu	345	50	38.9	13.4	For welded structures. Atmospheric corrosion resistance twice that of A36
ASTM A242	Multiple alloy, weathering steel	345	50	38.9	38.3	Weathering steel. Atmospheric corrosion resistance 4 times that of A36 (often 5 to 8 times that for type 1). Type 2 has excellent toughness.

Table 9-3 EX Steels and equivalent standard grades

EX No.	Composition, %					Equivalent SAE grade
	C	Mn	Cr	Mo	Other	
10	0.19–0.24	0.95–1.25	0.25–0.40	0.05–0.10	0.20–0.40 Ni	8620
15	0.18–0.34	0.90–1.20	0.40–0.60	0.13–0.20	...	8620
16	0.20–0.25	0.90–1.20	0.40–0.60	013–0.20	...	8622
17	0.23–0.28	0.90–1.20	0.40–0.60	0.13–0.20	...	8625
18	0.25–0.30	0.90–1.20	0.40–0.60	0.13–0.20	...	8627
19	0.18–0.23	0.90–1.20	0.40–0.60	0.08–0.15	0.0005 B min	94B17
20	0.13–0.18	0.90–1.20	0.40–0.60	0.13–0.20	...	8615
21	0.15–0.20	0.90–1.20	0.40–0.60	0.13–0.20	...	8617
24	0.18–0.23	0.75–1.00	0.45–0.65	0.20–0.30	...	8620
30	0.13–0.18	0.70–0.90	0.45–0.65	0.45–0.60	0.70–1.00 Ni	4815
31	0.15–0.20	0.70–0.90	0.45–0.65	0.45–0.60	0.70–1.00 Ni	4817
32	0.18–0.23	0.70–0.90	0.45–0.65	0.45–0.60	0.20–0.40 Ni	4820
33	0.17–0.24	0.85–1.25	0.45–0.65	0.05 min	...	4027
34	0.28–0.33	0.90–1.20	0.20 min	0.13–0.20	...	8630
36	0.38–0.43	0.90–1.20	0.40–0.60	0.13–0.20	...	8640
38	0.43–0.48	0.90–1.20	0.45–0.65	0.13–0.20	...	8645
39	0.48–0.53	0.90–1.20	0.45–0.65	0.13–0.20	...	8650
40	0.51–0.59	0.95–1.20	0.45–0.65	0.13–0.20	...	8655
54	0.19–0.25	0.70–1.05	0.40–0.70	0.05 min	...	4118
55	0.15–0.20	0.70–1.00	0.45–0.65	0.65–0.80	1.65–2.00 Ni	4817
56	0.08–0.13	0.70–1.00	0.45–0.65	0.65–0.80	1.65–2.00 Ni	9310
57	0.08 max	1.25 max	17–19	1.75–2.25	...	30303
58	0.16–0.21	1.00–1.30	0.45–0.65	...	...	4118
59	0.18–0.23	1.00–1.30	0.70–0.90	...	...	8620
60	0.20–0.25	1.00–1.30	0.70–0.90	...	...	8622
61	0.23–0.38	1.00–1.30	0.70–0.90	...	...	8625
62	0.25–0.30	1.00–1.30	0.70–0.90	...	...	8627
63	0.31–0.38	0.75–1.10	0.45–0.65	...	0.0005–0.003 B	...
64	0.16–0.21	1.00–1.30	0.70–0.90	...	...	...
65	0.21–0.26	1.00–1.30	0.70–0.90	...	...	...

All steels contain 0.035% P max, except EX 57 (0.040% P max); 0.040% S max, except EX 57 (0.15–0.35% S); and 0.20–0.35% Si, except EX 54 (0.33% Si max), EX 57 (1.00% Si max), EX 58 to 62 (0.15–0.30% Si), and EX 63 to 65 (0.15–0.35% Si).

contoured (radiused) edges, instead of sharp edges that are difficult to protect against wear and corrosion.

Table 9-4 Principal specification-writing groups in the United States

Name	Designation
Aerospace Material Specification (of SAE)	AMS
Aluminum Association	AA
American Bureau of Shipping	ABS
American Iron and Steel Institute	AISI
American National Standards Institute	ANSI
American Petroleum Institute	API
American Railway Engineering Association	AREA
American Society of Mechanical Engineers	ASME
American Society for Testing and Materials	ASTM
Association of American Railroads	AAR
General Services Administration	FED
Society of Automotive Engineers	SAE
Unified Numbering System	UNS
U.S. Department of Defense	MIL and JAN

Appearance

In selecting materials, it is important to remember that a product should be in harmony with its environment. Both the line and construction materials used will influence the appearance of any product. For example, a domestic appliance must be designed for both functionality and ease of mass production, and at the same time present a pleasing appearance for sales appeal. Appearance may not be vital when a product is hidden from view.

Standards and Specifications

Throughout the design process, engineers responsible for material selection must be aware of and make reference to pertinent performance and product specifications and standards. Conformance indicates that the product has been properly designed to cope with environmental

Table 9-5 AISI-SAE system of designations

Numerals and digits(a)	Type of steel and/or nominal alloy content
Carbon steels	
10*xx*	Plain carbon (Mn 1.00% max)
11*xx*	Resulfurized
12*xx*	Resulfurized and rephosphorized
15*xx*	Plain carbon (max Mn range — 1.00 –1.65%)
Manganese steels	
13*xx*	Mn 1.75
Nickel steels	
23*xx*	Ni 3.50
25*xx*	Ni 5.00
Nickel-chromium steels	
31*xx*	Ni 1.25; Cr 0.65 and 0.80
32*xx*	Ni1.75; Cr1.07
33*xx*	Ni 3.50; Cr 1.50 and 1.57
34*xx*	Ni 3.00; Cr 0.77
Molybdenum steels	
40*xx*	Mo 0.20 and 0.25
44*xx*	Mo 0.40 and 0.52
Chromium-molybdenum steels	
41*xx*	Cr 0.50, 0.80 and 0.95; Mo 0.12, 0.20, 0.25, and 0.30
Nickel-chromium-molybdenum steels	
43*xx*	Ni 1.82; Cr 0.50 and 0.80; Mo 0.25
43BV*xx*	Ni 1.82; Cr 0.50; Mo 0.12 and 0.25; V 0.03 min
47*xx*	Ni 1.05; Cr 0.45; Mo 0.20 and 0.35
81*xx*	Ni 0.30; Cr 0.40; Mo 0.12
86*xx*	Ni 0.55; Cr 0.50; Mo 0.20
87*xx*	Ni 0.55; Cr 0.50; Mo 0.25
88*xx*	Ni 0.55; Cr 0.50; Mo 0.35
93*xx*	Ni 3.25; Cr 1.20; Mo 0.12
Nickel-chromium-molybdenum steels (continued)	
94*xx*	Ni 0.45; Cr 0.40; Mo 0.12
97*xx*	Ni 0.55; Cr 0.20; Mo 0.20
98*xx*	Ni 1.00; Cr 0.80; Mo 0.25
Nickel-molybdenum steels	
46*xx*	Ni 0.85 and 1.82; Mo 0.20 and 0.25
48*xx*	Ni 3.50; Mo 0.25
Chromium steels	
50*xx*	Cr 0.27, 0.40, 0.50, and 0.65
51*xx*	Cr 0.80, 0.87, 0.92, 0.95, 1.00, and 1.05
50*xxx*	Cr 0.50, C 1.00 min
51*xxx*	Cr 1.02, C 1.00 min
52*xxx*	Cr 1.45, C 1.00 min
Chromium-vanadium steels	
61*xx*	Cr 0.60, 0.80, and 0.95; V 0.10 and 0.15 min
Tungsten-chromium steel	
72*xx*	W 1.75; Cr 0.75
Silicon-manganese steels	
92*xx*	Si 1.40 and 2.00; Mn 0.65, 0.82, and 0.85; Cr 0.00 and 0.65
High-strength low-alloy steels	
9*xx*	Various SAE grades
Boron steels	
*xx*B*xx*	B denotes boron steel
Leaded steels	
*xx*L*xx*	L denotes leaded steel

(a) "*xx*" in the last two (or three) digits of these designations indicates that the carbon content (in hundredths of a percent) is to be inserted.

and mechanical stresses encountered in service. The design engineer must ensure that the standard selected applies to the product being fabricated and must be certain the standard selected provides assurance that the end product will be able to perform its assigned tasks.

Specifications (also known as *standards*) are written for essentially two purposes:

- As a means for the individual purchaser of goods and services to define their requirements for the instruction of suppliers
- As a means for organizations such as trade associations and technical societies to establish national standards for their constituencies (suppliers of goods and services)

User specifications may or may not be included in purchase orders, and may or may not reference national standards written and published by organizations such as:

- American Iron and Steel Institute (AISI)
- Society of Automotive Engineers (SAE)
- American Society for Testing and Materials (ASTM)
- American Society of Mechanical Engineers (ASME)

Several standards-writing groups in the United States are listed in Table 9-4. Government agencies, such as the Department of Defense, also develop standards. However, the government has indicated that it wants to discontinue this activity.

Standards and Specifications for Carbon and Alloy Steels

AISI-SAE Designations. The most widely used system for designating carbon and alloy steels is that of AISI and SAE. Technically, they are separate systems, although the standards are nearly identical and carefully coordinated by both organizations. The numerical designations in Table 9-5 are used by both AISI and SAE. Limits and ranges for chemical composition are the same. Any differences in listing are the result of differences in eligibility for listing. The basis for AISI is production tonnage, while that of SAE is significant usage by two major users for whom the steel has unique engineering characteristics.

AISI standards are available from AISI. SAE standards are published in the annual *SAE Handbook*. These standards are listings of SAE designations and include limits and ranges for chemical compositions. Neither AISI nor SAE provides enough information necessary to describe a steel product for procurement purposes.

ASTM specifications, by comparison, are complete and generally adequate for specification purposes. Many ASTM specifications apply to specific products; for example, A574 is for alloy steel socket-head cap screws. More typically, these specifications are oriented to the performance of the fabricated end product and allow considerable latitude in the chemical composition of the steel used.

ASTM specifications represent a consensus among producers, specifiers, fabricators, and users of steel mill products. In many instances, dimensions, tolerances, limits, and restrictions in ASTM standards are the same as those in the AISI *Steel Product Manual*.

AISI-SAE designations for the composition of carbon and alloy steels normally are incorporated into ASTM specifications for bars, wires, and billets for forging. Some ASTM specifications for sheet products include AISI-SAE designations for composition. ASTM specifications for plates and structural shapes generally specify limits and ranges of chemical composition, without the AISI-SAE designation.

Many ASTM standards have been adopted by ASME with little or no change. ASME uses the prefix "S" and the ASTM designation for these specifications. For example, ASME SA-213 and ASTM A 213 are identical. Major categories in the ASTM system are listed in Table 9-6.

Table 9-6 Generic ASTM specifications

Specification	Material
A6	Rolled steel structural plate, shapes, sheet piling, and bars, generic
A20	Steel plate for pressure vessels, generic
A29	Carbon and alloy steel bars, hot rolled and cold finished, generic
A505	Alloy steel sheet and strip, hot rolled and cold rolled, generic
A510	Carbon steel wire rod and coarse round wire, generic
A568	Carbon and HSLA hot-rolled and cold-rolled steel sheet and hot rolled strip, generic
A646	Premium-quality alloy steel blooms and billets for aircraft and aerospace forgings
A711	Carbon and alloy steel blooms, billets, and slabs for forging

AMS Designations

These standards, also referred to as Aerospace Materials Specifications and published by SAE, are complete specifications and generally adequate for procurement purposes. Most are intended for aerospace applications. For this reason, mechanical property requirements usually are more severe than those for less demanding applications. Also, raw material processing requirements, such as the requirement for consumable electrode melting, are common in AMS specifications.

UNS Designations

The Unified Numbering System (UNS) has been developed by ASTM, SAE, and several other technical societies, trade associations, and U.S. government agencies. A UNS number, which is a designation of chemical composition and not a specification, is assigned to each chemical composition of a metal alloy.

The UNS designation consists of a letter and five numerals. Letters indicate the broad class of alloys; numerals define specific alloys within that class. Existing systems of designation, such as the AISI-SAE system for steels, have been incorporated into the UNS designation. The Unified Numbering System is described in SAE Recommended Practice J1086 and ASTM E 527.

Aluminum Standards

ANSI H35.1-1982 is the national standard for aluminum and its alloys in wrought and cast forms, together with their tempers. The designation system for wrought aluminum and its alloys is based on four digits, with the first digit indicating the group, as follows:

Aluminum, 99.00 minimum and greater	1*xxx*
Aluminum alloys, grouped by major alloying elements:	
Copper	2*xxx*
Manganese	3*xxx*
Silicon	4*xxx*
Magnesium	5*xxx*
Magnesium and silicon	6*xxx*
Zinc	7*xxx*
Other elements	8*xxx*

The system for cast aluminum and aluminum alloys also has four digits. A decimal point after the third digit identifies aluminum and its alloys in the form of castings and foundry ingot. Alloy group is indicated by the first digit, as follows:

Aluminum, 99.00% and greater	1*xx.x*
Aluminum alloys, grouped by major alloying elements:	
Copper	2*xx.x*
Silicon, with added copper and/or magnesium	3*xx.x*
Silicon	4*xx.x*
Magnesium	5*xx.x*
Zinc	7*xx.x*
Tin	8*xx.x*
Other elements	9*xx.x*

UNS numbers correlate numbering systems used by individual users and producers of aluminum and its alloys with those used in other numbering systems.

Other Resources

The designer has a variety of tools available to aid in the selection of engineering materials. Engineering societies such as ASM International publish reference books and handbooks that contain data on the properties of materials for engineering applications. In addition, electronic databases on material properties are accessible through data disks and CD-ROMs for networks and personal computers. In large companies, the services of material specialists often are available.

Chapter 10

Failure of Metals under Service Conditions

In the material selection process, the different conditions that can cause metals to fail must be taken into account. Rupture, wear, temperature effects, and corrosion are the major causes for these failures. This chapter discusses failures due to rupture, wear, and temperature effects; corrosion and preventive measures for dealing with it are discussed separately in Chapter 11.

Rupture, Wear, and Temperature Effects

Rupture. Metals rupture, or break, in service primarily because of applied mechanical forces that generate high tensile stresses. Brittle, ductile, and fatigue fracture are discussed in this chapter. A dramatic example of brittle fracture is shown in Fig. 10-1.

Wear can be defined as the unintended and undesired removal of metal from contacting surfaces by mechanical action. Mechanisms include abrasive wear, erosive wear, grinding wear, gouging wear, and adhesive wear. Adhesive wear is illustrated in Fig. 10-2.

Temperatures well above or well below room temperature can cause failures in service. Causes include creep, elevated-temperature fatigue, and thermal fatigue. An example of high-temperature fracturing of tubing in service is shown in Fig. 10-3.

Brittle Fracture

Little or no permanent deformation (plastic type) is involved in brittle fracture. Many nonmetals—such as chalk, rock, and brick—are subject to this type of fracture because they are lacking in ductility. However, brittle metals—such as hardened tool steels and gray cast iron—are used daily in engineering applications. Properly handled, their performance is very satisfactory in many types of service. However, if a tool such as a metal cutting file is bent, for example, it can break suddenly due to brittle fracture; there is little or no permanent deformation, and the pieces can be put back in perfect alignment.

In general, it is characteristic of very hard, strong, notch-sensitive metals to be brittle. Conversely, it is generally true that softer, weaker metals usually are ductile. Gray cast iron is the exception: It is brittle because it contains a large number of internal graphite flakes. These flakes act as internal stress concentrations and limit the ability of the metal to flow or deform, which is necessary for ductile behavior.

These and certain other metals known to be brittle are normally suitable for applications where there is little or no danger of fracture. Certain other common metals—especially low-carbon and medium-carbon steels—normally are considered to have ductile properties. Under certain circumstances, however, these normally ductile steels can fracture in a totally brittle manner. The tanker pictured in Fig. 10-1 was one of 19 ships that failed by brittle fracture (broke totally in two) during World War II. More than 200 other ships had partial brittle fractures of the hull. When tested, fractured plates had normal ductility as specified. In some cases fractures occurred in ships that were still being outfitted and had never been to sea.

Explanation. Brittle fracture of normally ductile steels has occurred primarily in large, continuous, boxlike structures such as ships, box beams, pressure vessels, tanks, pipes, bridges, and other restrained structures. Four factors must be present simultaneously to cause brittle failure of a normally ductile steel:

- The steel itself must be susceptible to brittle fracture.
- Stress concentrations must be present—such as a weld defect, fatigue crack, stress-corrosion crack, or a designed notch (such as a sharp corner, thread, or hole). The stress concentration must be large enough and sharp enough to be a critical flaw in terms of fracture mechanics.
- A tensile stress must be present. One of the major complexities is that the tensile stress need not be an applied stress on the structure but can be a residual stress completely within the structure. In fact, a part or structure can be completely free of an external or applied load, and fail suddenly and catastrophically while sitting on a bench or on

Fig. 10-1 SS Schenectady at its outfitting dock in 1941. This tanker was one of 19 ships to fail by brittle fracture during World War II.

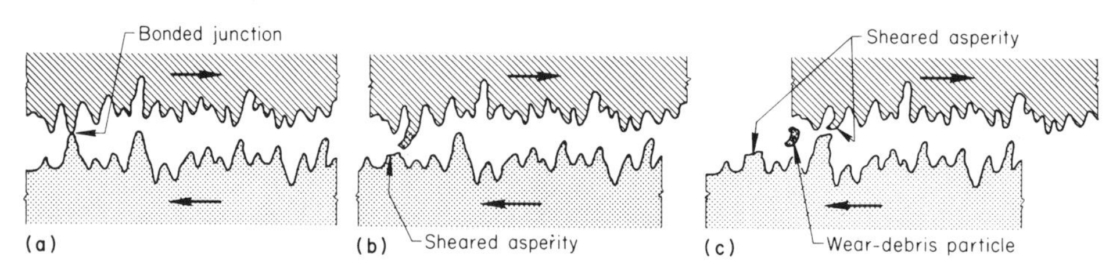

Fig. 10-2 Schematic illustration of one process by which a particle of wear debris is detached during adhesive wear. As the surfaces slide across each other, a bonded junction (a) is torn from one peak, or asperity (b), then is sheared off by impact with a larger, adjacent peak to form a particle of wear debris (c). The peaks are greatly exaggerated in this sketch, but the principle is accurate; metal also may be transferred from one surface to another by the microwelding process. Arrows indicate direction of sliding.

the floor. Welded, torch-cut, or heat-treated steels are particularly susceptible to this type of failure.

- Temperature must be relatively low for the steel involved. As a rule, the lower the ductile-brittle transition temperature of a steel, the greater the possibility of brittle fracture. Under certain circumstances the transition temperature may be above room temperature.

Preventive measures include:

- Stress concentrations such as sharp corners, threads, and grooves should be minimized during design.

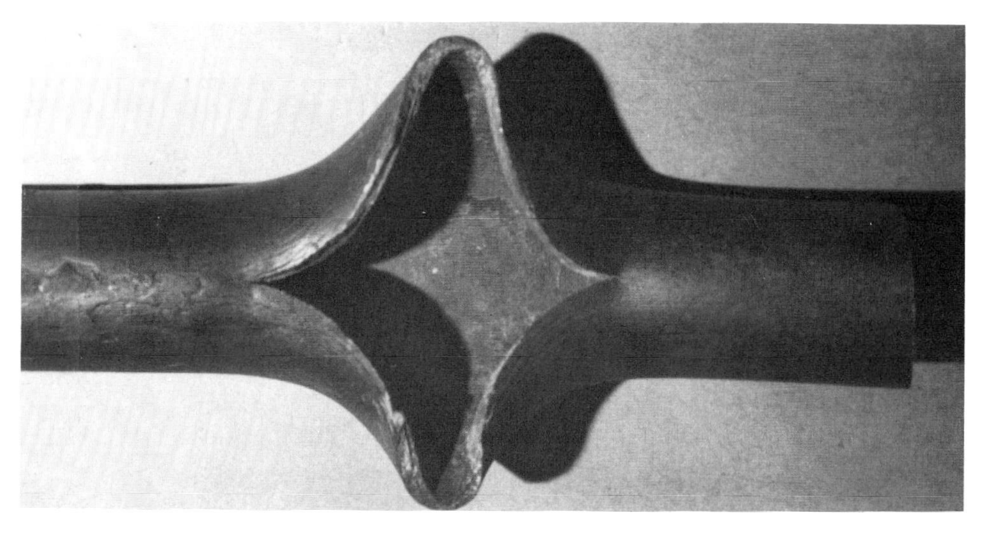

Fig. 10-3 Thick-lip "fishmouth" failure of a 50 mm (2 in.) diam superheater tube. The tube bent away from the fracture because of the reaction force of the escaping steam.

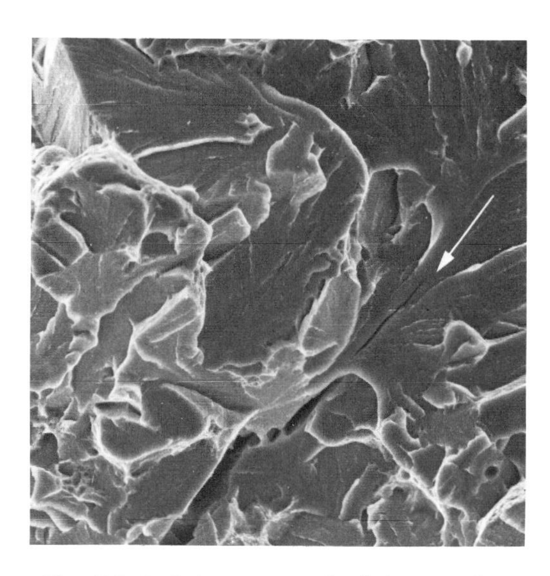

Fig. 10-4 SEM micrograph of cleavage fracture in hardened steel. Note progression of "river" marks in the direction of arrow. Grain boundaries were crossed without apparent effect. 2000×, shown here at 75%

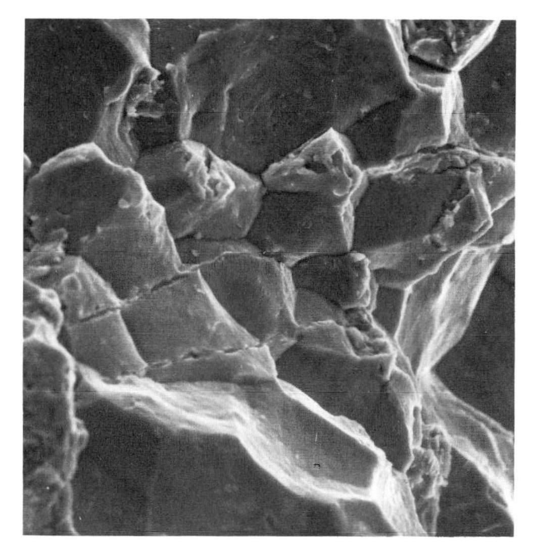

Fig. 10-5 SEM micrograph of intergranular fracture in hardened steel. Note that fracture takes place between the grains; thus, the fracture surface has a "rock candy" appearance that reveals the shapes as part of the individual grains. 2000×, shown here at 75%

- Tensile stresses usually are inevitable during service loading, but care can be taken to minimize damaging residual stresses, especially when shrinkage stresses from welding are involved.
- Temperatures can be controlled in some applications. In some instances, temperature may not be a problem. Some processing equipment normally is operated continually at elevated temperatures. In this case, brittle fracture may not be a consideration unless there is an undesirable environmental factor, such as absorption of hydrogen or hydrogen sulfide.

Brittle fracture of a normally ductile steel is a real possibility when service temperatures are relatively low. However, it is possible to prevent the problem by taking advantage of recently developed grades of steel with increased notch

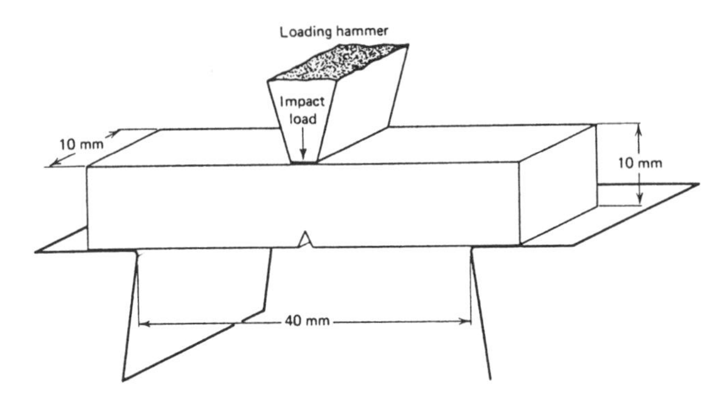

Fig. 10-6 Schematic of Charpy impact specimen and test arrangement

Fig. 10-7 Brittle versus ductile fracture in two 1038 steel bolts deliberately heat treated to have greatly different properties when pulled in tension. The brittle bolt (left) was water quenched with a hardness of 47 HRC, but had no obvious deformation. The ductile bolt (right) was annealed to a hardness of 95 HRB (equivalent to 15 HRC) and shows tremendous permanent deformation.

Fig. 10-8 Type 302 stainless steel tensile specimen with the typical cup-and-cone fracture characteristic of ductile metals fractured in tension. In this case the slant fracture at the surface of the test specimen was in both directions; in other instances it may be in only one direction, forming a perfect cup and cone.

toughness. Carbon and alloy steels along with some nonaustenitic stainless steels are susceptible to brittle fracture: all have body-centered cubic (bcc) crystal structures. Most nonferrous metals and austenitic stainless steels have face-centered cubic (fcc) structures and are not susceptible. Brittle fractures usually propagate by either or both of two fracture modes: cleavage and/or intergranular. *Cleavage fractures* are characterized by the splitting of crystals, as shown in Fig. 10-4. In this instance, the steel has been hardened. The fracture, as viewed under a scanning electron microscope (SEM), goes through the grains. In *intergranular fracture,* as shown in Fig. 10-5, breakage takes place along grain boundaries, analogous to failure through the mortar in a brick wall. Again, the steel is in a hardened condition.

Ductile Fracture

Ductility is defined as the ability of a metal to flow or deform with the application of force. Fracture may or may not be the end result, depending on the magnitude of the force. Ductility and toughness are related. Toughness usually is measured in the presence of a notch or other stress concentration. The Charpy test (Fig. 10-6) is used to measure impact strength. A notched specimen, supported at both ends in the tester, is struck by a hammer behind the notch and the energy absorbed in fracture determined. The ability to absorb energy and to deform plastically prior to fracture is characteristic of both ductility and toughness.

Impact toughness is a valuable property of ductile metals. For example, many cars are designed so that in the event of a front-end collision, the vehicle will not stop instantly but rather will fold like an accordion. This absorption of energy serves to slow the rate of deceleration and lessen the impact on the occupants.

Characteristics. Ductile and brittle fractures have some common characteristics. The characteristics unique to ductile failure include:

- Plastic deformation in the region of failure is considerable, as shown in Fig. 10-7. Brittle failure (bolt on left) and ductile failure (bolt on right) are compared. The brittle bolt was water quenched and was high in hardness. Failure but no deformation occurred. The ductile bolt was softer (had greater ductility) and underwent obvious permanent deformation. Both bolts were pulled in tension (as in tensile testing). Both were made from the same grade of low-carbon steel.
- Ductile fracture occurs when shear stress exceeds the shear strength of the metal. Fracture is in the shear direction.
- The surface of a ductile fracture is dull and fibrous in appearance. A classic example of ductile fracture is shown in Fig. 10-8. The stainless steel test specimen illustrates the typical cup-and-cone fracture characteristic of ductile metals in tension. The narrowing or necking indicates extensive stretching of the grains in the metal in the reduced area.
- The shear impact aspect of tensile failure is illustrated in Fig. 10-9. The diagonal ridges are called *Lüders lines,* or *stretcher strains.* In this instance, the tensile test specimen was a low-carbon cast steel. *Shear* is defined as force that causes two contiguous parts of the same body to slide relative to each other in a direction parallel to the plane of contact.

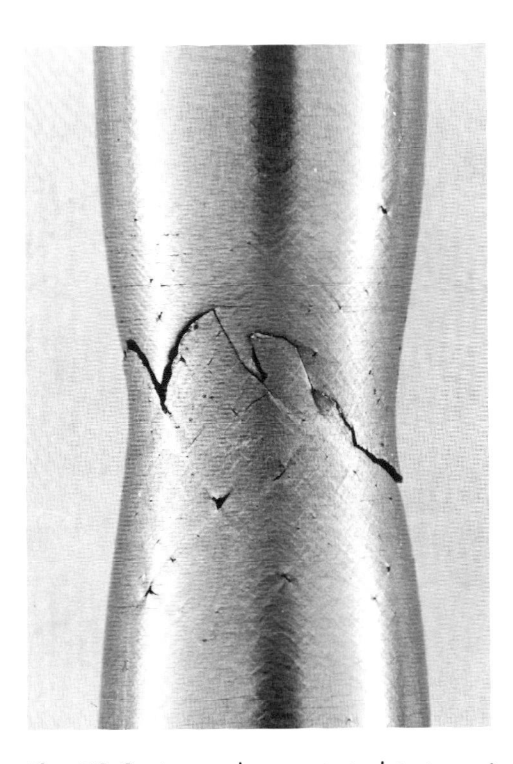

Fig. 10-9 Low-carbon cast steel test specimen emphasizing 45° shear aspect of tensile fracture of a ductile metal. Diagonal ridges are Lüders lines; porosity in steel shows many localized fractures.

The mechanism of ductile fracture is summarized in Fig. 10-10.

Fatigue Fracture

Machine parts can fail from fatigue in normal service and in the absence of excessive over-

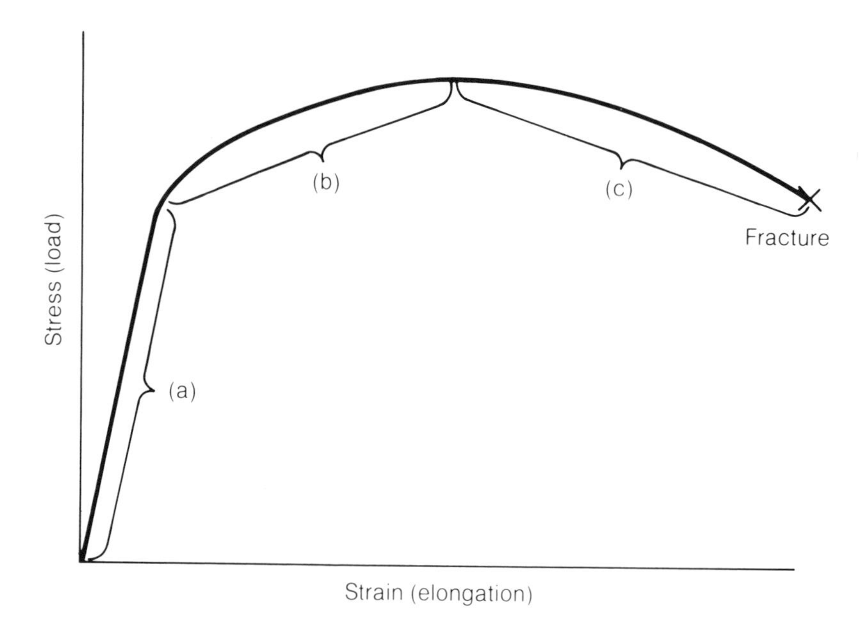

Fig. 10-10 Typical stress-strain diagram showing different regions of elastic and plastic behavior. (a) Elastic region in which original size and shape will be restored after release of load. (b) Region of permanent deformation, but without localized necking. (c) Region of permanent deformation with localized necking prior to fracture at X

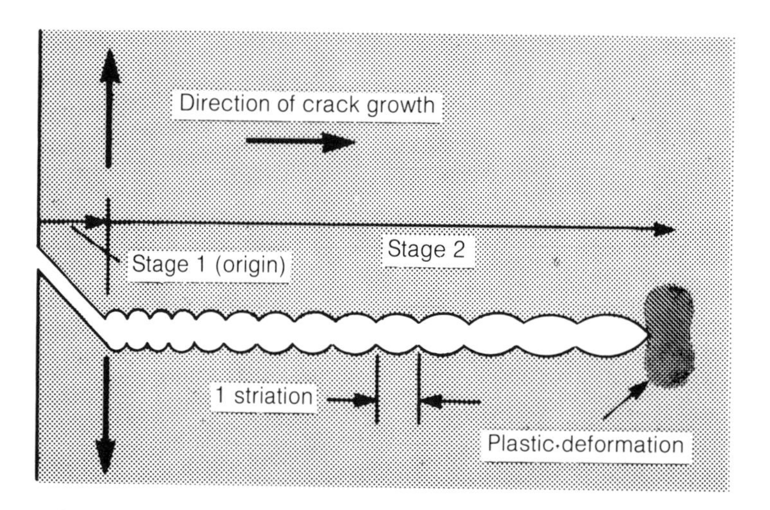

Fig. 10-11 Highly enlarged schematic cross-sectional sketch of stage 1 and stage 2 fatigue. The edge of the metal is at left. When tensile forces repeatedly act on the surface, the microstructural changes of stage 1 cause a submicroscopic crack to form. With each repetitive opening, the crack jumps a small distance (one striation). Note that the spacing of each striation increases with the distance from the origin, assuming the same opening stress. The metal at the tip of the fatigue crack (right) is plastically deformed on a submicroscopic scale.

loading. Such failures are particularly serious because they can happen without warning.

Explanation. Many submicroscopic changes take place in the crystal structure of a metal due to relatively low-level repetitive loading applications. These minute changes may progress gradually to form tiny cracks that may grow to become large cracks under continued cyclic loading and can lead to fracture of the part or structure. The final fracture may be brittle or ductile in nature, depending on the metal and the circumstances of the stress and the environment.

Stages. Fatigue fracture takes place in three stages:

Step 1: Initiation. Failure is the end result of cumulative microchanges in structure, representing thousands or millions of load applications. Shearing (sliding) stresses repetitively imposed on the crystal lattice structure of a metal cause irreversible microstructural changes, each of which is a potential initiation site for fatigue failure.

Step 2: Propagation. As repetitive loading continues, the direction of each tiny crack changes from parallel to the shear stress direction to perpendicular to the tensile stress direction. After the original crack is formed, it becomes an extremely sharp stress concentration that tends to drive a crack ever deeper with each application of tensile stress.

Fig. 10-12 Fatigue striations showing the result of spectrum loading in a laboratory test of aluminum alloy 7075-T6. In this test, the specimen was loaded 10 cycles at a high stress, then 10 at a lower stress, and alternated with these stresses as the fatigue crack continued to propagate. This produced 10 large striations, then 10 small striations, alternately across the fracture surface. 4900×

Figure 10-11 is a schematic of stages 1 and 2. Note that the direction of shear stress is indicated by the horizontal arrow; the vertical arrow indicates the direction of tensile stresses. *Striations* (Fig. 10-11 and 10-12) are microscopic lines that show the location of the tip of a fatigue crack.

Beach marks and *ratchet marks* are macroscopic characteristics of fatigue fracture (Fig. 10-13). These features are visible to the unaided eye or at magnifications of 25× to 50×. Beach marks, which are oval in shape and have the appearance of high-water marks left behind on a beach after the tide has gone out, show the location of the tip of a fatigue crack. Ratchet marks (small, shiny areas on the periphery of the part illustrated in Fig. 10-13) are ridges on a fatigue fracture that indicate where two adjacent fatigue areas have grown together.

Stage 3: Final Rupture. As the propagation of the fatigue crack continues, gradually reducing the cross-sectional area of the part, it eventually weakens the metal so greatly that final, complete fracture can occur with a single additional application of load. The fracture mode may be ductile (with a dimpled fracture surface) or brittle (with a cleavage or intergranular fracture surface, or combination).

In analyzing a failure, careful attention must be paid to the size, shape, and location of the final fracture area. The information obtained may be helpful in understanding the relationship between stresses on the part and the strength of the part, or it can indicate imbalance and nonuniform stresses.

The Many Faces of Wear

Wear can be defined two ways: (1) the undesirable removal of material by mechanical action, and (2) wear not caused directly by sliding action, such as fatigue that produces cavities or pits. Both types will be discussed here, including abrasive wear, erosive wear, grinding wear, gouging wear, adhesive wear, fretting wear, and cavitation fatigue.

Wear is an enormously expensive problem but, comparatively speaking, it is not as serious as fatigue and is usually foreseeable. Various rubbing surfaces in any machine are expected to wear out eventually; in many instances, the size of the problem can be minimized by lubrication, filtering, materials engineering know-how, and remedial design. In many respects, wear is similar to corrosion. Both have many types and subtypes, both are somewhat predictable, and both are difficult to test and to evaluate in lab or service tests.

Abrasive Wear

Abrasive wear can be characterized by one word: cutting. Cutting (Fig. 10-14) is caused when (1) hard particles suspended in liquid contact a material surface or (2) projections from one surface roll or slide under pressure against another surface. When a hard metal under pressure comes into contact with a softer metal, it usually causes microscopic distortion of the surface structure and a chip or fragment of the softer metal is removed; in the process, heat is generated by friction between the two metals.

Abrasive wear can be dealt with in several ways:

- *Increase surface hardness.* This is a remedy with plus and minus sides. On the plus side, making tools such as knives and blades harder adds resistance to dulling. However, higher hardness can make the cutting tool susceptible to brittle fracture.
- *Remove* hard, adhesive foreign particles with filters for air, water, and oil. The automotive engine is an example. Air and oil filters are used to prevent the entry of external foreign particles and to trap and collect internal foreign particles before they damage the engine. In some instances, particularly where many high-speed particles slide and roll across a metal surface, foreign particles cannot be removed.
- *Replace* worn parts. One of the common strategies is to facilitate replacement in the design stage.

Erosive Wear

Erosive wear takes place when particles in a fluid or other carrier slide and roll at relatively high velocity against a metal surface. Each par-

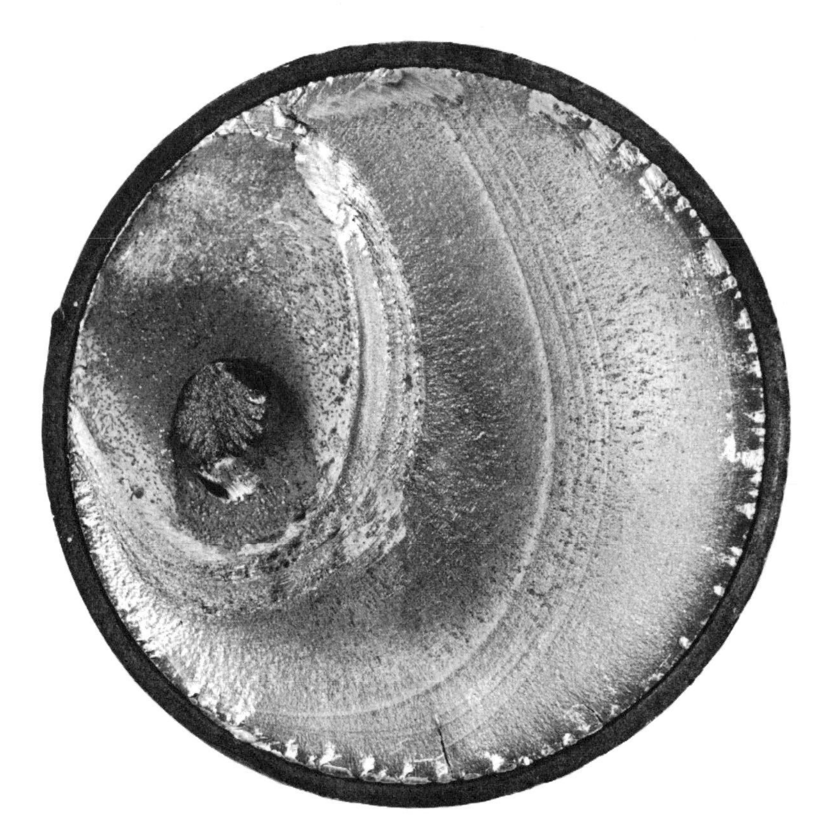

Fig. 10-13 Surface of a fatigue fracture in a 1050 steel shaft, with hardness of about 35 HRC, that was subjected to rotating bending. Presence of numerous ratchet marks (small, shiny areas at surface) indicates that fatigue cracks were initiated at many locations along a sharp snap-ring groove. The eccentric pattern of oval beach marks indicates that the load on the shaft was not balanced; note final rupture area (stage 3) near left side.

ticle cuts a tiny particle from the surface; when this is repeated over a long period of time, a significant amount of erosion will result. Erosive wear is common in pumps and impellers, fans, steam lines and nozzles, on the inside of sharp bends in tubes and pipes, and in sandblasting and shotblasting equipment.

Characteristics of erosive wear include:

- *General removal of soft surface coatings* or soft material. Fan and turbine blades are susceptible to erosive wear. In automotive applications, for example, the paint on the rear (concave) side of the blade usually is removed by the scouring or cutting action of dust and dirt particles in the air. The concave side of the fan has a positive pres-

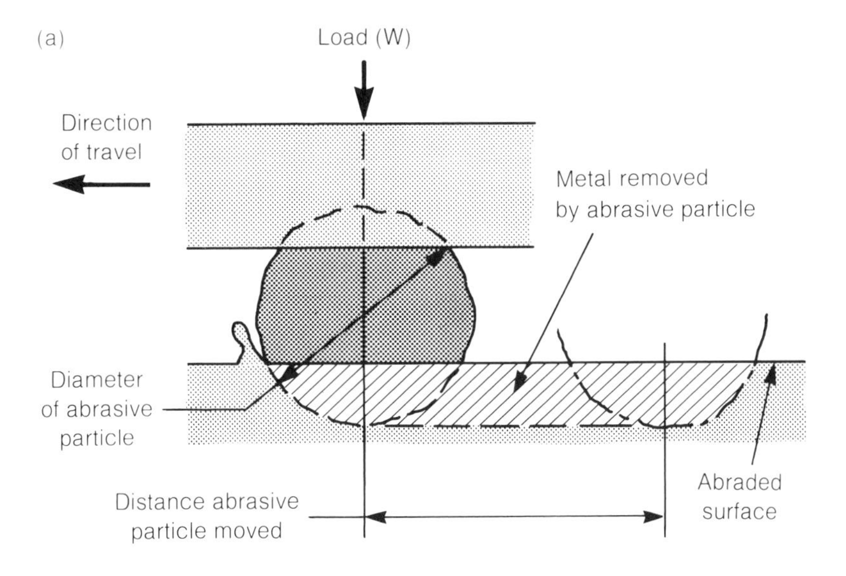

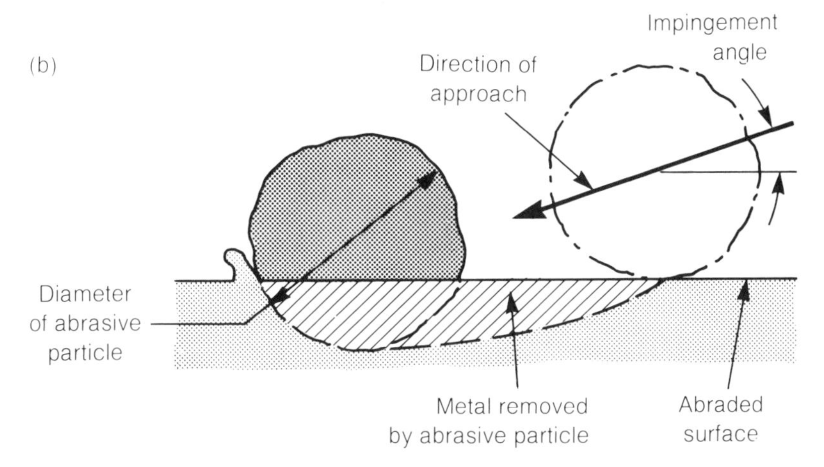

Fig. 10-14 Idealized representations of the two types of force applications on abrasive wear particles. (a) Represents the cutting or plowing action of a contained particle under pressure. That is, the particle is not free but is under pressure from other particles or a solid object. This is characteristic of grinding and gouging abrasion, in which the hard particles are forced to scratch or cut the metal surface. (b) Represents the cutting or plowing action of a loose particle flowing across the metal surface after impinging on the surface. This is characteristic of erosive wear, in which free particles strike the surface at an angle, then slide across the surface.

sure, while the convex side has negative pressure; the positive pressure forces particles against the concave surface and leads to erosive wear.

- *Grooving or channeling* of the material. This kind of wear is common in assemblies involving fluids (liquids or gases) where, due to design, fluids flow faster or change direction in certain locations. Examples are pumps or impellers in which vanes push particle-laden fluid into various passages. Sharp curves or bends cause more erosion than gentle curves. In textile machines, thread or yarn moving at high speed can cause erosion. In Fig. 10-15, a sudden change in direction of the yarn caused grooving and erosive wear in the eyelet.
- *Rounding of corners.* Erosive wear can change the shape of impellers, turbine blades, and vanes, reducing their efficiency. A before-and-after example of such damage is shown in Fig. 10-16. With continued service, the vanes would have been totally destroyed by erosion.

Fig. 10-15 Erosive wear of a yarn eyelet made of hardened and tempered 1095 steel. Grooving was caused by a sharp change in direction of the yarn as it came out of the hole. Service life was improved by changing the eyelet material to M2 high-speed tool steel, which contains spheroidal carbides in a matrix of martensite. Service life probably could have been improved also by changing the angle of exit or by rounding the corner to make a bell-mouth hole.

Grinding Wear

Particles under high stress that cut or plow many very small grooves across a metal surface at relatively low speed are the primary cause of grinding wear. High-stress, low-speed operation is characteristic of tillage tools (plows, cultivators, rakes, etc.) and other ground-contact parts such as track shoes for bulldozers and cutting edges of blades. Cutting edges become dull and their shape is changed, making them less efficient or totally inefficient. Grinding wear can be identified if wear occurs at high-stress locations, particularly at points and edges, causing a general change in part shape.

Several preventive measures can be taken:

- *Higher hardness* is a way of coping with this problem, but the downside is the potential for creating great problems.
- *Surface coatings* are another option. Hardfacing by welding, metal spraying, or other means is frequently used to improve resistance to grinding wear. Deposits usually contain large quantities of carbides, such as tungsten, titanium, chromium, molybdenum, and vanadium types. In some applications oxides, borides, or nitrides may be more satisfactory.
- *Diffusion treatments* are yet another possibility. These include carburizing, carbonitriding, nitriding, chromiding, or boronizing.
- *Controlled grinding wear* also can be used to give certain cutting tools self-sharpening capabilities. This technique is based on what is called the *rat's tooth principle* (Fig. 10-17). The front tearing teeth of rats have very hard enamel on their convex surface and relatively soft dentine on their concave side. When the rat is eating, the high stress or concave surface is at the rear of the teeth. This concave area also wears much faster. However, the front or convex side experiences little or no wear.

This same principle is applied to certain cutting tools—plowshares, for example. Cutting edges can be made self-sharpening. In service the soft, high-stress surface is worn, while the hard surface remains relatively undamaged. Note in Fig. 10-18 that in this instance the softer steel is on the front side of the tool, and the hardfacing alloy is applied to the rear or low-stress side of the tool. In the mining industry, digging tools are sometimes hardfaced on

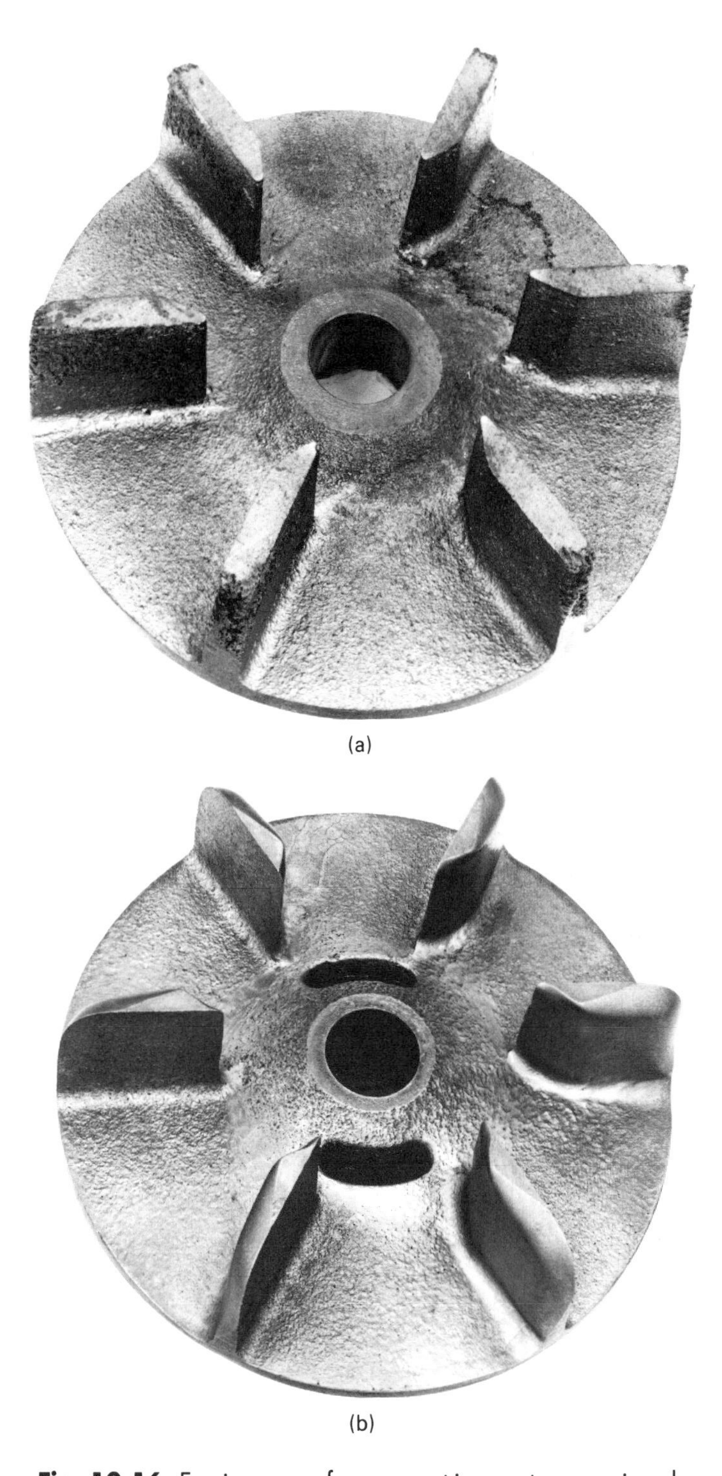

(a)

(b)

Fig. 10-16 Erosive wear of a gray cast iron water pump impeller. (a) New impeller. (b) The sharp corners of the impeller have been completely rounded off by sand in the cooling system. The change in shape of the vanes reduces the efficiency of the pump; if erosive wear were to continue long enough, the vanes—and efficiency—would be completely gone.

only one side to maintain self-sharpening capabilities (Fig. 10-19).

Gouging Wear

When hard, abrasive products are crushed, battered, or pounded under extremely high stress, rapid deterioration of contact surfaces can be expected. The cause of this kind of wear is extremely high-stress battering or impact, which tends to cut or gouge large fragments from the surface of a metal. Such conditions are encountered, for example, in earthmoving, mining, quarrying, oil well drilling, steelmaking, cement and clay product manufacture, railroading, dredging, and lumbering. In some instances it can be more economical to use replaceable parts, such as teeth for backhoe buckets (Fig. 10-20).

Adhesive Wear

Adhesive wear also can be characterized by a single word: welding. Other terms used include scoring, scuffing, galling, and seizing.

The adhesive wear process has been depicted earlier in Fig. 10-2. Two surfaces sliding with respect to each other may or may not be separated by lubricant. When a peak from one surface comes in contact with a peak from the other surface, instantaneous microwelding may take place due to the heat of friction generated, as shown in Fig. 10-2(a). Continued relative sliding between the two surfaces fractures one side of the welded junction (Fig. 10-2b), making the peak on one side higher than the peak on the other side. The higher peak is now available to contact another peak on the opposite side (Fig. 10-2c). The tip may be fractured by the new contact or rewelded to the opposite side, and the cycle will be repeated. In either case, adhesive wear starts on a small scale but rapidly escalates as the two sides alternately tear and weld metal from the surface of the other.

Several preventive measures can be taken:

- *Keep the bulk temperature of the lubricant relatively cool:* Adhesive wear is caused by locally high temperatures.
- *Use contacting metals that are insoluble to each:* There can be no adhesive wear if the two metals are not weldable to each other.
- *Use smooth surfaces* with no projections to penetrate the lubricant film.
- *Use chemical films such as phosphate coatings and special lubricants* to prevent the metal-to-metal contact that leads to adhesive wear. Phosphate coatings, used in addition to lubricant, help separate metal surfaces of machinery during the wear-in period. Projections are removed from mating surfaces; phosphate crystals help to retain lubricant. With the passage of time, phosphate coatings gradually wear away.

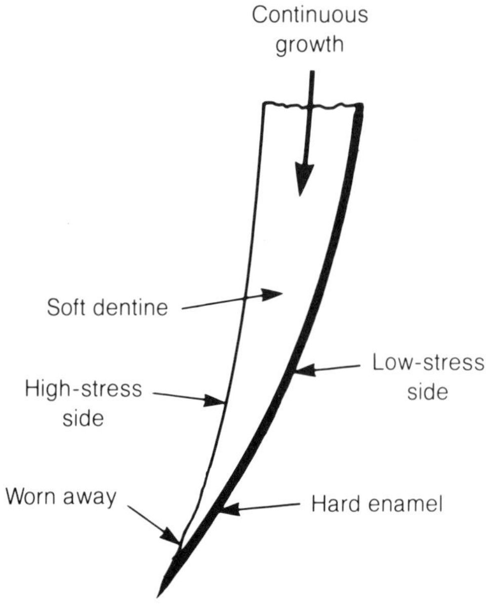

Fig. 10-17 Diagram of a self-sharpening rat tooth

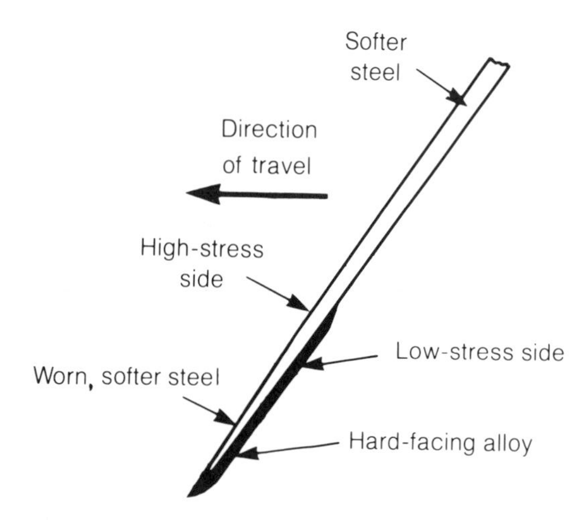

Fig. 10-18 Diagram of a self-sharpening plowshare using the same principle as in the rat tooth shown in Fig. 10-17. As the plowshare cuts through the soil from right to left, the relatively soft steel on the forward high-stress side is gradually worn away, but the hardfacing applied to the rear low-stress side is continually exposed at the sharp tip. Eventually, of course, the part must be replaced, but service life may be very long in certain types of soil, particularly soil without rocks.

- *Use extreme-pressure (EP) lubricants* to coat metal surfaces exposed to high sliding velocities, such as hypoid gear sets in automotive axles. These lubricants form extremely thin compounds on surfaces, preventing metal-to-metal contact.

Fretting Wear

Fretting wear is similar to adhesive wear in that microwelding occurs on mating surfaces. In adhesive wear, however, facing metals slide across each other, while in fretting wear metal-

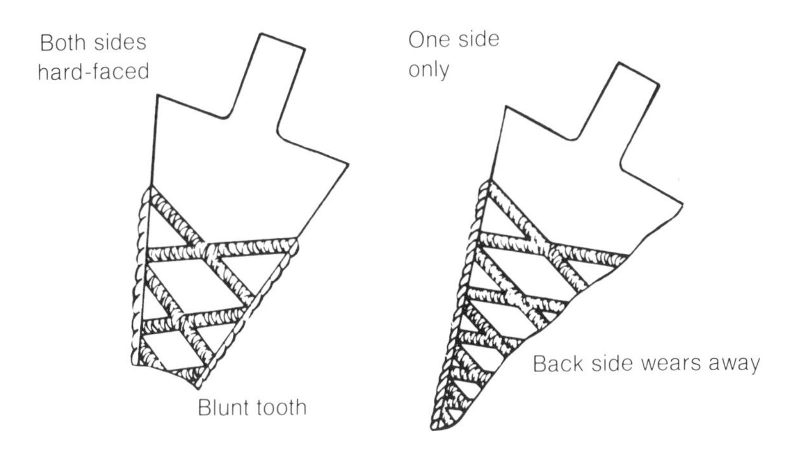

Fig. 10-19 Self-sharpening of a digging tooth from ground-contact equipment by controlled wear through selective hardfacing. The pattern of hardfacing can be varied to suit the conditions, but note that the blunt tooth is hardened on both sides, while the self-sharpening tooth is hardened on only one side.

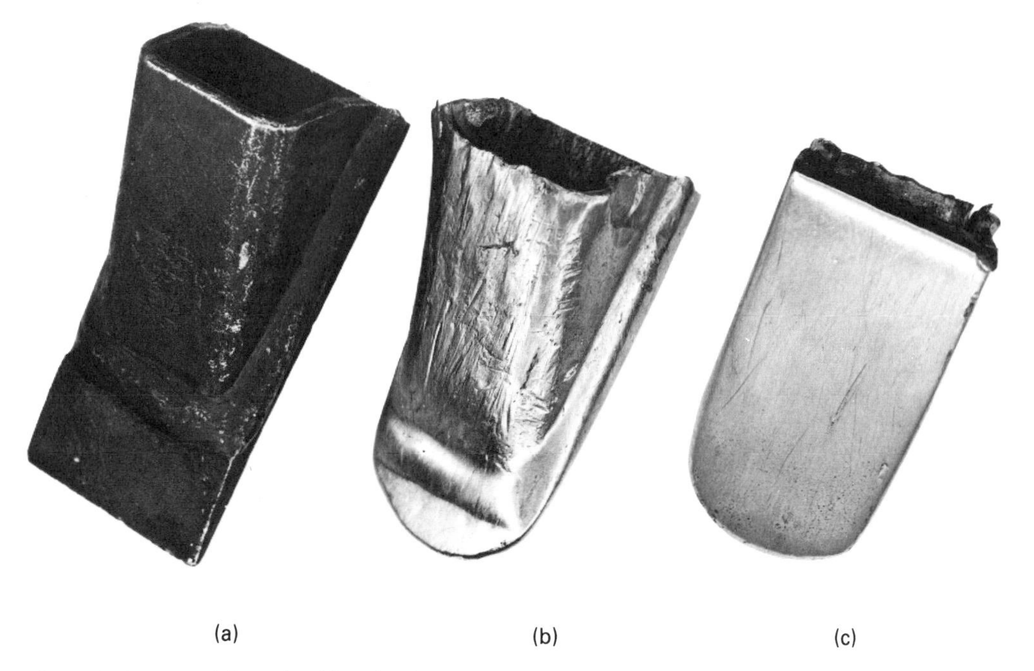

Fig. 10-20 Tooth for a backhoe bucket. (a) Original condition. (b) and (c) Both sides of a tooth worn by long operation in rocky, frozen soil. The soft top of the tooth (b), made of 1010 steel, wore considerably more than did the flat opposite side (c) of 8640 medium-hard steel. The tooth is a replaceable part that is pinched over a stub to hold it in position.

to-metal interfaces are essentially stationary. Microwelding is made possible by minute elastic deflections in the metals. Cyclic motions of extremely small amplitude are sufficient to cause microwelding on both metal surfaces. This kind of wear also is known as fretting corrosion, false brinelling, friction oxidation, chafing fatigue, and wear oxidation. An example of fretting wear is shown in Fig. 10-21.

Because fretting wear is essentially a stationary phenomenon, debris is retained at or near the locations where it was formed originally. This debris usually consists of oxides of the metals in contact. Ferrous metal oxides are brown, reddish, or black; aluminum oxides form a black powder.

Preventive measures include:

- *Eliminate or reduce vibration* by using vibration damping pads or by stiffening certain parts of a structure to increase the natural frequency of vibration. In some instances, neither measure will prove successful.
- *Use an elastomeric bushing or sleeve* in a joint. Any motion will be absorbed by the elastomeric material, preventing metal-to-metal contact.
- *Lubricate the joint.* The problem here is that because the joint is essentially stationary, liquid lubricant cannot flow through the interface (as is the case with continual sliding motion). Therefore, some greases, solid-film lubricants (e.g., molybdenum disulfide), and oils are used to reduce or delay fretting.

Contact Stress Fatigue

So far in this chapter the root causes of two types of wear have been analyzed: one type characterized by the undesirable removal of material from contacting surfaces due to mechanical action, and a second type in which the wear is not caused directly by sliding action. Presented here is a third scenario for metal removal. In this instance, wear is caused by con-

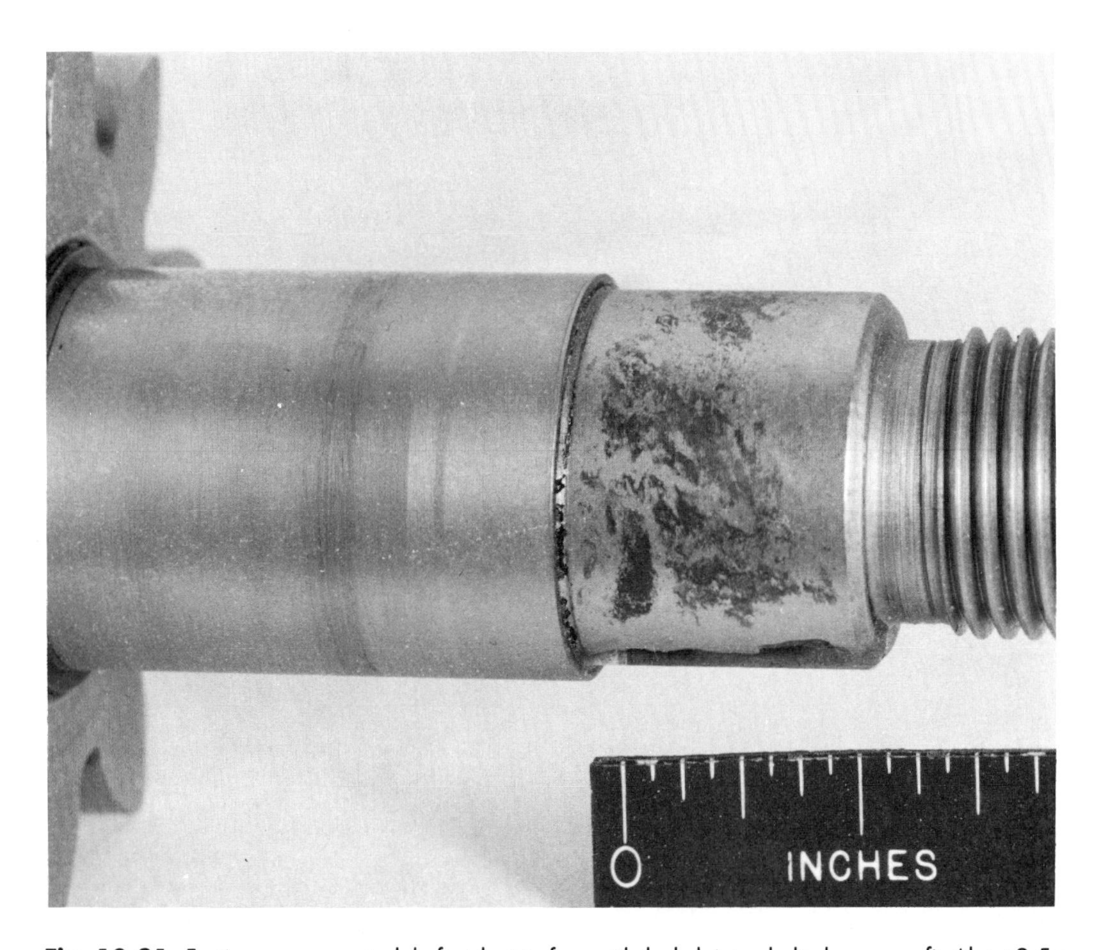

Fig. 10-21 Fretting wear on a steel shaft at the interface with the hub intended to be a press fit. About 2.5×

tact stress fatigue that results from a variety of mechanical forces and environments.

This type of fatigue involves the same mechanism resulting from cyclic slip under repetitive load applications for many thousands or millions of load cycles. However, instead of gross fracture of parts, initially only fragments are removed from the surface. In time, this results in pits or cavities in the surface. Contact stress fatigue frequently is the limiting factor in load-carrying capability.

Pits appearing on the surfaces of contacting parts as a result of this type of fatigue exhibit three different types of behavior:

- *Initial microscopic cavities.* Some pits start as microscopic cavities and may stay at that level for the life of the part. This condition is evidenced by a dull, frosted appearance on an otherwise bright surface.
- *Development of larger cavities.* Although initially these cavities may be microscopic, under continued service (rolling and sliding under load) gradually they may become larger.
- *Growing cavities.* These cavities are large at the outset and just continue to grow.

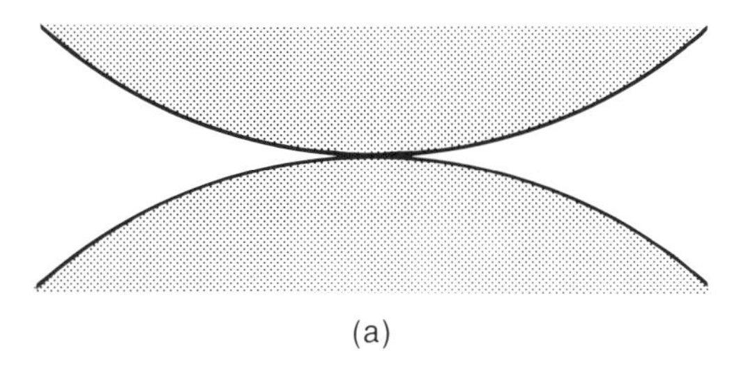

(a)

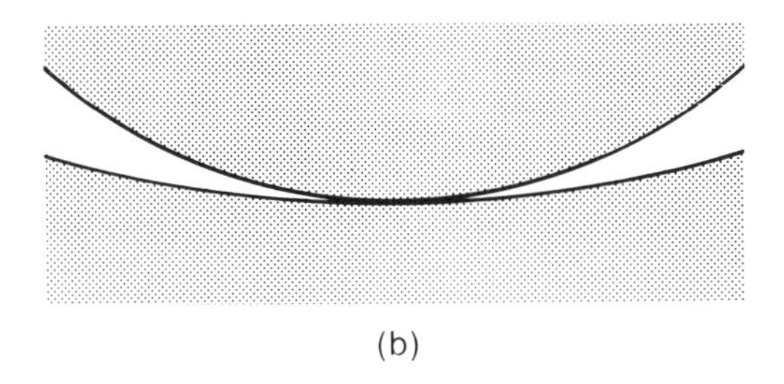

(b)

Fig. 10-22 Counterformal, or convex, surfaces in contact. Examples include gear teeth and roller or needle bearings rolling on a shaft, an inner raceway, or a flat surface. (b) Conformal surfaces, where a convex surface is in contact with a concave surface. Examples are ball bearings in contact with an inner or outer raceway, roller or needle bearings in contact with an outer raceway, and a shaft in contact with a sliding bearing or on a flat surface.

Types 2 and 3 can completely destroy the surfaces of hardened steel gears, rolling-element bearings, roller cams, and other parts or assemblies where there is a combination of rolling and sliding.

Parts subject to this type of failure generally have two convex or counterformal surfaces in contact under load. Typical are gear teeth and various types of antifriction bearings. This same type of failure can occur when a convex shape fits within a concave shape, such as a shaft within a sliding bearing, or balls in a ball-bearing race. Figure 10-22 shows both types of conditions.

Cavitation Fatigue

Cavitation fatigue is caused by vibration and movement in various liquids, with water being the most common. Because many liquids are corrosive to most metals, the problem of cavitation fatigue is entwined with the problem of contact stress fatigue.

Cavities frequently act as stress concentrations and can cause fracture of a part, especially where gear teeth are involved. Also, metal removed from cavities usually is very hard and brittle, and is readily crushed and fragmented into much smaller particles. These, in turn, can cause abrasive wear as well as other damage when carried by lubricant to other parts of a mechanism. This type of fatigue can be a serious problem for marine propellers of all sizes, diesel engine cylinder liners, pump impellers, hydraulic pumps and equipment, turbines, torque converters, and other parts that contact or vibrate in various liquids.

Pits can vary in size from very small to very large—from pinheads to golf balls, or even larger—and can completely penetrate the thickness of a metal. Damage to the structure can be catastrophic and losses in functional efficiency can be substantial. Typical examples of cavitation fatigue are shown in Fig. 10-23.

Methods of dealing with this problem include:

- *Increasing the stiffness of the part.* This should reduce its amplitude of vibration, thereby increasing its natural vibration frequency. It may be possible to increase wall thickness or add stiffening ribs to change vibration characteristics.
- *Increasing the smoothness of the surface.* Cavities tend to cluster in certain low-pressure areas. It may be possible to eliminate surface peaks and valleys by dispersing the cavities.

- *Increasing the hardness and strength of the metal.* However, this may only delay the problem, rather than prevent it.

Temperature-Induced Failures

Heat speeds the movement of atoms, and cold temperatures slow it; metals expand with heat and contract with cold. Repetitive heating and cooling cycles can cause the failure of metals in service. Elevated-temperature fatigue, thermal fatigue, creep rupture, and scaling are examples of failures that occur at high temperatures. Some failures, such as the brittle type, occur at low temperatures (i.e., below room temperature).

Elevated Temperature Fatigue

In this instance, fatigue strength decreases with increasing temperature. Take Astroloy, a nickel-base superalloy used in forged jet engine blades; this alloy has a fatigue strength of 620 MPa (90 ksi) when exposed to a temperature of 650 °C (1200 °F) for 1000 hours. However, when exposed to 815 °C (1500 °F) for 1000 hours, fatigue strength drops to 210 MPa (30 ksi);

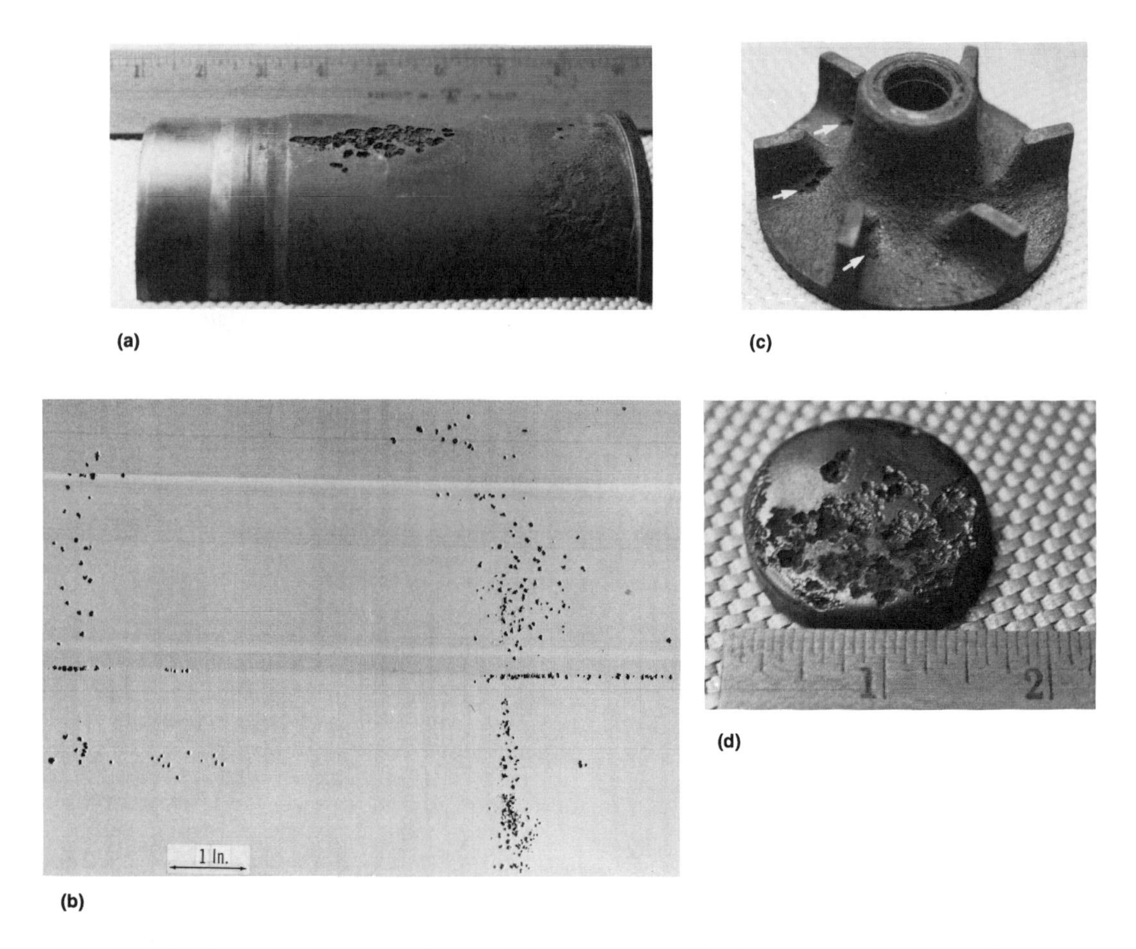

Fig. 10-23 Examples of cavitation fatigue. (a) Gray cast iron diesel-engine cylinder sleeve. The pitted area is several inches long, and the pits nearly penetrated the thickness of the sleeve. Note the clustered appearance of the pits at preferred locations. (b) Another gray cast iron diesel-engine cylinder sleeve unwrapped by a special photographic process. Again, note the clustered locations, with the most severe pitting on the thrust side of the sleeve, against which the piston slides on the power stroke of the combustion cycle. The lighter pitting at left is on the opposite, or antithrust, side of the sleeve. (c) Pitting at preferred locations on the vanes of a gray cast iron water-pump impeller. This impeller rotated in a clockwise direction; the arrows show some of the pits formed in the metal on the suction side of the vanes. (d) Pitting perforated this steel freeze plug from a gasoline engine, causing coolant to leak. Vibration of the wall of the engine block at this location caused this type of damage on the coolant side.

and with exposure to 1095 °C (2000 °F) for 1000 hours, strength plummets to 55 MPa (8 ksi).

Thermal Fatigue

Cyclic thermal stressing (expansion and contraction) caused by repetitive heating and cooling is the sole source of thermal fatigue. On cooling, residual tensile stresses are produced if the metal is prevented from moving (contracting) freely. Fatigue cracks form as cycling continues. A new crack develops each time the metal is cooled. An example of a thermal fatigue crack in an engine exhaust valve is shown in Fig. 10-24.

Remedies are available. Thermal failure of many parts can be prevented by designing parts with curves rather than straight lines. Expansion loops and bellows in elevated-temperature piping and tubing systems take advantage of this principle.

Creep Rupture

Creep is defined as a time-dependent strain occurring under stress. *Creep rupture* can be de-

Fig. 10-24 Thermal fatigue crack in the hardfacing alloy on an exhaust valve from a heavy-duty gasoline engine. About 2.5×

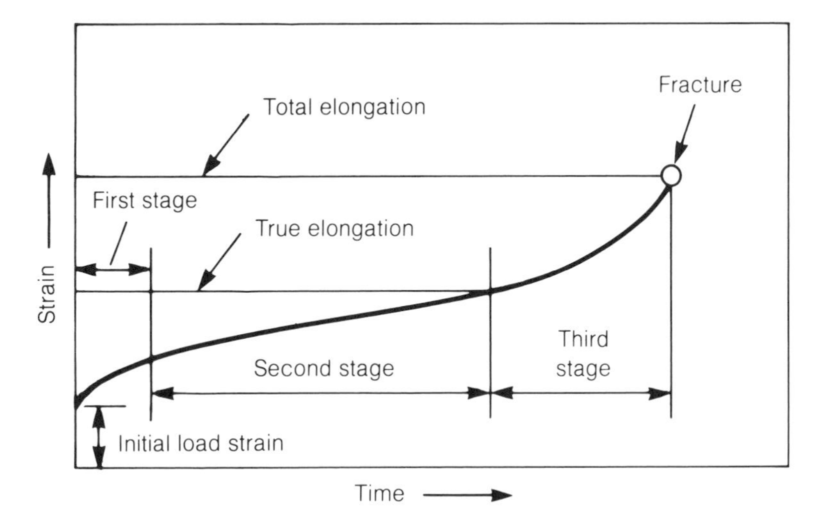

Fig. 10-25 Schematic of a tension-creep curve, showing the three stages of creep

fined as failure at a given time and temperature. The schematic of a tension-creep curve is shown in Fig. 10-25. In this instance, strain or elongation is due to tensile stress at specific times. Values for temperature and stress are fixed.

Creep is a three-stage process (Fig. 10-25):

- *Stage 1*: Changes shown take place during the first few moments after a load is applied. Following initial elastic strain, the metal undergoes increasing plastic strain at a decreasing strain rate.
- *Stage 2:* Metal is still stretching under tension, but at a slower rate than in Stage 1.
- *Stage 3:* A gradual increase in strain will be shown prior to fracture.

Figure 10-26 shows an example of creep fracture of stainless steel tubing.

Development work on the jet engine started in England and Germany in the prewar stages of World War II, although the available high-temperature alloys used initially for turbine blades fell short of being equal to the task. In Germany, for instance, a service life of about 25 hours was reported; turbine inlet temperatures were about 595 °C (1100 °F). In time, new alloys with greater high-temperature strength were developed, but they failed frequently in a brittle manner in a relatively short time span. The problem was solved by improving stress rupture strength through the use of fine-grained, adherent oxide coatings. Today, Inco 718, a nickel-base superalloy, is a popular choice for these high-temperature applications.

Oxidation Failure

Scaling, an oxidation product, can cause failure at elevated temperatures, particularly in combination with repetitive heating/cooling cycles. Scale forms during exposure to high temperatures in an oxidizing atmosphere, usually air. Scale flakes when metal cools or contracts because of differences in the thermal expansion/contraction characteristics of the scale and the metal to which it is, or was, attached. Scale is not a metal, but rather a compound consisting of scale and the parent metal.

Ferritic stainless steels in the 400 series have good resistance to oxidation at high temperatures. For example, type 409 stainless steel is often the choice for automotive exhaust systems and catalytic converters.

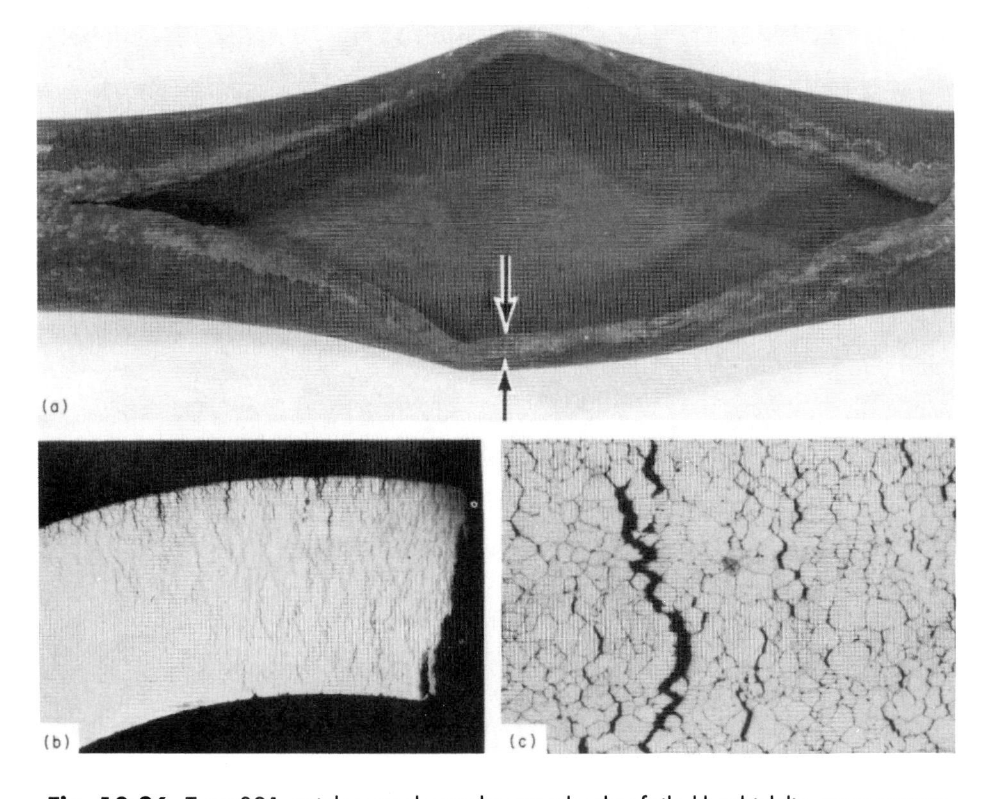

Fig. 10-26 Type 321 stainless steel superheater tube that failed by thick-lip stress rupture

Low-Temperature Failure

Some carbon steels have good ductility in room-temperature tensile tests, but fail in a brittle manner in the same test at a lower temperature. This behavior is believed to be related to microstructure: Ductility is dependent on the slippage of atoms, which seems to be impeded in steels with a bcc structure. Susceptibility to this type of failure can be reduced by variations in composition, steelmaking practices, and heat treatment.

Chapter 11

Coping with Corrosion

How big is the corrosion problem?

You can get an idea by studying Fig. 11-1, which spells out the entire life cycle of a typical metal. Energy requirements are plotted on the vertical axis, time on the horizontal axis.

In effect, metals in end products are on temporary loan from nature, and any countermeasures amount to delaying tactics. In tracking the life of a metal from ore to refining to refined metal to end product, corrosion appears to be inevitable; and by inference, any relief is temporary.

Supporting evidence goes like this:

- Ores in their natural, stable state usually contain oxide or sulfide impurities (corrosion products, also known as rust).
- In the refining process, impurities are stripped from the ore; relatively pure metals are produced. The price of refining includes a great expenditure in energy and a metallurgical trade-off: The refined metal is less stable than the ore, and is destined to deteriorate due to corrosion and revert to its prior chemically stable, orelike condition. For example, iron is found in the ore as iron oxide, or rust. With usage, the iron or steel product eventually will revert to iron oxide, or rust. Alternatives to deterioration via corrosion are remelting and recycling.

Corrosion is defined as deterioration of a metal due to chemical or electrochemical reactions with the environment. Example of a chemical reaction is dissolving a metal in an acid. Example of an electrochemical reaction: a reaction caused by the passage of current through a medium containing moving ions.

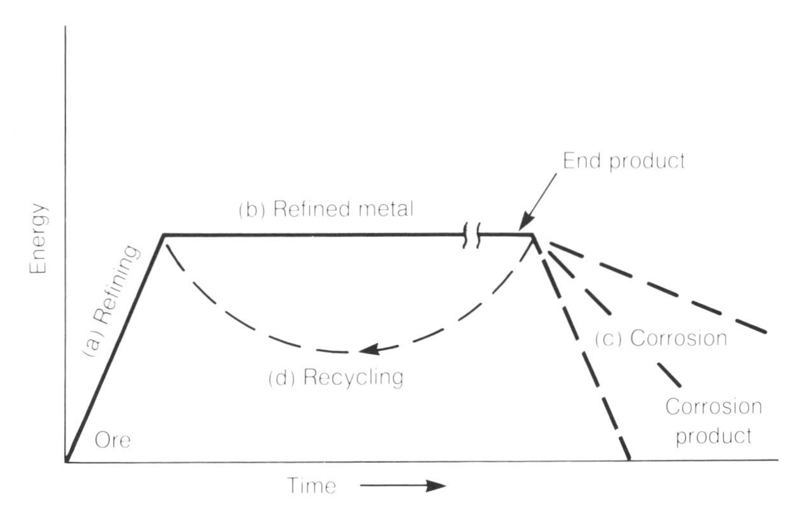

Fig. 11-1 Life cycle of a typical metal

Ions are produced when an atom gains an electrical charge (+ or –) through the gain or loss of one or more of its electrons.

Six types of corrosion are discussed in this chapter:

- Galvanic corrosion
- Uniform corrosion
- Crevice corrosion
- Stress-corrosion cracking
- Corrosion fatigue
- Selective leaching

In each instance, possible remedies are reviewed.

Galvanic Corrosion

The basic principles of electrochemical reactions that cause *galvanic corrosion* are identical to those of a simple battery, as shown in Fig. 11-2. Three components are needed: two different metals (or a metal and the nonmetal graphite) in either physical or electrical contact in an *electrolyte* (an electrically conductive liquid or paste). If these conditions are met, one of the two materials will be corroded; the other will be protected, will release hydrogen, and an electric current will be generated. The material that corrodes is called the *anode,* and the one that does not corrode is known as the *cathode*.

Metals are usually rated in a descending order, ranging from the most easily corroded (or anodic), to the least easily corroded (or cathodic). The galvanic series in seawater is listed in Table 11-1. Note that magnesium and magnesium alloys are most easily corroded and appear on the anodic end of the list; gold and platinum are at the cathodic end of the list, as they are the least easily corroded metals. Gold, platinum, silver, and other precious metals are also known as *noble* metals because of their strong resistance to corrosion.

When two metals make electrical contact in an electrolyte, the farther apart they are in the galvanic series, the more likelihood exists that the more anodic metal will corrode. An electrolyte is a chemical compound, or mixture of compounds, that conducts an electrical current when molten or in solution. An example of galvanic corrosion is shown in Fig. 11-3, and a preventive measure (use of an insulating washer and gasket) is shown in Fig. 11-4.

Galvanic corrosion can be prevented or minimized by any or all of the following measures:

- Prevent the flow of electric current by physical separation, or insulate the dissimilar metals from each other by using nonconductive, nonabsorbent materials such as plastic, waxy coatings, certain heavy greases, or paint.

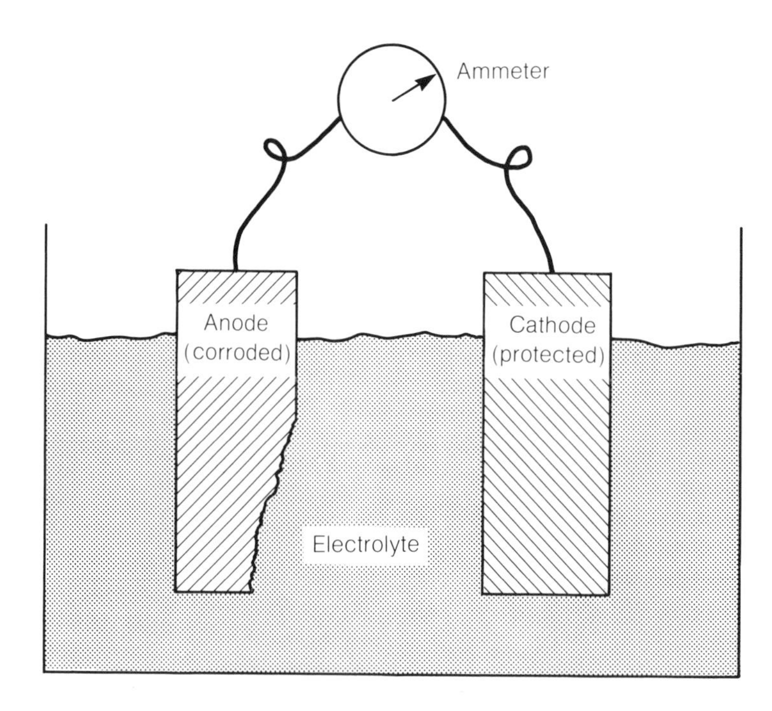

Fig. 11-2 Galvanic cell showing the basic principles of the electrochemical nature of corrosion

- Eliminate the electrolyte. There can be no galvanic corrosion without an electrolyte, which explains why there is little or no galvanic corrosion in dry desert atmospheres. When humidity is below about 30 to 35%, atmospheric corrosion usually cannot occur.
- When it is necessary to use different metals together, choose those metals that are close together in the galvanic series. For example, contact between aluminum and steel in an electrolyte may cause gradual pitting and deterioration of the aluminum. However, using brass or copper with aluminum under the same conditions will result in much faster deterioration of the aluminum.
- Use a large anode metal and a small cathode metal. For example, plain steel rivets (small cathode) through aluminum sheet or plate (large anode) may prove to be satisfactory in service, but aluminum rivets through steel will corrode rapidly in an electrolyte. Also, copper rivets in steel plate may be satisfactory in service, but steel rivets through copper plate will corrode rapidly in an electrolyte.
- Use corrosion inhibitors in the electrolyte in a closed system. The principle is applied by using antifreeze in cars. Cooling systems contain dissimilar metals such as gray cast iron, aluminum, copper, brass, tin-lead solder, and steel.
- Protect a structural metal by contact with a sacrificial (expendable), more anodic metal. Use of zinc to protect iron or steel is the most common example. Galvanized steel can be produced by dipping the metal into molten zinc, and by electroplating zinc onto steel. Alternatives are coating with a zinc-rich primer or a zinc-rich plating. Zinc is gradually corroded to protect the steel.

Uniform Corrosion

Rust on iron or steel is the most common type of corrosion and is known as *uniform corrosion.* Other metals that corrode uniformly include aluminum, copper, and brass. Aluminum, for one, forms a film on its surface during the corrosion process. This film provides protection against corrosion. However, similar films formed on other metals may not provide such protection.

Uniform actually is a misnomer. Corrosion is uniform only on a macroscale, not on a microscale. Corrosion occurs as a result of microscopic galvanic cells on the surface of metals. If an electrolyte is present, corrosion takes place because of chemical differences and impurities in the metal in microscopic anodic and cathodic areas.

Uniform corrosion is fairly predictable and relatively easy to cope with, provided other types of corrosion are not present. Preventive measures include use of a more suitable material, such as a noble metal or a stainless steel, or use of a coating to protect the metal. Coatings

Table 11-1 Galvanic series in seawater

Anodic end (most easily corroded)
Magnesium
Magnesium alloys
Zinc
Galvanized steel or galvanized wrought iron
Aluminum alloys 5052, 3004, 3003, 1100, 6053, in this order
Cadmium
Aluminum alloys 2117, 2017, 2024, in this order
Low-carbon steel
Wrought iron
Cast iron
Ni-Resist (high-nickel cast iron)
Type 410 stainless steel (active)
50-50 lead-tin solder
Type 304 stainless steel (active)
Type 316 stainless steel (active)
Lead
Tin
Copper alloy 280 (Muntz metal, 60%)
Copper alloy 675 (manganese bronze A)
Copper alloys 464, 465, 466, 467 (naval brass)
Nickel 200 (active)
Inconel alloy 600 (active)
Hastelloy B
Chlorimet 2
Copper alloy 270 (yellow brass, 65%)
Copper alloys 443, 444, 445 (admiralty brass)
Copper alloys 608, 614 (aluminum bronze)
Copper alloy 230 (red brass, 85%)
Copper 110 (ETP copper)
Copper alloys 651, 655 (silicon bronze)
Copper alloy 715 (copper nickel, 30%)
Copper alloy 923, cast (leaded tin bronze G)
Copper alloy 922, cast (leaded tin bronze M)
Nickel 200 (passive)
Inconel alloy 600 (passive)
Monel alloy 400
Type 410 stainless steel (passive)
Type 304 stainless steel (passive)
Type 316 stainless steel (passive)
Incoloy alloy 825
Inconel alloy 625
Hastelloy C
Chlorimet 3
Silver
Titanium
Graphite
Gold
Platinum
Cathodic end (least easily corroded)

include paint, oxide films, plating, and cladding:

- *Paint* is effective as long as the paint film remains intact. However, if only one side of a sheet or steel is painted, this will provide only a short-term solution. Corrosion will start on the unpainted side.
- *Anodizing* is an example of an oxide coating. Anodized aluminum, for example, has essentially a relatively thick aluminum oxide film formed on its surface. A thin oxide film forms naturally on aluminum exposed to air. Oxides, and some other oxide coatings, are ineffective if they are damaged and are subject to pitting in certain environments, particularly those containing chlorides and other halides such as fluorine, bromine, and iodine.
- *Plating* with a less active or more cathodic metal can be very effective if the base metal contains no pinholes, scratches, or abrasions and is completely and uniformly covered.
- In *cladding,* a metal is protected by using another metal as the "bread" for a sandwich—an open-face sandwich if only one other metal is involved and a regular sandwich if the "bread" is on both sides of the metal to be protected. An example of the former is an alclad (aluminum) coating on aluminum alloys used in airplanes. A nickel coin is an example of the latter; a copper alloy is sandwiched between two slices of nickel alloy bread. Inner and outer coatings or single metals are applied by rolling.

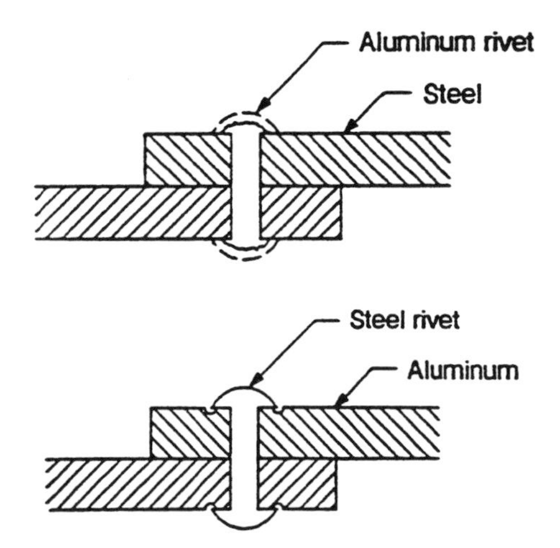

Fig. 11-3 Example of galvanic corrosion

Allowing the part to corrode is another possible solution. Railroad rails are not protected in any manner. In service they become covered with an oily or greasy coating that tends to resist rusting on the sides and bottom of the rails. With sufficient usage a shiny surface is maintained on the top of the rails, which are subjected to heavy contact stress.

Crevice Corrosion

A variation of galvanic corrosion, *crevice corrosion* is difficult to fight without careful control of design, construction materials, engineering, and quality. There are two types of crevice corrosion:

- One involves a crevice or joint between two surfaces.
- The other involves a moist accumulation of dirt or debris, is anodic, and sees little oxygen.

The site of the moist deposit, or *poultice,* is more likely to corrode than an area that is cathodic and exposed to a high oxygen level. An example is

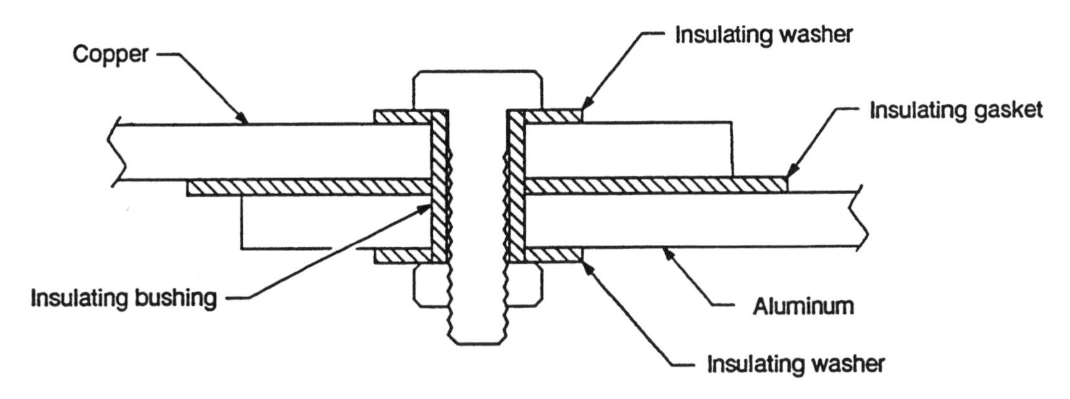

Fig. 11-4 Method of avoiding galvanic corrosion between dissimilar metals

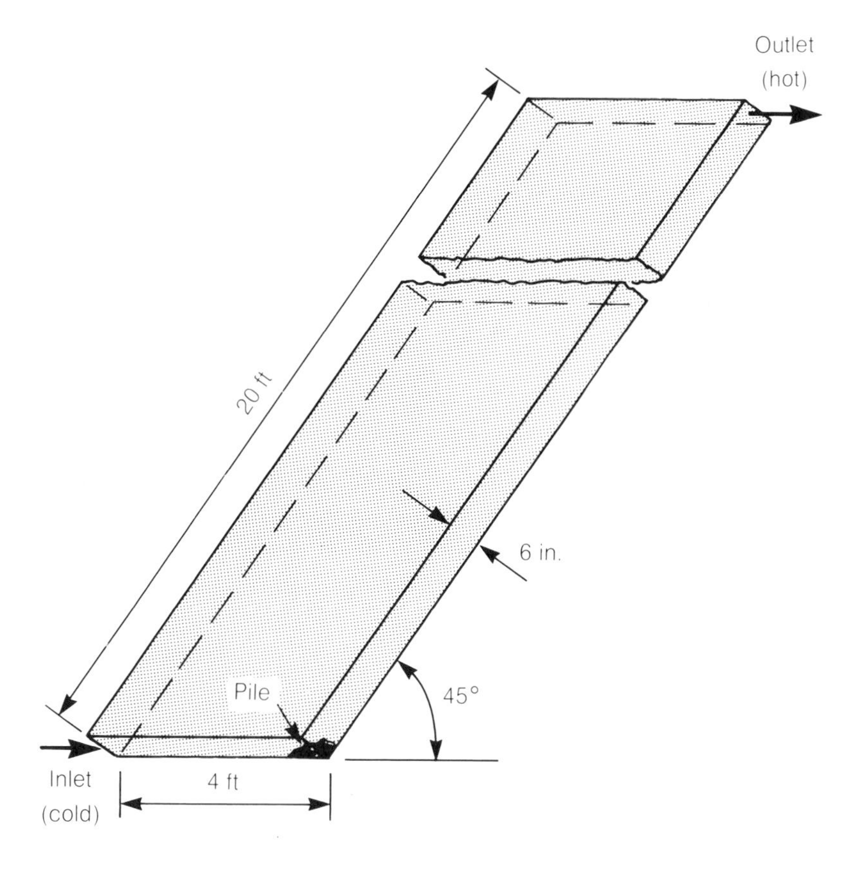

Fig. 11-5 A 115 mm (4.5 in.) long hole caused by crevice corrosion in 6.5 mm (0.25 in.) thick steel plate. The photograph (⅓×) shows the inside of the lower right corner of a large steel box (see sketch) that acted as a heat exchanger or panel to cool extremely hot exhaust gases in a steel mill. Cold water entered at the lower left corner, and the heated water left at the upper right corner of the panel, tilted about 45°. The recirculating mill water contained many particles of rust, dirt, and so on, which gradually accumulated in the dead corner at lower right, opposite the inlet. The pile of particles apparently caked together to form a large cementlike plug (dark area around hole in photograph) over the gradually corroding steel underneath. The corroded hole, with very thin edges, grew to the very large size shown without leaking. Eventually, however, the plug could not support the water over the hole; it then burst, releasing a large volume of water.

shown in Fig. 11-5; note that the large hole in the 6.5 mm (0.25 in.) steel plate was caused by poultice corrosion. Concealed metal under the poultice tends to pit and will eventually perforate the metal. An example of a joint where crevice corrosion can occur is shown in Fig. 11-6, which also shows a preventive measure: eliminating crevices by welding them shut.

Car bodies, with concealed joints and panels, are particularly susceptible to both types of crevice corrosion. When a brown, rusty stain appears on a rocker panel or other part, it is already too late. Corrosion is at work in crevices or joints and under poultices in the presence of moisture, which may be laden with deicing salts or sea air. When corrosion first hits exterior surfaces, it stains the paint and eventually penetrates the metal, leaving holes. Painting the outside stain does not stop the underlying corrosion. In addition, structural members such as box frames and cross members can be severely weakened.

As shown in Fig. 11-6, corrosion may also occur under fasteners, such as bolted or riveted joints. This can occur even if both metals are the same. The problem is greater when dissimilar metals are in contact, because both crevice and galvanic corrosion can be involved, as shown in Fig.11-7. In this instance, corrosion pits are formed, resulting in fatigue fractures.

Possible preventive measures include:

- Avoid bolted or riveted joints unless the metals are coated, ideally both before and after joining. However, be aware that if welding is an alternative to bolting and/or riveting, the heat of welding can destroy coated surfaces. Further, when any fastening method is used it may be necessary to spray or dip the completed assemblies with paint or a waxy or greasy corrosion-resistant material that flows to cover any uncoated areas. A sacrificial anodic metal may be used to protect the structure.
- Close or seal crevices; if moisture or another electrolyte cannot breach the crevice, corrosion cannot occur. Drain holes must be provided if moisture cannot be blocked out.
- Inspect for and remove deposits frequently. An alternative is to use filters, traps, or settling tanks to remove particles from the system.
- Use solid, nonabsorbent gaskets or seals, such as those made of solid rubber or plastics. Surfaces must be smooth to promote sealing, and clamping forces must be suitable for the application.

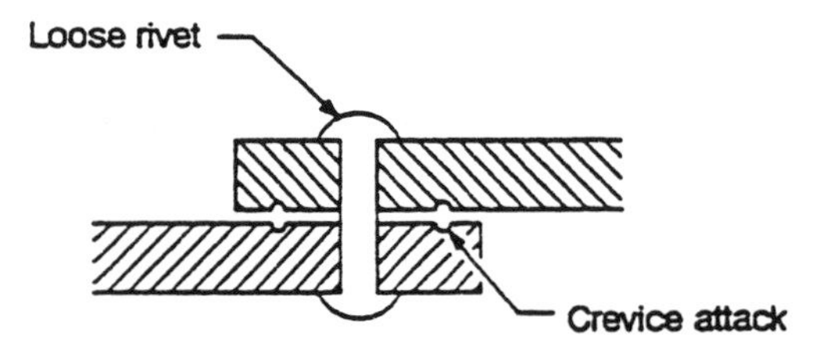

Fig. 11-6 Welding as a method of preventing crevice corrosion

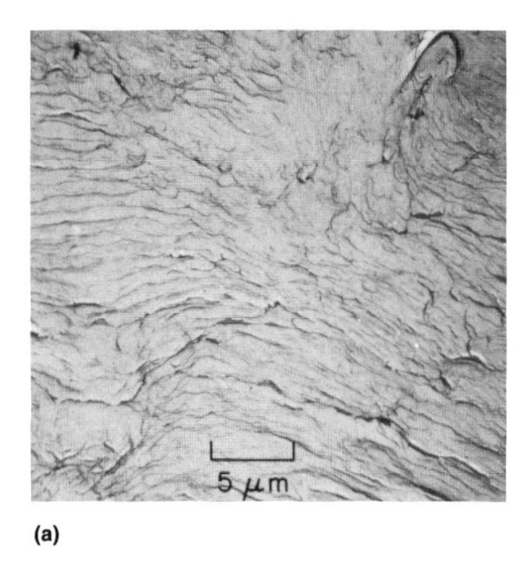

(a)

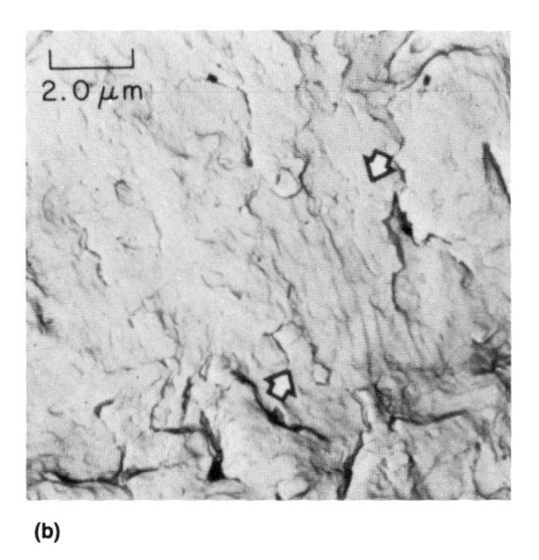

(b)

Fig. 11-7 (a) Fatigue striations in a fracture surface of soft aluminum alloy 1100. 2000×. (b) Poorly formed striations (between arrows) on a fatigue-fracture surface of D-6ac steel with a tensile strength of 1795 to 1930 MPa (260 to 280 ksi). 4900×

Stress-Corrosion Cracking

Stress-corrosion cracking (SCC) is defined as cracking under the combined action of corrosion and tensile stresses; stresses may be applied externally or internally (residually). Cracks may be either transgranular or intergranular, and are perpendicular to the tensile stress. Usually there is little or no obvious evidence of corrosion.

A classic example of SCC is shown in Fig. 11-8. Failure of the brass cartridge case was due to what is sometimes called *season cracking*. This problem was originally encountered by the British army during its campaigns in India in the 1800s. Thin-walled necks of the cartridge cases cracked spontaneously during the monsoon season. The source of the problem was traced to a combination of high temperature and humidity, and traces of ammonia in the air. Now we are aware that most zinc-containing copper alloys, such as 70% copper and 30% zinc, are susceptible to SCC when surfaces are under tensile stress (pulled) in the presence of certain chemicals, such as moist ammonia, mercurous nitrate, and the amines.

Somewhat similar to fatigue, SCC is a progressive type of fracture. Over a period of time, a crack or series of cracks grows gradually until a critical size is reached; then stress concentration can cause a sudden brittle fracture.

Supporting theory has not been developed. However, several unique characteristics of the process have been identified:

- For a given metal or alloy, only certain specific environments contribute to this type of failure, with no apparent general pattern.
- Pure metals are less susceptible than impure metals, although binary alloys such as copper-zinc, copper-gold, and magnesium-aluminum alloys generally are susceptible.
- Cathodic protection has been used to prevent the initiation of SCC.
- Microstructural features such as grain size and crystal structure influence susceptibility in a given environment.

Fig. 11-8 Stress-corrosion crack in the thin neck of a cartridge case

Nearly all metals are susceptible in the presence of static tensile stresses in specific environments. For example, carbon and alloy steels are subject to caustic embrittlement with exposure to sodium hydroxide at relatively low tensile stresses. Table 11-2 lists environments that may cause SCC in many common metals.

Stress-corrosion cracking is the result of a combination of static tensile stress and a particular environment. Residual tensile stresses, including those originating during assembly, generally are thought to be more often the cause of SCC than applied tensile stresses. Tensile stresses also are generated in other ways: shrinking, fitting, bending, or torsion during assembly. The only requirement for SCC is the presence of a combination of a tensile stress on the surface of the metal and a critical environment. The stress need not exceed the yield strength of the metal.

Prevention sounds deceptively simple: Eliminate the tensile stress or the corrosive environment. Out in the cruel world, however, this remedy belongs in the "easier said than done" category. Self-help measures include shot peen-

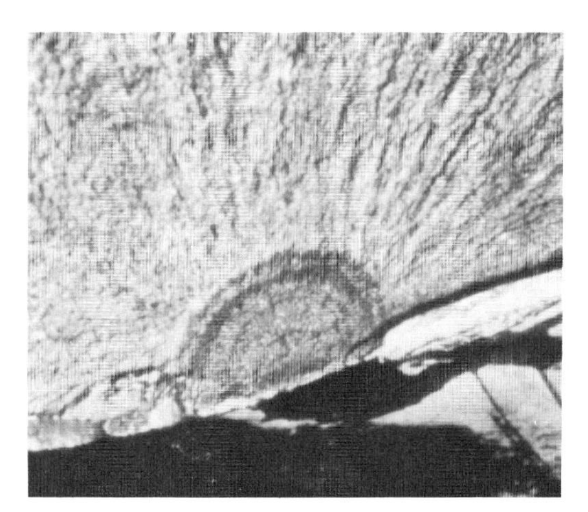

Fig. 11-9 Stress-corrosion crack in a high-strength steel part. The fracture surface appears to have the characteristic beach-mark pattern of a fatigue fracture. However, this was a stress-corrosion fracture in which the pattern was caused by differences in the rate of corrosion penetration. Final fracture was brittle. 4×

ing and surface rolling, which put compressive residual stresses on the surface of a part. Heat treatment can be effective when tensile residual stresses are high. Stresses are relieved by warming parts to subcritical temperatures and temperatures at which a change in phase occurs.

Identification of SCC is challenging, because it can be confused with another type of fracture. For example, Fig. 11-9 shows a stress-corrosion crack in a high-strength steel part that has a fracture pattern that could be mistaken for fatigue fracture. However, the part had not been cyclically stressed, and fatigue could not be the problem. The beach-mark pattern is the result of differences in the rate of penetration of corrosion on the surface as the crack advanced. Progression of the crack was relatively slow until it reached critical size. Then the part failed in a brittle manner. Stress-corrosion cracking also is frequently confused with hydrogen embrittlement cracking due to exposure in service to hydrogen gas or hydrogen sulfide.

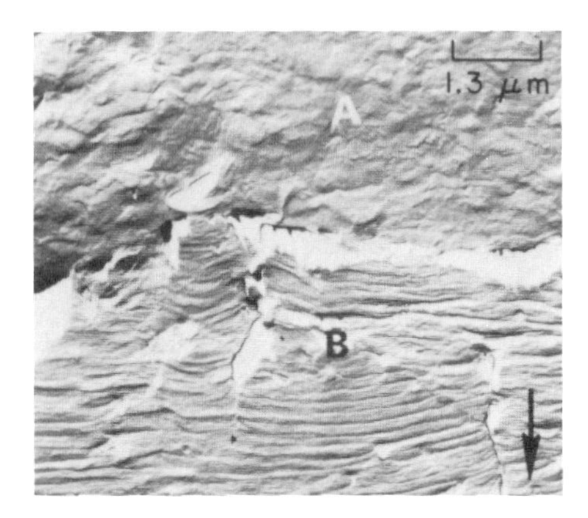

Fig. 11-10 Fatigue fracture in aluminum alloy 2024-T3 tested first in vacuum (region A) and then in air (region B). Arrow at lower right indicates direction of crack propagation. Note the flat, featureless fracture surface A with no correlation between fracture appearance and the cyclic nature of the imposed loading. In contrast, the regular fatigue striations formed while testing in air (region B) correlate with the crack advance of each loading cycle. 7500×

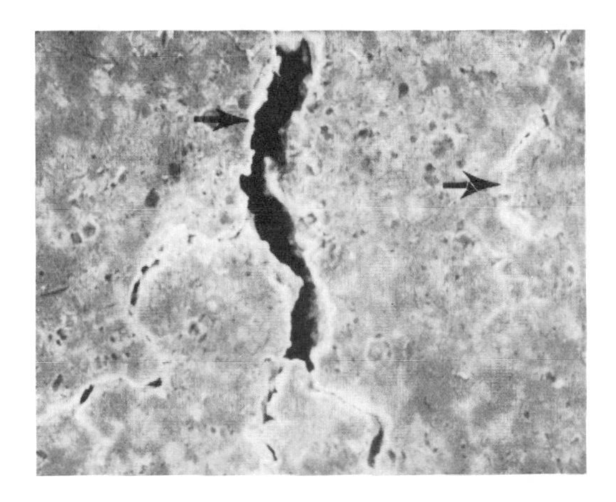

Fig. 11-11 Scanning electron micrograph of corrosion pits on the surface of a gas turbine airfoil showing both large and small pits (arrows) that led to fatigue fractures. The material is the precipitation-hardening stainless steel 17-4 PH. 330×, shown here at 75%

Table 11-2 Environments that may cause SCC under certain conditions

Material	Environment
Aluminum alloys	Na-Cl-H_2O_2 solutions
	NaCl solutions
	Seawater
	Air, water vapor
Copper alloys	Ammonia vapors and solutions
	Amines
	Water, water vapor
Gold alloys	$FeCl_3$ solutions
	Acetic acid-salt solutions
Inconel	Caustic soda solutions
Lead	Lead acetate solutions
Magnesium alloys	Na-Cl-K_2CrO_4 solutions
	Rural and coastal atmospheres
	Distilled water
Monel	Fused caustic soda
	Hydrofluoric acid
	Hydrofluosilicic acid
Nickel	Fused caustic soda
Carbon and alloy steels	NaOH solutions
Carbon and alloy steels (continued)	NaOH-Na_2SiO_2 solutions
	Calcium, ammonium, and sodium nitride solutions
	Mixed acids (H_2SO_4–HNO_3)
	HCN solutions
	Acidic H_2S solutions
	Moist H_2S gas
	Seawater
	Molten Na-Pb alloys
Stainless steels	Acid chloride solutions such as $MgCl_2$ and $BaCl_2$
	NaCl-H_2O_2 solutions
	Seawater
	H_2S
	NaOh-H_2S solutions
	Condensing steam from chloride waters
Titanium	Red fuming nitric acid

The complexity of SCC is indicated by the host of pertinent factors involved: the alloy, heat treatment, microstructure, stress system, part geometry, time, environmental conditions, and temperature. It is obvious that suspected cases of environment-related fractures must be studied and analyzed carefully.

Corrosion Fatigue

Corrosion fatigue is caused by repeated or fluctuating stresses in a corrosive environment. Fatigue life is shorter than would be expected if the sole cause were either repeated or fluctuating fatigue or a corrosive environment. Environment is the key factor.

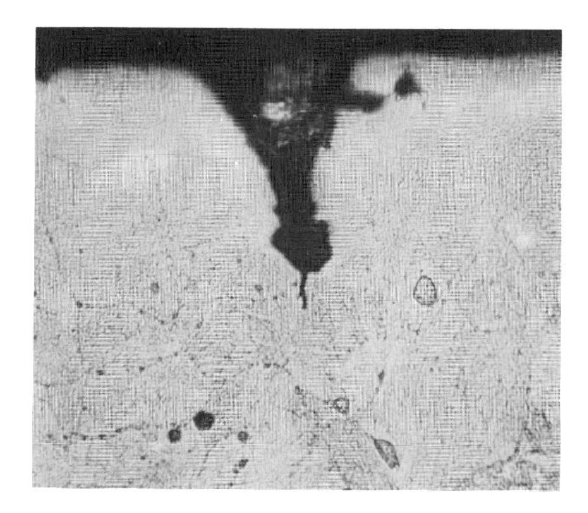

Fig. 11-12 Micrograph of a polished and etched section through a pit like those shown in Fig. 11-11. Note the start of a fatigue crack growing from the bottom of the pit. Pitting was caused by continued operation in a seawater environment. 500×, shown here at 75%

Even when the environment is air, its influence is evident, as demonstrated in Fig. 11-10. In this instance, aluminum alloy 2024-T3 formed normal fatigue striations when it was tested in air (area B). These features are absent in area A when the part was tested in a vacuum. The longer and more frequently a fatigue crack is opened to a corrosive environment, the more severe will be the effect of the environment on shortening fatigue life.

In many cases, fatigue is initiated from small pits on a corroded surface that act as stress concentrations (Fig. 11-11 and 11-12). In other cases, it appears that a fatigue crack initiates first and is made to grow faster either by moisture or by another corrodent that enters a crack by *capillary action* (movement of a liquid into an opening in a solid).

Identification of corrosion fatigue fractures is complicated by the effect of the environment. The site where a fatigue fracture started is more likely to be the most severely corroded because it has been exposed to the environment for the longest time.

As usual, prevention is easier in theory than it is in practice. The most effective measures include these actions:

- Reduce or eliminate corrosion by any of the conventional means, such as painting or plating. It may be possible to reduce the aggressiveness of an environment by adding inhibitors or by changing the concentration of solutions in closed systems.
- Reduce the tensile stress causing the fatigue problem. It may be possible to reduce the applied stress by reducing the load applied to the part.
- Redesign to increase section size.
- Reduce tensile stresses by increasing compressive residual stresses on critical surfaces by shot peening or surface rolling.

Table 11-3 Combinations of alloys and environments subject to selective leaching

Alloy	Environment	Element removed
Brasses	Many waters, especially under stagnant conditions	Zinc (dezincification)
Gray iron	Soils, many waters	Iron (graphitic corrosion)
Aluminum bronzes	Hydrofluoric acid, acids containing chloride ions	Aluminum
Silicon bronzes	Not reported	Silicon
Copper nickels	High heat flux and low water velocity (in refinery condenser tubes)	Nickel
Monels	Hydrofluoric and other acids	Copper in some acids, and nickel in others
Alloys of gold or platinum with nickel, copper, or silver	Nitric, chromic, and sulfuric acids	Nickel, copper, or silver (parting)
High-nickel alloys	Molten salts	Chromium, iron, molybdenum, and tungsten
Cobalt-tungsten-chromium alloys	Not reported	Cobalt
Medium-carbon and high-carbon steels	Oxidizing atmospheres, hydrogen at high temperatures	Carbon (decarburization)
Iron-chromium alloys	High-temperature oxidizing atmospheres	Chromium, which forms a protective film
Nickel-molybdenum alloys	Oxygen at high temperature	Molybdenum

Shot peening puts a relatively shallow layer—usually several thousandths of an inch deep—on a surface into compression. This protection may not be adequate if corrosion makes pits that penetrate the peened area. Deeper penetration may be obtained by surface rolling or by pressing grooves in critical areas. However, both remedies are limited by part geometry. Propeller shafts for ships, for example, are surface rolled where the propeller hub fits the shaft. This location is critical because the rotating propeller hub must be separated from the stationary rear support or bearing by a sealing system, which often does not work properly. As a result, water—often seawater—corrodes the steel shaft, making corrosion fatigue a possibility.

- Change the material to one that is more resistant to the environment. This is regarded as a last resort as it could open the door to new problems relating, for example, to economics and material availability, along with general engineering and manufacturing suitability.

Selective Leaching

In *selective leaching,* an element is removed from an alloy by corrosion. The most common example is *dezincification,* or the removal of zinc from brass. However, many alloys are subject to selective leaching under certain conditions. Elements in an alloy that are more resistant to the environment remain behind.

Two mechanisms are involved:

- Two metals in an alloy are dissolved; one metal redeposits on the surface of the surviving elements.
- One metal is selectively dissolved, leaving the other metals behind.

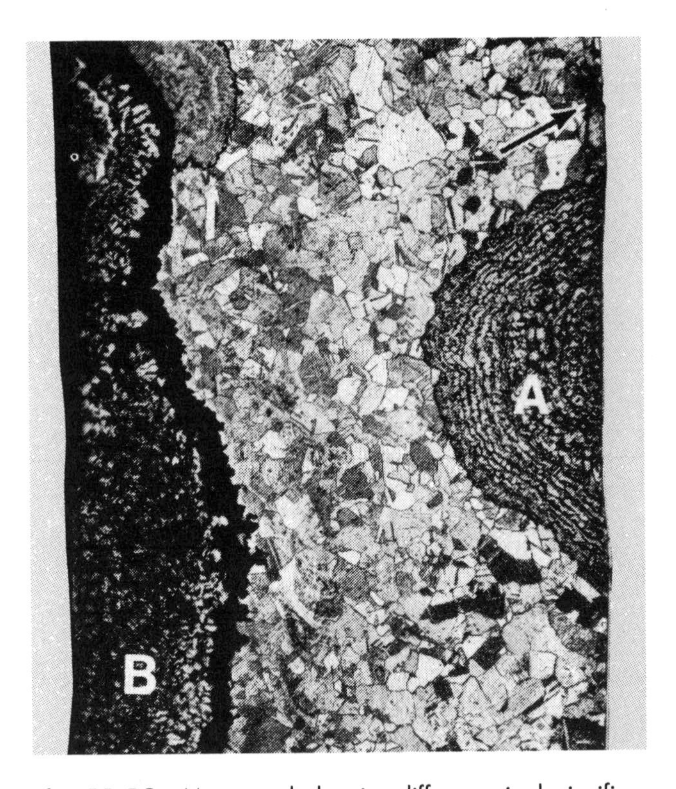

Fig. 11-13 Micrograph showing difference in dezincification of inside and outside surfaces of a plated copper alloy 260 (carriage brass, 70%) pipe for domestic water supply. Area A shows plug-type attack on the nickel-chromium-plated outside surface of the brass pipe that initiated below a break in the plating (at arrow). Area B shows uniform attack on the bare inside surface of the pipe. Etched in $NH_4OH{\cdot}H_2O_2$. 85×

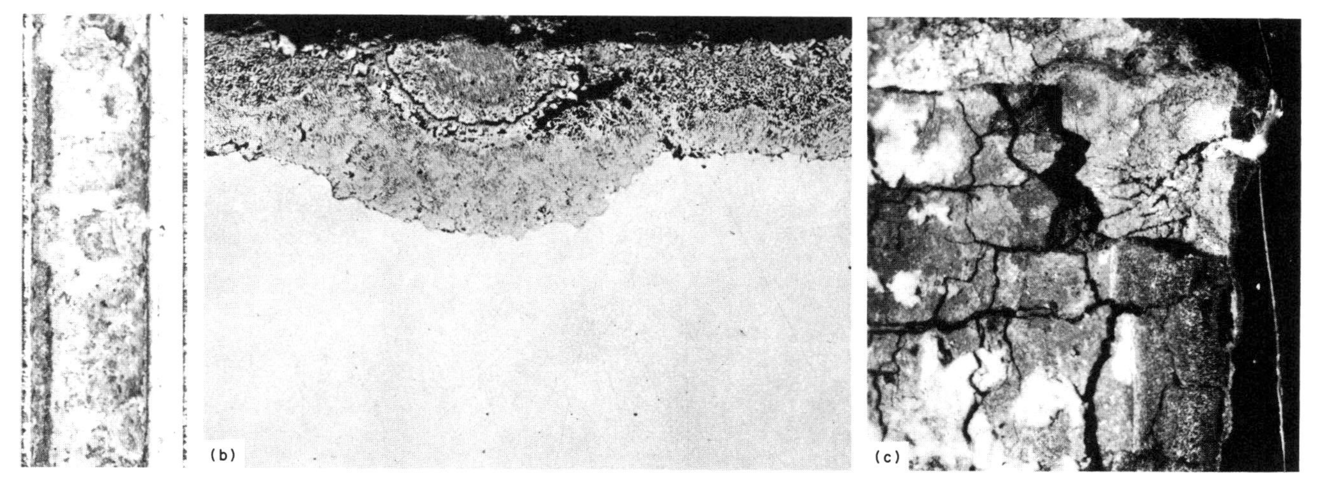

Fig. 11-14 Copper alloy 270 (yellow brass, 65%) inner-cooler tube in an air compressor that failed by dezincification. (a) Unetched longitudinal section through the tube. (b) Micrograph of an unetched specimen showing a thick uniform layer of porous, brittle copper on the inner surface of the tube and extending to a depth of about 0. 25 mm (0.01 in.) into the metal, plug-type dezincification extending somewhat deeper into the metal, and the underlying sound metal. 75× (c) Micrograph of an unetched specimen showing complete penetration to the outside wall of the tube and the damaged metal at the outside wall at a point near the area shown in (b). 9×

The first system is involved in the dezincification of brasses, and the second system is involved when molybdenum is removed from nickel alloys in molten sodium hydroxide. Alloys susceptible to selective leaching in specific environments, and the elements removed, are listed in Table 11-3.

Dezincification

Brasses containing less than 85% copper are subject to selective leaching. *Leaching* is defined as the removal of an element or compound from a solid alloy or mixture by preferential dissolution in a suitable liquid. In this instance, zinc corrodes preferentially, leaving a porous residue of copper and corrosion products. Alpha brass containing 70% copper and 30% zinc is particularly susceptible when exposed to an aqueous electrolyte at elevated temperatures.

The Process. Brass dissolves; zinc ions stay in solution; copper plates back on. The process can take place in the absence of oxygen, as evidenced by the fact that zinc corrodes slowly in pure water. However, the presence of oxygen increases the rate of attack. Dezincified areas usually show 90 to 95% copper, including some copper oxide.

Two Types. Dezincification may be either uniform or plug-type attack. Examples of the latter are shown in Fig. 11-13 and 11-14. High zinc content favors uniform attack, whereas low zinc content favors the plug type. Slightly acidic room-temperature water that is low in salt content is likely to produce uniform attack; neutral or alkaline water high in salt content and above room temperature often produces plug-type attack.

Access of corrodent, flow rate, and other factors sometimes are involved. In Fig. 11-13, plug-type attack occurred on the exterior of a nickel-chromium-plated brass pipe for domestic water supply. Untreated municipal water leaking from the faucet packing gland ran down the pipe and attacked the brass through a break in the plating caused by mechanical damage. Water inside the pipe produced fairly uniform, layer-type dezincification on the pipe, which had an initial wall thickness of 0.76 mm (0.030 in.). Only about one-third of the original wall remained as sound metal in the corroded region, which was 0.61 mm (0.024 in.) thick.

In Fig. 11-14, yellow brass intercooler tubes began to leak. They were sectioned longitudinally as shown in Fig. 11-14(a), so that inside surfaces could be examined. Visual examination disclosed a thick layer of porous, brittle copper. A uniform layer of spongelike copper that penetrated to a depth of about 0.25 mm (0.01 in.) was found on the inside surface (Fig. 11-14b), along with a plug-type deposit that penetrated somewhat deeper into the tube wall. Sound metal underlying the damaged surface layer is also visible. Damaged metal at the outside wall of the tube at a point where the wall had been completely penetrated is shown in Fig. 11-14(c).

On further examination it was learned that the tubes had been fabricated from a 65% copper yellow brass. Only a trace of the 35% zinc component remained in the brittle layer. An uninhibited brass with high zinc content should not have been used, and the original alloy was replaced by arsenical aluminum brass.

Chapter 12

Quest for Quality

The quest for quality is a continual pursuit of a moving target. Granted, in this century great strides have been made in the science and practice of metallurgy, in the technology used in the manufacture of end products, and in quality control technology (refer to the chronology of achievement in Chapter 2, as well as that in Chapter 13.

But there is a "however"—a huge one.

At this point in time, total quality control does not seem to be any more possible than total prevention of the corrosion of metals. The challenge then boils down to:

- Controlling the myriad factors that account for variations in the properties of engineering materials, which signal degradation in quality
- Controlling the myriad factors that account for variations in the performance of machines and processes involved in the manufacture of end products, which again signal degradation in quality
- Controlling the myriad factors that account for variations in the performance of people involved in the above activities, which also signals degradation in quality

The metallurgist is supremely aware of these challenges. In all matters relating to quality this person exhibits a relentless skepticism. He or she is a hard sell, not buying anything sight unseen; "the best we can do" is not good enough. In other words, the metallurgist is quick to question and slow to accept.

The scope of the challenge is mapped out by the following excerpts from a lesson in "Metallurgy for the Non-Metallurgist™," the MEI course offered on a continuing basis by ASM International:

- The objective of all manufacturers is to assure satisfaction with the product.
- The usefulness of quality control as a decision-making tool is based on the recognition of variability as a major factor in the production of goods and the performance of services.
- All products and services have variability because it is impossible for all units of a product or service to be made exactly alike. . . . Variability may be difficult to measure when a product is made by precision equipment, but variability is always present. At times it is blatantly obvious, as evidenced by excessive scrap, rework, returned goods, or service calls. There usually is an unacceptably high cost associated with obvious variability.
- No statistical methods are used to separate variability due to uncontrollable causes. In this respect, quality control is a marriage of the techniques of manufacturing engineering and statistical mathematics.
- If quality is defined as fitness for use, then quality control is the broad task of maintaining a state of product quality that assures the customer of a useful product and the manufacturer of a viable business enterprise. This has a particular meaning in metalworking, metal finishing, and metal treating companies. Too often, for example, it is assumed that quality is automatically achieved when a practice such as "This is the way we've always done it" is followed . . . instead of asking, "Will this really result in customer satisfaction?"

A Potpourri of Variability

Item: The inevitability of variability in quality is given recognition by the U.S. Air Force in its official damage tolerance concept, which is based on the ability of a product such as a jet engine to resist failure due to flaws, cracks, and other damage in the materials of construction for a specified period.

Item: It seems reasonable to assume that an automated machine has the ability to cut metal to the same length every time. Not so, according to Fig. 12-1. In this instance, the desired length was 50 mm (2 in.), but in the example, lengths varied from about 48 to 53 mm (1.9 to 2.1 in).

Item: Is stainless steel really stainless—that is, does it rust? As evidenced in Fig. 12-2, stainless steel does rust; the dark areas in the micrograph indicate intergranular corrosion of a cast stainless steel pump that has been in contact with a corrosive gas.

Item: Figure 12-3 gives lie to "What you can't see won't hurt you." In this instance, the rough surfaces pictured often will appear smooth to the naked eye. The purpose of these drawings is to illustrate a kind of adhesive wear, which is characterized by a type of welding. Contacting peaks, or aspirates, are sheared off, and the resulting metal debris is subsequently welded by contact to the tip of another asperity. The progression is indicated in Fig. 12-3(a) to (c). The caption to Fig. 10-2 provides details.

Item: With the help of available testing technology, it is possible to measure the severity of a variable—in Fig. 12-4, the location and depth of a crack on a crane hook. The crack was located with the use of magnetic particle testing (top photo). Subsequently, the part was sectioned and the same technique was used to determine the depth of the crack (bottom micrograph).

Item: It pays to ask probing questions, as evidenced by the following example from a metallurgist. In investigating the cause of a weld failure, he went back to the welder and asked "The specification calls for preheating the part before it is welded. Did you do this?" The answer was, "Yes." The metallurgist followed up with, "When did you preheat?" The answer elicited, "The evening before I welded." Problem solved; preheating should have been followed immediately by welding. The incident suggests the importance of asking enough questions.

Item: In the pursuit of quality, suppliers to the manufacturing industry are routinely audited by their customers. Both customers and suppliers conduct self-audits of their facilities. Certain branches of the government, such as the Department of Defense, audit suppliers. And some organizations provide auditing services to all parties concerned; these are characterized as third-party audits. One organization offering such services has audited several hundreds of suppliers over the last few years. To date, no supplier has passed with flying colors, and a small minority have flunked first audits. In the majority of audits, findings of breakdowns in quality typically range from three or four on the low side to a couple of dozen on the high side. Breakdowns tend to be simple in nature:

- Suppliers have failed to keep tabs on customer satisfaction, or lack thereof.
- People on the shop floor do not always follow instructions—of either their managers or their customers.
- People on the shop floor sometimes fail to inform their managers of problems in a timely manner, and may even elect to solve problems on their own.
- Suppliers do not always have their machines calibrated by outside services to provide some assurance that the machines are operating as they should.
- Nationally and internationally accepted standards of performance are not always followed.
- Suppliers do not always have programs for training shop personnel.

Item: The pursuit of quality in the manufacturing industry is international, as evidenced by the following incident—which highlights two universal problems: occasional breakdown in

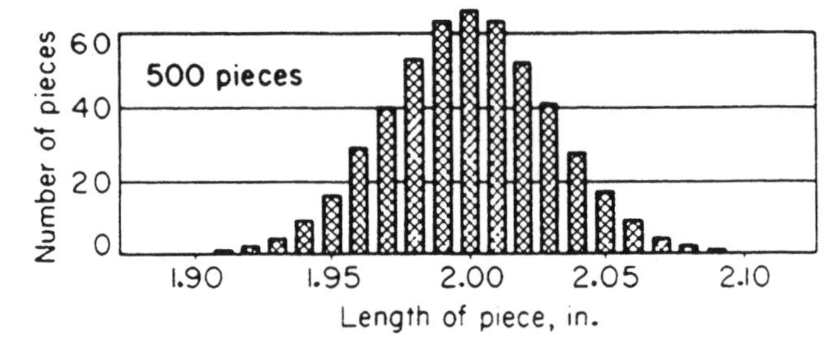

Fig. 12-1 Frequency distribution for length measurements of 500 pieces of bar stock that were cut to 50 mm (2 in.) lengths in an automatic machine

the performance of both people and machines. In this particular incident, a visitor from the West was on a guided tour of a model auto plant in a Far Eastern country noted for its all-out dedication to quality.

The guest and his guide were on a walkway above what to the guest looked like acres of automated stamping machines. The guest also noted an almost total absence of people below; only one person was busy doing something at a bench, and his back was turned to both guide and guest.

The guest commented on this almost peopleless plant to his host, who pointed to a number of large, rectangular panels suspended from the ceiling at the walkway level. Each panel was attached to an individual machine, and red and green lights on the panel indicated whether or not the machine was operating properly. According to the guide, these panels were a convenience for the operator and for managers on the walkway. What followed concerned the only operator in sight.

The guest noted that one of the lights on a panel had changed from green to red and pointed this out to his host, who apparently had not seen the red light but immediately rushed down a stairway and practically ran toward the operator, who still had his back turned.

It was apparent to the guest watching from above that one of the machines had jammed and stopped the operation. The operator, a young man, removed the part that had jammed and threw it into a nearby trash can. The machine went back into operation immediately.

The operator went directly back to his bench without looking to see whether the machine had resumed operations, and continued to do whatever it was that commanded his total attention. The two men on the catwalk continued their tour. The guest looked back at the panel only to notice that the red light was on yet again. Obviously, the machine had jammed once more. The operator continued to focus on his work at the bench.

The guest did not mention the second breakdown to his guide.

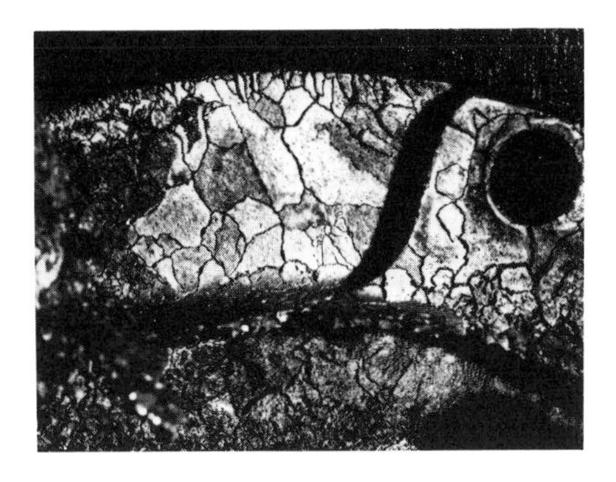

Fig. 12-2 Intergranular corrosion of a cast stainless steel pump component that came into contact with $HCl\text{-}Cl_2$ gas fumes

Overview of Testing and Inspection Technology

A variety of technology is available to ascertain the quality of metal parts, or lack thereof. Commonly used techniques include:

- *Mechanical Testing.* Strength, hardness, ductility, and toughness are determined by destructive testing methods. This means that the part or specimen being tested will be destroyed during the testing process. When individual parts are expensive and only a few are made, destructive testing becomes an economic consideration. Topics discussed in this section include tensile testing, hardness testing (both macrohardness and microhardness types), impact testing, and fatigue testing.
- *Nondestructive Testing.* As the name implies, parts or specimens tested during this process are not damaged and may be put into service after testing. Five of these testing methods are discussed: liquid penetrant, magnetic particle, radiographic, ultrasonic,

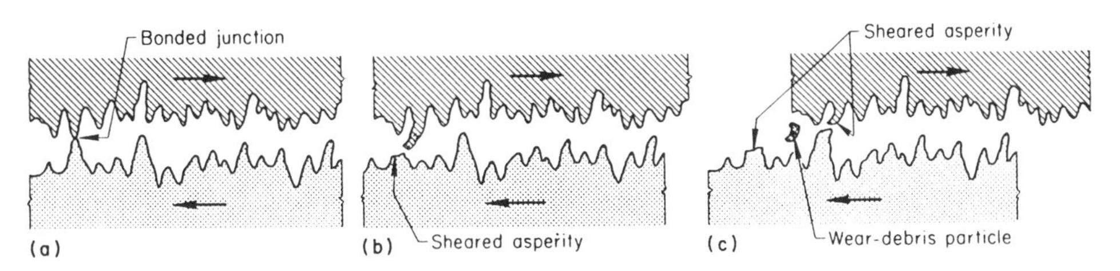

Fig. 12-3 Schematic illustration of a process by which a particle of wear debris is detached during adhesive wear. Such rough surfaces often appear smooth to the naked eye.

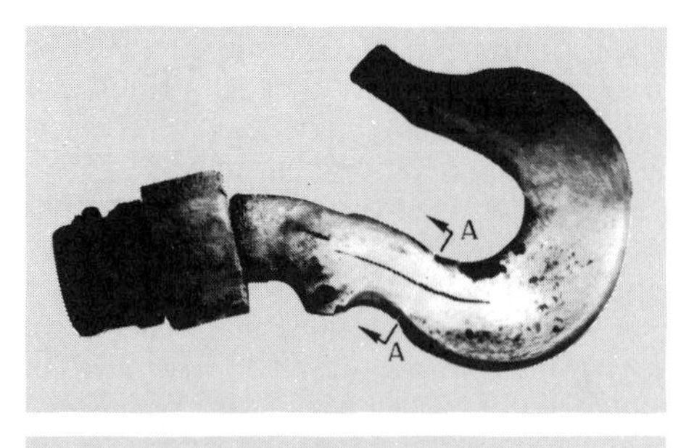

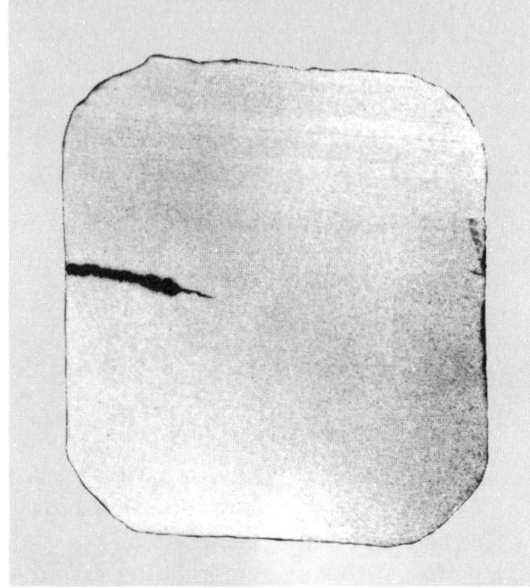

Section A-A

Fig. 12-4 Crane hook showing magnetic particle indication of a forging (top) and section through hook showing depth of lap (bottom)

and eddy current. Each method has advantages and disadvantages.

- *Metallography*. Here, quality is determined by slicing samples of parts or specimens and taking a much closer look under the microscope. At the same time, a permanent record of findings is created by taking a photograph of the area(s) of interest. These photos are called photomicrographs, or micros for short.

Mechanical Testing

Tensile Testing

Tensile testing concerns the stress-strain relationship of a material in both the elastic and plastic strain regions of deformation under load. *Elastic deformation,* in which a part returns to its original shape when the load is removed, is demonstrated in Fig. 12-5. *Plastic deformation,* where a part does not return to its original shape when the load is removed, is illustrated in Fig. 12-6.

A standard tensile specimen is pictured in Fig. 12-7. Ends of the test bar are either threaded or have a dog-bone shape so that they can be secured in a tensile testing fixture.

A universal tensile testing machine is shown in Fig. 12-8. In this instance, a tensile bar is located in the lower third of the equipment on the left. Adjoining equipment on the right includes a computer keyboard, a monitor, a recorder, and a printer.

Typical properties evaluated in tensile testing include yield strength or yield point, tensile strength, and ductility. *Yield point* is unique to materials such as low-carbon steel that show an increase in strain with no in-

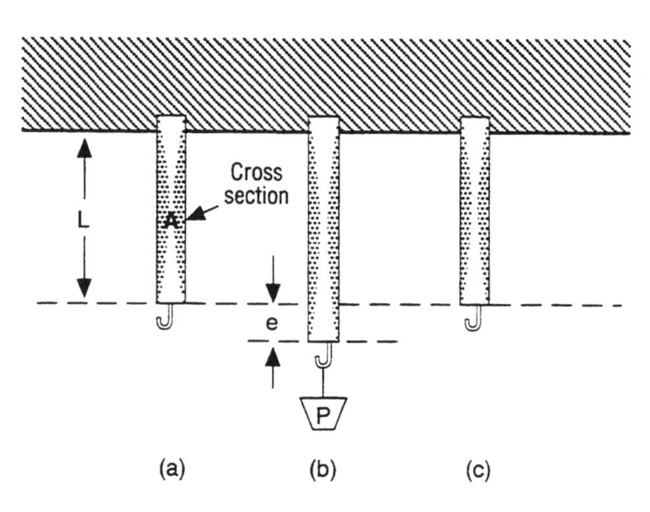

Fig. 12-5 Elastic deformation. (a) Before loading. (b) Loaded. (c) After removal of load

crease in stress. *Tensile strength* or *ultimate tensile strength* is the highest stress level reached on an engineering stress-strain plot. *Ductility* is usually evaluated in terms of percentage elongation or reduction in area.

Macrohardness Testing

In *macrohardness testing,* loads exceed 1 kg (2.2 lb), and the resulting indentation is visible to the naked eye. Tests in this category include the Brinell, Rockwell, Rockwell superficial, and Vickers (heavy-load) tests. The first three will be discussed here.

Brinell Test. The setup for a Brinell test is shown in Fig. 12-9(a). The indentation is measured in millimeters, as shown in Fig. 12-9(b). A hardened steel or carbide ball 10 mm (0.4 in.) in diameter normally is used. The force applied is 3000 kg (6600 lb) or less, depending on the material; the diameter of the spherical impression is measured with a microscope to an accuracy of 0.05 mm (0.002 in.).

A Brinell hardness number can be related to tensile strength. For a homogeneous steel in which all elements are alike, the hardness number is multiplied by 500 to obtain an approximate tensile strength in pounds per square inch (psi); 1000 psi is equivalent to 1 kip per square inch (ksi). To convert ksi to the metric designation of megapascals (MPa), ksi is multiplied by 6.89. For example, a Brinell hardness number of

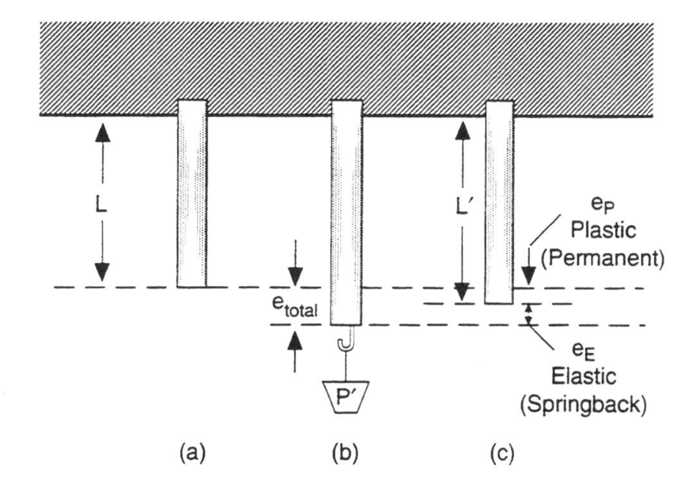

Fig. 12-6 Plastic deformation. (a) Before loading. (b) Loaded. (c) After removal of load

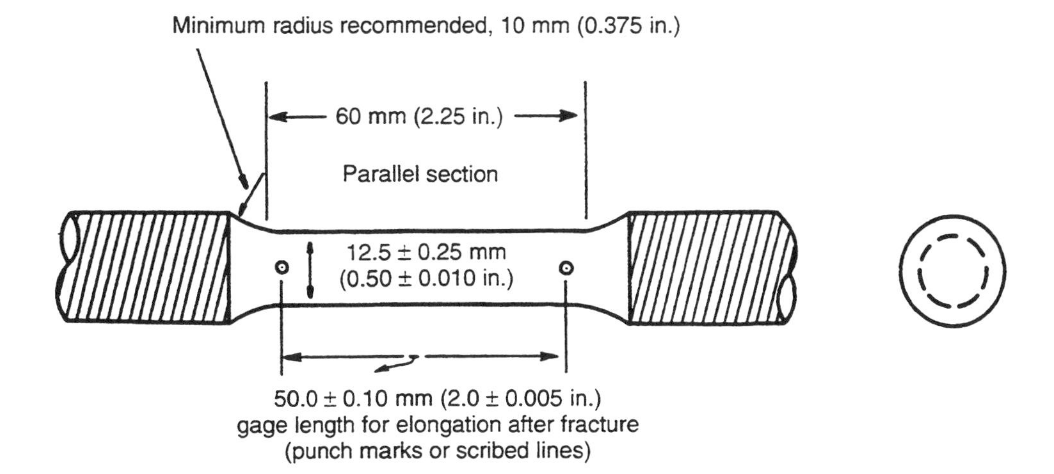

Fig. 12-7 Dimensions for standard round tensile specimen

350 is equivalent to approximately 175,000 psi (350 × 500), or 175 ksi; 175 ksi converts to approximately 1205 MPa (175 × 6.89).

As Brinell testers are often used to test parts such as forgings, castings, bar stock, pipe, tubing, plate, and other heavy duty components, their construction is often more rugged than that of the other types of hardness testers.

The Rockwell hardness test comes in two varieties: regular Rockwell and superficial Rockwell. The superficial Rockwell test applies lighter loads than the regular Rockwell test and has a special N-brale indenter that is more precise than the regular brale indenter. Indentations produced by both light and heavy loads are shown in Fig. 12-10(a) and (b), respectively.

Regular Rockwell Hardness Test. In the regular Rockwell test, cone-shaped diamond indenters and hardened steel ball indenters are used. The diamond-cone indenter shown in Fig. 12-11 is called a brale indenter. Balls usually are 1.6 mm (0.063 in.) in diameter. Loads applied by either type of indenter are 150, 100, or 60 kgf. The abbreviation indicates the force applied in kilograms.

A schematic of a Rockwell tester is shown in Fig. 12-12. Note the indenter and the device or anvil that holds the specimen.

The Rockwell hardness number obtained in testing relates inversely to the depth of the indentation caused by the load force. For example, indentations are shallow when the material being tested is hard; they are deeper when the material is soft. An indentation of about 0.08 mm (0.003 in.) would be produced by a steel with a Rockwell number of 62, whereas a steel with a Rockwell number of 40 would have an indentation measuring about 0.13 mm (0.005 in.).

The regular Rockwell test is normally chosen when the area of the specimen being tested is not large enough to qualify for the Brinell test. However, the diamond indenter makes the Rockwell test suitable when hardness is above the range of the Brinell carbide ball.

Superficial Rockwell Hardness Test. As mentioned, the special N-brale indenter is more precise than the regular Rockwell brale indenter. However, ball indenters are the same for both types of tests. The minor load force is 3 kgf (Fig. 12-10a). Major loads in this instance are 45, 30, and 15.

Parts tested usually are small and thin, such as spring steel and light case-hardened parts. In indentation hardness tests, minimum material thickness usually is 10 times the depth of the indentation.

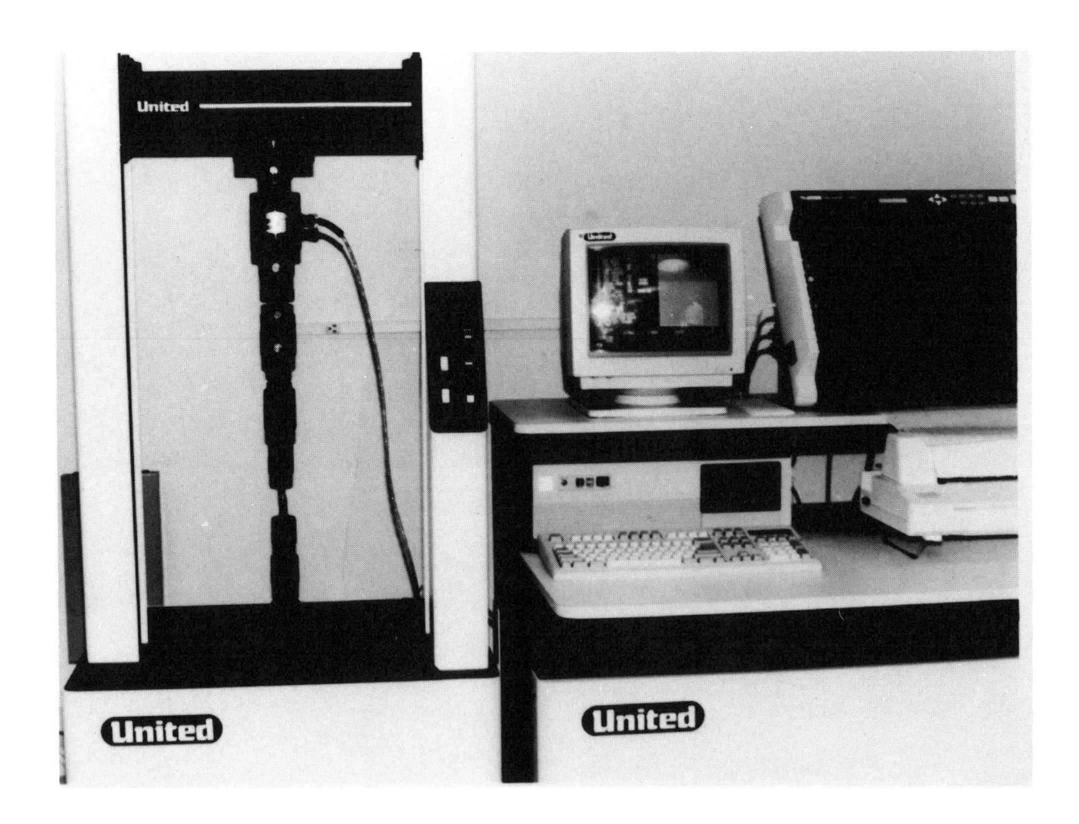

Fig. 12-8 Universal tensile testing machine

Microhardness Testing

In *microhardness testing,* loads are less than 1 kgf, and indentations can be seen only with the aid of a microscope. Knoop and Vickers (low-load) tests are in this category. Indenters used and the indentations they make are compared in Fig. 12-13.

Vickers Indenter. As shown in Fig. 12-13(a), the indenter is a square-base diamond pyramid. The applied load force generally is 1 kgf or less. A typical impression also is shown.

Test specimens must have a polished metallographic surface for viewing and measuring with a microscope at a magnification of about 200× to 400×. The specimen is placed in a fixture and the microscope focused on the area(s) of interest. As in other hardness tests, the higher the number, the harder the material.

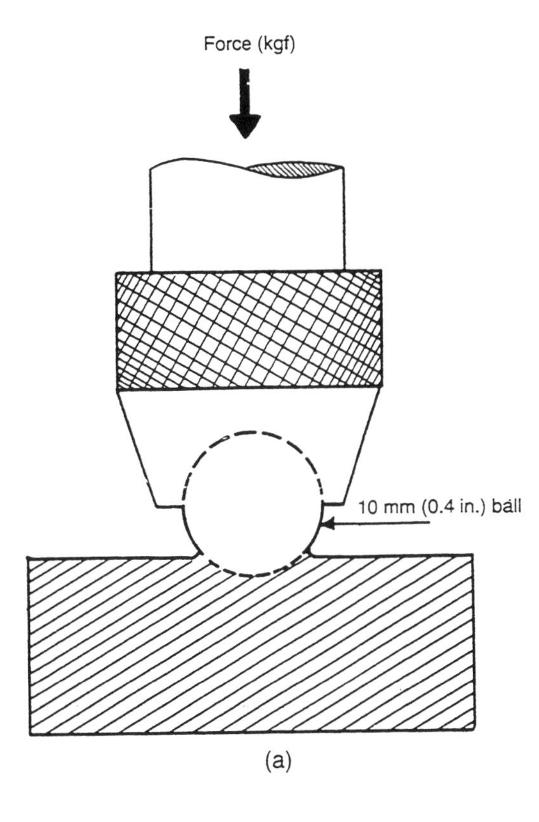

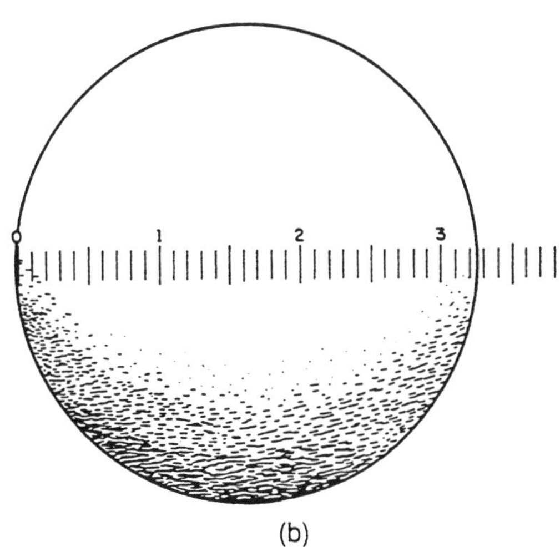

Fig. 12-9 (a) Schematic diagram of Brinell indenting process. (b) Brinell indentation with measuring scale in millimeters

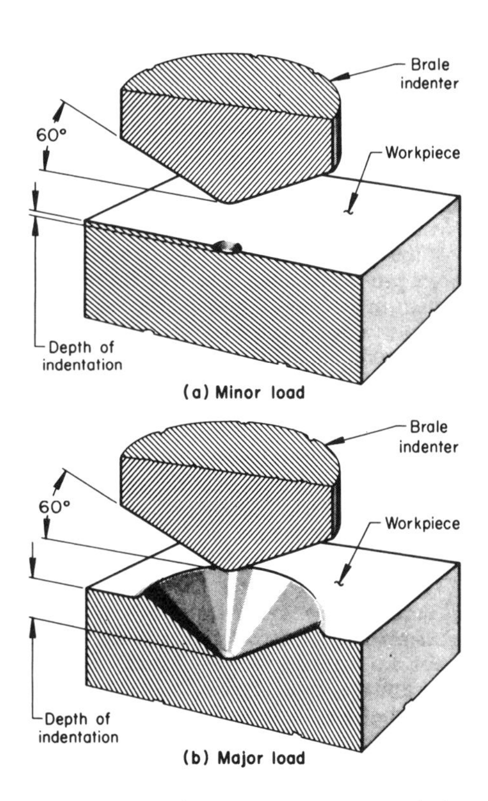

Fig. 12-10 Indentations in a workpiece made by application of minor load (a) and major load (b) from a diamond Brale indenter in Rockwell hardness testing

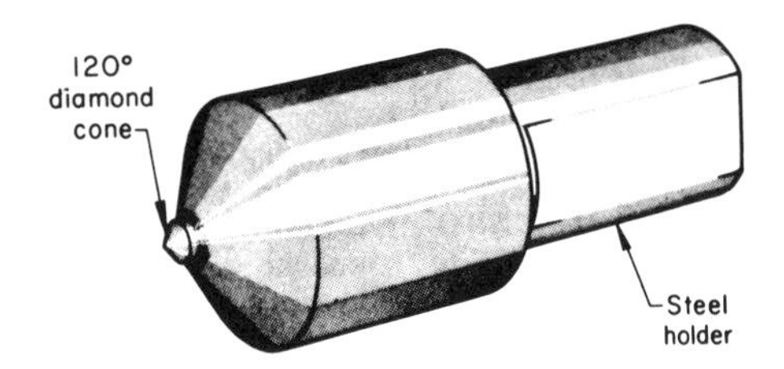

Fig. 12-11 Diamond-cone brale indenter used in Rockwell hardness testing. ~2×

Knoop Indenter. The tester and operating procedures for testing are the same as those for the Vickers test. Differences are in the geometry of the indenture and calculation of area. The Knoop indenter is a rhombus-base diamond pyramid that provides an impression of the shape shown in Fig. 12-13(b). Only the long diagonal is measured.

Because the Knoop impression has one short diagonal, it can be used to advantage where impressions must be made close together, as in the measurement of case depth. Also, the Knoop impression often is preferred when specimens are very thin. The depth of a Knoop indentation is only half that of a Vickers indentation.

Impact Testing

In an *impact test,* the specimen is subjected to dynamic shock loading to evaluate the response of a material to rapid changes in loading, such as that experienced by an axle shaft when a car hits a chuckhole. A standard impact tester and its operation are illustrated in Fig. 12-14. When a specimen is placed in the path of the swinging pendulum, the pendulum must bend and/or fracture the specimen if it is to continue its arc. The amount of energy required to break the specimen is absorbed by the pendulum, which reduces the height to which the pendulum rises. The difference between the height to which the pendulum can rise unimpeded (h) and the height to which the pendulum rises after breaking the specimen (h') represents the energy absorbed in breaking the specimen. This energy (h–h′), expressed in newton-meters (or joules, J) or foot-pounds (ft·lb), is called the *impact energy* of the material being tested.

The two most common impact tests are the Charpy and Izod tests. They differ mainly in the manner in which specimens are held and broken (Fig. 12-15).

The Charpy test often is done on specimens over a range of temperatures because some materials may exhibit a ductile-to-brittle transition at low temperatures. For example, only a small amount of energy is needed to fracture low-carbon steel specimens at lower temperatures.

Fatigue Testing

Parts subjected to repeated cyclic stresses often fail in service, even though the applied stress is considerably below the yield strength of the material. In service, stresses may cycle between tension and compression, or the magnitude of tensile stresses may vary. The combination of tension and compression stresses and their ultimate effect on failure in service for a simple beam is illustrated in Fig. 12-16.

Figure 12-17 shows a typical rotating-beam fatigue testing machine. In order to determine

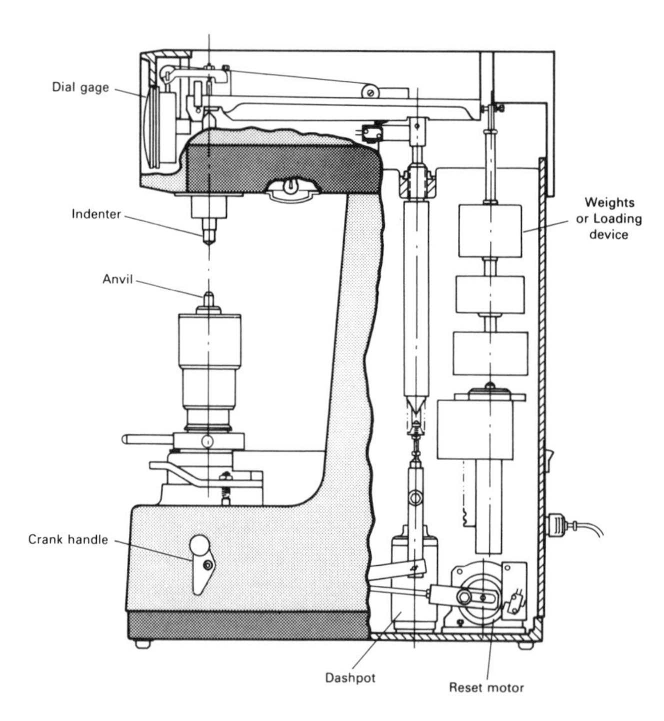

Fig. 12-12 Schematic of Rockwell testing machine

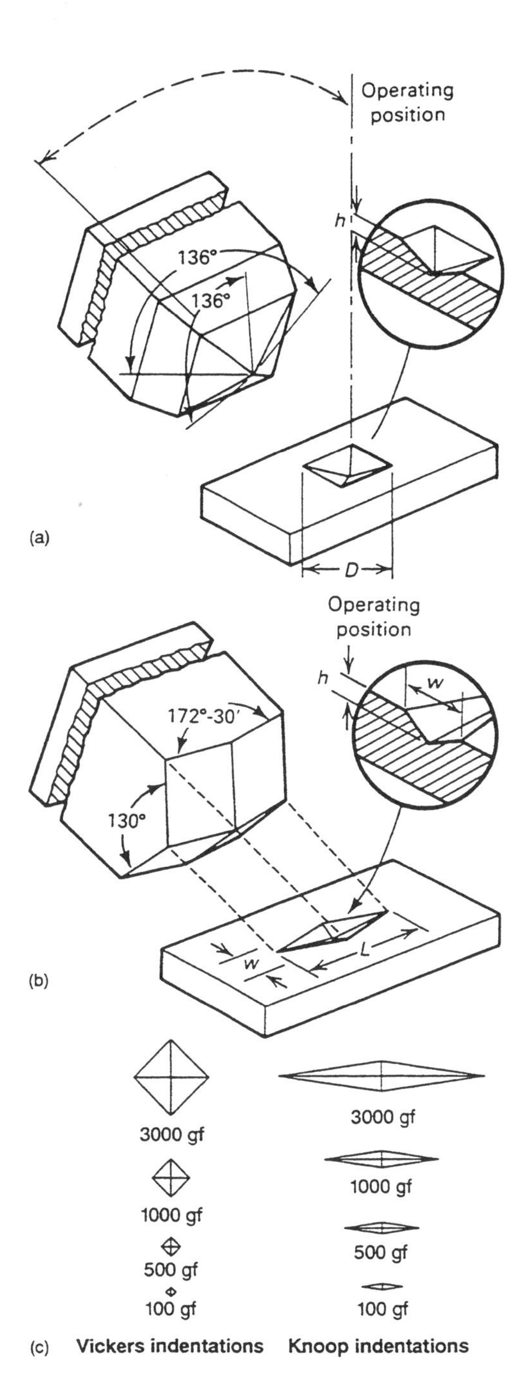

Fig. 12-13 (a) Indentation made by square-base diamond-pyramid indenter used in Vickers hardness test. *D*, mean diagonal of the indentation in millimeters. (b) Indentation made by rhombus-base diamond-pyramid indenter used in Knoop hardness test, along with scale. (c) Comparison of Knoop and Vickers indentations

the number of stress reversals that can be tolerated without failure, specimens are subjected to decreasing loads. This value is called the *endurance limit* or *fatigue limit*.

The endurance limit of steel is roughly related to tensile strength and varies with surface finish, notches, and other residual stresses. Residual tensile stress is lower than the fatigue limit; residual compressive stress, as in shot peening, raises the fatigue limit.

Nondestructive Testing

In *nondestructive testing* (NDT), defects can be detected and evaluated without damage to the part. Five different inspection methods are discussed here:

- Liquid penetrant
- Magnetic particle
- Radiographic
- Ultrasonic
- Eddy current

Liquid Penetrant Inspection

The liquid penetrant inspection process is shown in Fig. 12-18. A discontinuity open to the surface of a metal is located by allowing a penetrating dye or fluorescent liquid to infiltrate it. Excess penetrant is removed in the next step, which is followed by applying a developing agent. This agent causes the penetrant to seep back out of the discontinuity and register as an indication.

Advantages/Disadvantages. Equipment usually is simpler and less expensive than for most other NDT methods. Also, testing procedures usually are less difficult. However, only imperfections open to the surface can be detected, and rough or porous surfaces are likely to produce false indications.

Magnetic Particle Inspection

Magnetic particle inspection locates surface and near-surface discontinuities in ferromagnetic materials. Figure 12-4, discussed earlier in this chapter, provides an example: the detection of what is known as a forging lap in a crane hook.

The Principle. When a material being tested is magnetized, discontinuities in the material cause leakage in the magnetic field. The presence of the leakage field in combination with discontinuities makes it possible to detect defects by applying ferromagnetic particles over the surface of the part. Some of the particles are then drawn to low-level leakage fields that occur around discontinuities. This magnetically held collection of particles forms an outline of the discontinuity and generally indicates its location, size, shape, and extent.

Advantages/Limitations. Magnetic particle inspection provides a sensitive means of locating small, shallow surface discontinuities but will not produce a pattern if particles cannot bridge the surface opening, although these particles probably will be picked up in visual inspection. On occasion, discontinuities that have not broken through to the surface but are within 6.4 mm (0.25 in.) of it may be detected.

The process does have limitations. For example:

- The material being tested must be ferromagnetic.
- Thin coatings of paint and other nonmagnetic coverings such as plating reduce the sensitivity of the process.
- Demagnetization and cleaning are often necessary following inspection.
- Magnetic particle indications are easy to see, but their interpretation often requires skill and experience.

Radiographic Inspection

Radiography is based on differential absorption of penetrating radiation. Three basic elements (Fig. 12-19) are involved: (1) a radiation source, (2) a testpiece or object being evaluated, and (3) a recording medium, usually film. The testpiece in Fig. 12-19 is a plate of uniform thickness containing an internal flaw that has absorption characteristics different from those of the surrounding material. Radiation from the source is absorbed by the testpiece as the radiation passes through it, but the flaw and surrounding material will absorb different amounts. Thus, the amount of radiation reaching the film in the areas beneath the flaw is different from the amount of radiation that impinges on the adjacent areas, producing a latent image of the flaw on the film. When the film is developed, the flaw will be seen as a shadow having different photographic density than the image of the surrounding material. An example is shown in Fig. 12-20. The arrow points to gross shrinkage porosity in an aluminum alloy casting.

Applications. Castings and weldments are frequent applications, especially when freedom from internal flaws is critical. Forgings are another application. Radiography also can be used in the inspection of mechanical assemblies to determine the placement and condition of components, and in the inspection of semiconductor devices for cracks, broken wires, unsoldered

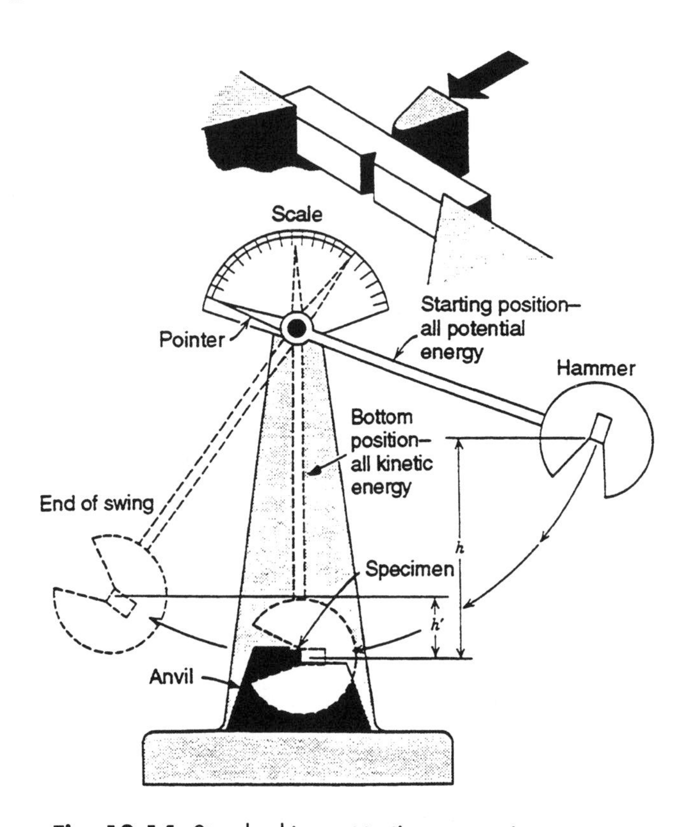

Fig. 12-14 Standard impact testing apparatus

connections, and the presence of foreign materials.

Limitations. Compared to other NDT processes, radiography is more expensive and is hazardous to the operator. Field inspection of thick sections is time consuming. Portable x-ray units are limited to steel sections about 75 mm (3 in.) thick. Large units are available for the inspection of sections up to 300 mm (12 in.) thick, but they are very expensive and very hazardous to operate.

Some flaws are difficult to detect with this method. Unless they are essentially parallel to the radiation beam, laminar defects such as cracks will present problems. Tight, meandering cracks in thick sections usually cannot be detected even when they are properly oriented, and minute discontinuities such as inclusions in wrought materials also are difficult to detect.

Safety requires special attention. Inspectors must be trained in government regulations, shielding, dosimetry, and other safety-related subjects.

Ultrasonic Inspection

In some respects, ultrasonic inspection works like radar. In C-scan inspection, for example, changes in ultrasonic waves passing across the surface of a structural component are displayed on a screen. The ultrasonic C-scan technique is used extensively to determine the initial integrity and void content of a manufactured part and to subsequently follow the progression of damage resulting from environmental loading.

An A-scan display for immersion testing of a plate containing a flaw is shown in Fig. 12-21(a). The test material is a 25 mm (1 in.) thick aluminum alloy plate containing a flaw at a depth of 11 mm (0.44 in.), or approximately 45% of the thickness of the plate. In this instance, straight-beam immersion testing was done in a water-filled tank. A normal oscilloscope display is shown in Fig. 12-22(b). Any changes to the normal signals that are picked up as parts are scanned will show on the oscilloscope screen. Note the differences in the length of the signals

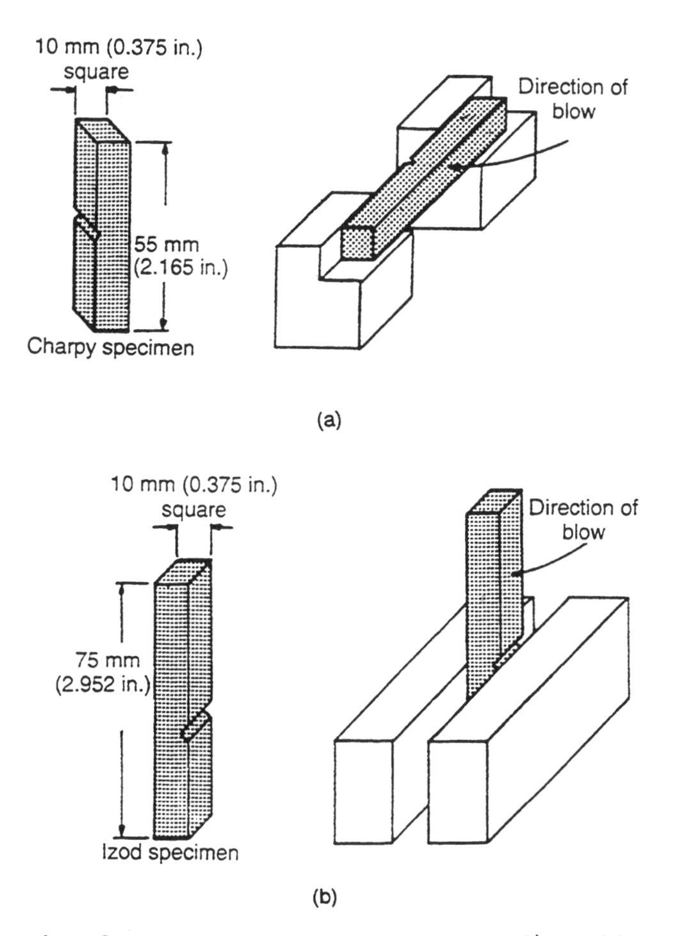

Fig. 12-15 Two major impact test specimens: Charpy (a) and Izod (b)

and that the length of the signal identifying the flaw is shorter than the signal to its left.

Advantages/Disadvantages. Ultrasonic inspection has a number of advantages, including:

- Flaws deep in a part can be detected. Inspection to depths of several feet is done routinely.
- Due to the high sensitivity of the process, it is possible to detect extremely small flaws.
- Positions of internal flaws can be determined with great accuracy, facilitating the estimation of flaw sizes, their nature, orientation, and shape.
- There is absolutely no exposure to radiation.

Major limitations include:

- Operators must be well trained and experienced.
- Parts that are rough, irregular in shape, very small or thin, or not homogeneous are difficult to inspect.
- Contact with the part surface or liquid or semiliquid couplants is necessary in order to provide effective transfer of the ultrasonic beam between search units and the part being inspected.
- Reference standards are required, both for operation of the equipment and for characterization of the flaws.

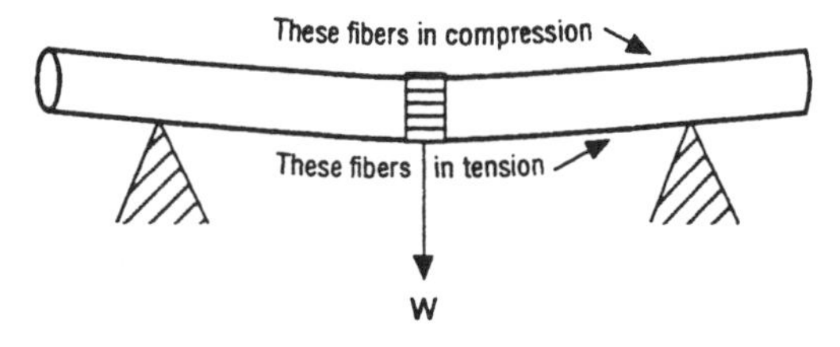

Fig. 12-16 Effect of loading on distribution of stress

Eddy Current Inspection

In the eddy current method, the part to be inspected is placed within, or adjacent to, an electric coil in which an alternating current flows. The alternating current, also known as the exciting current, causes eddy current to flow in the part as a result of electromagnetic induction.

Two common types of inspection coils and the patterns of eddy current flow generated by the exciting current in the coils are shown in Fig. 12-22. Figure 12-22(a) shows a solenoid-type coil used in the inspection of cylindrical and tubular parts. A *solenoid* is a tubular coil that produces a magnetic field. Figure 12-22(b) shows a pancake-type coil used for testing flat surfaces.

The nature of current flow is determined by the electrical characteristics of a part, the presence or absence of flaws or other discontinuities in the part, and the total electromagnetic field within the part. A flaw such as a crack in a pipe causes a change in the characteristics of the flow of the current.

As shown in Fig. 12-23, a pipe travels along the length of the inspection coil. In section A-A, it is apparent that a crack is not present because the flow of eddy current (indicated by arrows) was not impeded, or is symmetrical. In section B-B, however, the eddy current flow is not symmetrical; this is because it was impeded, causing it to change in pattern. Such changes cause changes in the associated electromagnetic field that can be detected and measured by instrumentation.

Applications. Eddy current inspection is used to identify differences in various properties of metals and metal parts. Examples include:

- Sorting dissimilar metals
- Measuring the thickness of a nonconductive coating on a conductive metal, or the thick-

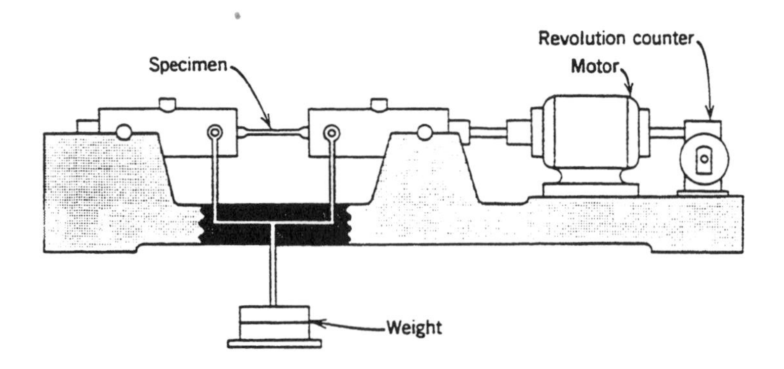

Fig. 12-17 Typical rotating-beam fatigue testing machine

ness of a nonmagnetic metal coating on a magnetic metal
- Detecting seams, laps, cracks, voids, and inclusions on or near the surface of a part

Advantages/Disadvantages. Main advantages include:

- Electrical contact with the part is not required.
- High-speed inspection is a suitable application.

Main disadvantages include:

- Accurate interpretation of instrument readings requires special skills and long experience.
- Instrument readings are affected by many properties of the material being inspected, other than flaws. Correlations between instrument readings and characteristics of the part being inspected must be carefully established and maintained.

Metallographic Examination

The metallurgist, armed with a microscope seeking clues to the properties of metals and their alloys, can be compared to the forensic expert. He knows, for example, that the properties of wrought materials in the longitudinal direction can vary considerably from those in the transverse direction, and that properties can vary considerably from the surface to the center of a cold-formed product.

Metallography deals with the structure of metals and alloys as revealed by the microscope. Knowledge of the relationship between the microstructure of a material and its chemical, physical, and mechanical properties is one of the main tools of the metallurgist. Microstructure is closely related to these properties.

Key steps in the process include:

- Selection of sample and area of interest
- Sectioning
- Mounting
- Grinding, polishing, and etching
- Examination under the microscope

Selection. Quality control procedures usually specify the number of samples to be examined and the stage of production and final inspection at which the examination should take place. These procedures generally also specify the area(s) of interest to be examined.

Sectioning. Because the structure of materials can be greatly altered by heat and by plastic deformation, considerable care must be taken in sectioning or cutting. The exercise of care extends to the selection of abrasive cutoff wheels and to the application of coolant. Ferrous materials usually are sectioned with aluminum oxide abrasives and nonferrous materials with silicon carbide abrasives. Coolants are generally water

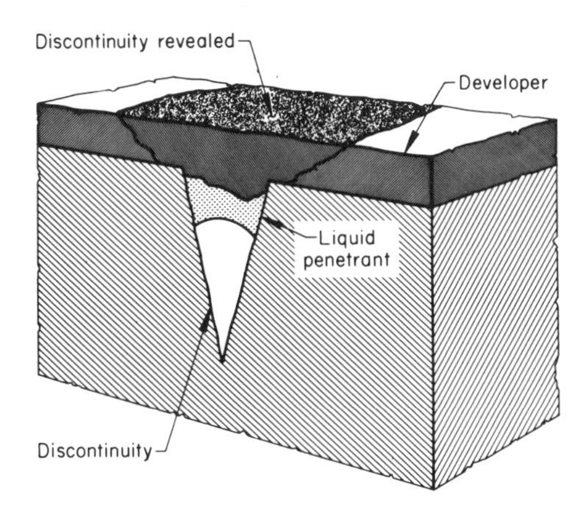

Fig. 12-18 Sectional view showing result of the migrating action of a liquid penetrant through a developer

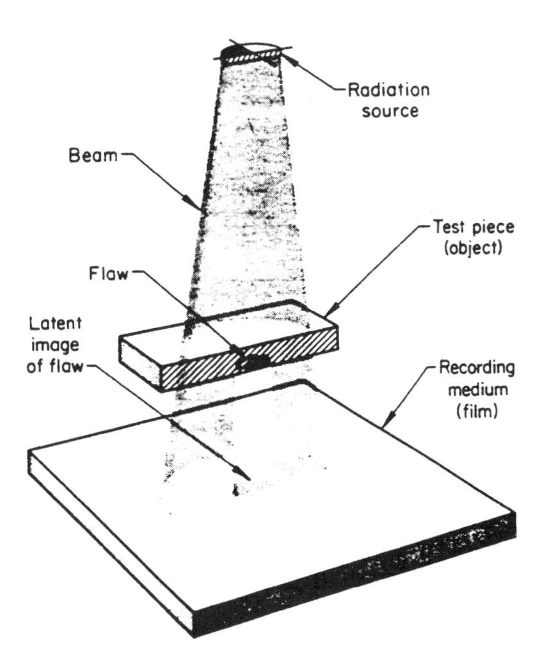

Fig. 12-19 Diagram of the basic elements of a radiographic system, showing method of detecting and recording an internal flaw in a plate of uniform thickness

based and contain additives such as soluble oil to minimize corrosion.

Mounting. In routine inspection, regularly shaped specimens are sometimes examined unmounted. However, most specimens are small and odd in shape, and are mounted in a mold with cold-mounting polymers such as acrylics, polyesters, or epoxies. Cold mounted specimens in an epoxy resin are shown in Fig. 12-24. Alternatively, some specimens may be mounted with polymers using a heat and press method.

Grinding, Polishing, and Etching. After mounting, specimens are ground in a series of steps using successively finer abrasives. Most polishing is done with cloths attached to rotating wheels and impregnated with a selected abrasive and lubricant. Etching is required to reveal the structure of the specimen. Many prepared etchants are available.

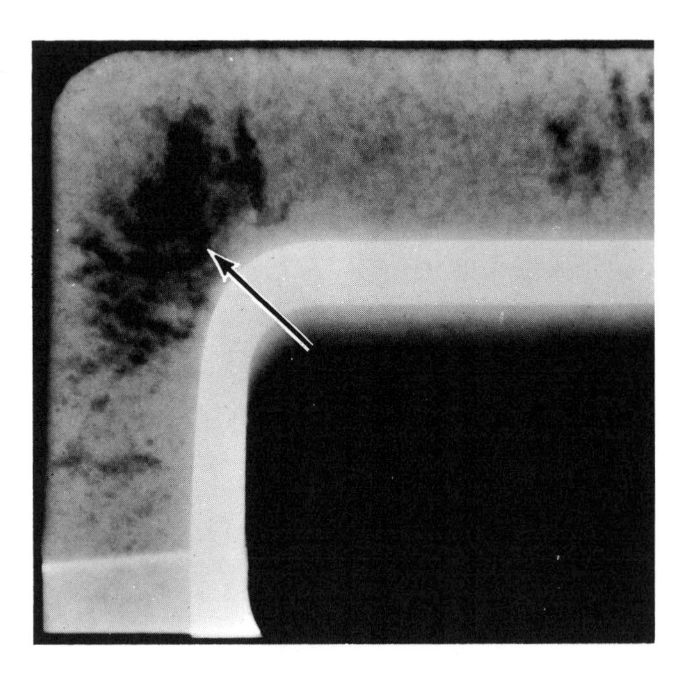

Fig. 12-20 Radiographic appearance of gross shrinkage porosity (arrow) in an aluminum alloy 319 manifold casting

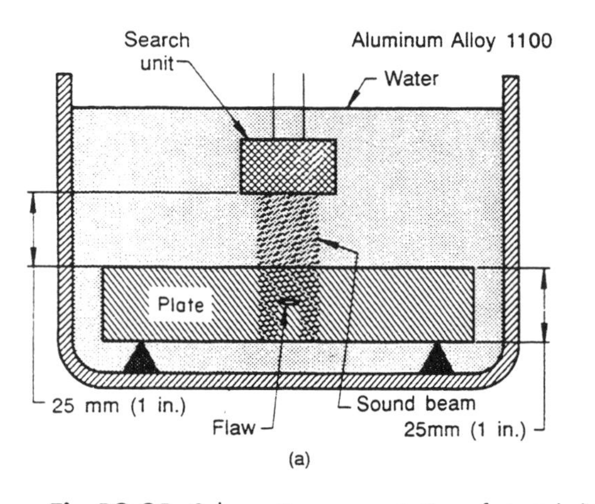

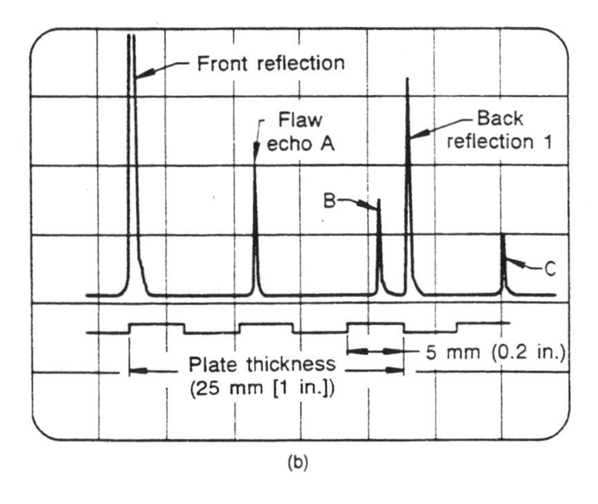

Fig 12-21 Schematic representation of straight-beam immersion inspection of a 25 mm (1 in.) thick aluminum alloy 1100 plate containing a planar discontinuity. (a) Inspection setup. (b) Normal oscilloscope display

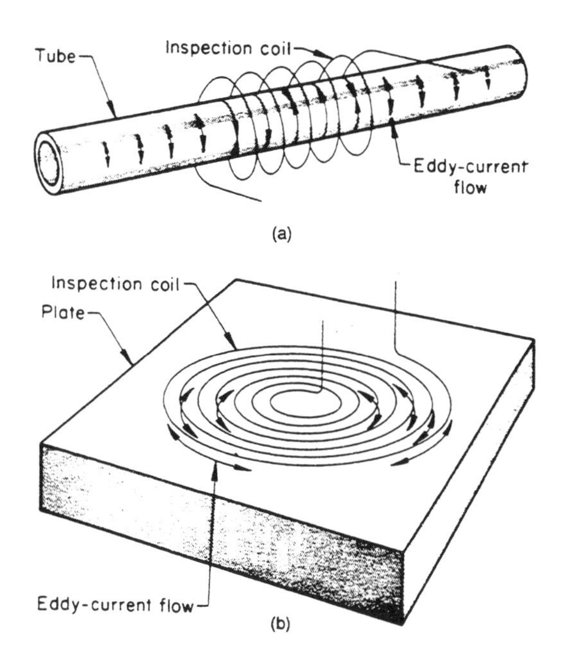

Fig. 12-22 Two common types of inspection coils and the patterns of eddy current flow generated by the exciting current in the coil. (a) Solenoid-type coil, applied to cylindrical or tubular parts. (b) Pancake-type coil, for flat surfaces

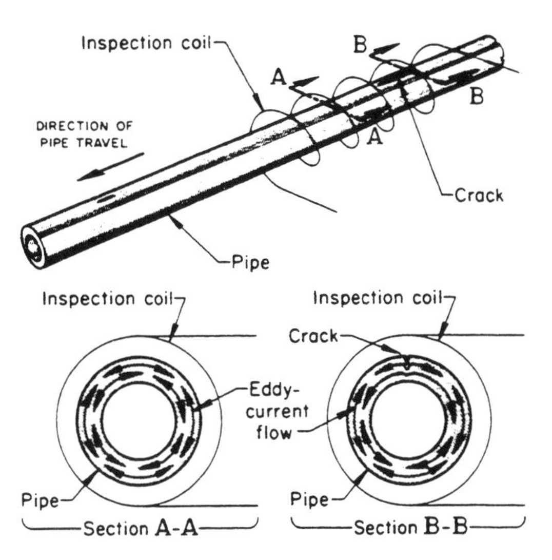

Fig. 12-23 Effect of a crack on the pattern of eddy current flow in a pipe

Fig. 12-24 Specimens cold mounted in epoxy resin

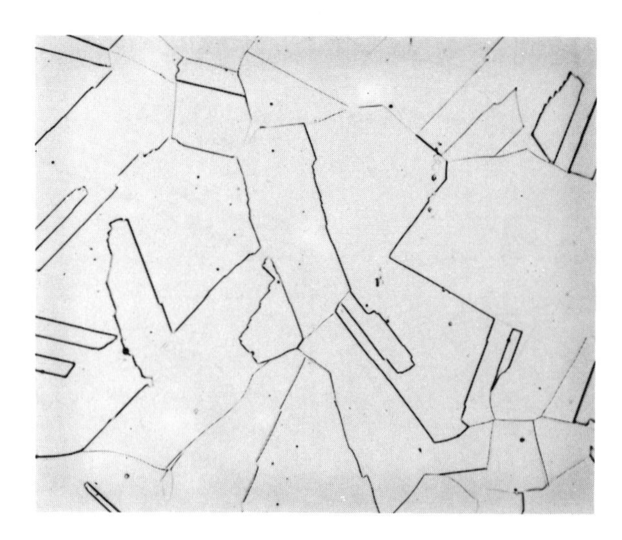

Fig. 12-25 Photomicrograph of type 304 austenitic stainless steel. 500×

Mechanical grinding and polishing cause some plastic deformation of the prepared surface, which can be eliminated by electrolytic metal removal (the reverse of plating). The result is a highly polished, scratch-free surface.

Examination. The equipment used to view the specimen under the microscope usually is referred to as a *metallograph*. The microscope depends on light and a lens system to reproduce the polished and etched specimen surface. The equipment includes a camera to make pictures—called photomicrographs, or micros for short—of the etched surfaces.

Photomicrographs provide a permanent record of the structure of a material and are published extensively in all types of documents. When used in conjunction with a microhardness test, a photomicrograph is useful in relating structure to mechanical properties. A micro of type 304 stainless steel is shown in Fig. 12-25.

Chapter 13

Progress by the Decade

The record of accomplishment in the science and application of metals and their alloys is notable in both scope and variety. Taken decade by decade, it is possible to trace a technology or product from its beginning to its status today. In fact, there is something for just about everyone.

1910–1920

- Around 1910 in France, Philipp Monnarts discovered the "stainlessness" of stainless steel. A fellow countryman, one Léon Guillet, reputedly had made the alloy a few years earlier.
- Charles Kettering of Delco (Dayton Electrical Laboratory Company) designed a small motor for the automobile self-starter. Earlier, as an engineer at National Cash Register, Kettering had developed a similar small motor that replaced the hand crank on the cash register. The self-starter was introduced in the 1912 model Cadillac. The idea for the device came from Henry Leland, then head of Cadillac. A friend had died as a result of injuries suffered in cranking a stalled Cadillac.
- Using x-rays, Max von Laue proved the existence of crystals in metals.
- Sir Ernest Rutherford proposed the theory of the atomic nucleus. Niels Bohr proposed the solar system model of the atom in 1913.
- A metal composite—a 42% nickel-iron core clad with 20% copper, called Du-Met—was developed by B.E. Eldred for use in sealing metal to soft glass. The process was to become the standard in fabricating lamps and radio tubes.
- The buffalo nickel, an alloy of 75% copper and 25% nickel, was introduced in the United States. It was replaced by the Jefferson nickel in 1938.
- British scientist Frederick Soddy deduced the existence of isotopes—atoms of an element with the same properties but with heavier-than-normal atoms.
- In 1913 Henry Ford experimented with the moving production line when demand for cars exceeded manufacturing capacity. The magneto line came first. Before that, each worker assembled all parts; the best pre-magneto production time was 18 minutes. On the moving line with 29 workers each handling one function, time was trimmed to 13 minutes. With refinements in the system, time was cut to 5 minutes.
- In 1914 the Union College Radio Club in the United States built a radio receiver using recently developed pilotron vacuum tubes. In July of that year, the radio picked up a German message announcing the start of World War I.
- During World War I, the name of German silver, a copper-nickel-zinc alloy, was changed to nickel silver. Major application: electroplating of silverware.
- The Panama Canal was opened in 1914. Its 500 electric motors operated locks, while 500 other electric motors were installed in other parts of the system, making it the largest electrical system in the world. Total horsepower was close to 30,000.
- Dr. Wheeler P. Davey of General Electric Company was the first to use x-rays to inspect steel castings. He presented a paper, "The Radiography of Metals," at a joint meeting in San Francisco of AIME and the American Electrochemical Society.
- The battleship USS *Arizona* contained about 345 gross tons (315 metric tons) of steel castings. Also, at the time, anchor chains for battleships were converted from forgings to castings. Single links, first cast

separately, were assembled in molds into which connecting links were poured.

- The Iron and Steel Institute of Japan was founded to promote the exchange of technological and economic information. Its first president was Kageyoshi Norro.
- The Chicago, Milwaukee & St. Paul railroad was electrified by General Electric, including 42 electric locomotives rated at 3000 volts and weighing 280 tons (255 metric tons) each. Braking problems on steep grades were reduced with regenerative brakes.
- In 1916, a two-way radio telephone linking Schenectady, New York, and Pittsfield, Massachusetts, was demonstrated. In the same year, a multiple tuned antenna was demonstrated. Its success suggested the possibility of a worldwide communication network.
- Heating and melting with high-frequency induction currents was invented by Edwin F. Northrup of Princeton University.
- An automatically controlled, continuous heat treating system combining hardening, quenching, and drawing was built by the Electric Furnace Company for the Sharon Works of the National Malleable and Steel Castings Company.
- In Germany, Professor Hugo Junkers built the J4, forerunner of the all-metal airplane. It had Duralumin cantilever median wings and tubular multispars joined by multiple sets of drawn sections that were supported by a thin corrugated sheet skin shaped to the wing profile.
- In 1917 riveting was still being used in ship construction in the United States. However, the British had started to use electric arc welding in place of rivets. Their practice was studied and brought back to the United States by a committee headed by Professor Comfort Adams of Harvard University.
- By 1917 stainless steel made by Stavanger Electro-Stalverk AS, in Norway, was being used in airplane engines. The alloy was forged into poppet valves by Hughes Johnson Stamping Company Ltd. in England.
- In 1917 General Electric Company went into limited production of home refrigerators at its plant in Ft. Wayne, Indiana.
- In 1918 the U.S. War Production Board ordered cutbacks in auto production to accommodate the production of trucks and other military vehicles. The automotive industry also produced shells, tractors, gun carriages, and aircraft engines. Ford made subchasers called Eagles. At this time, the White Sewing Machine Company got into the truck business.
- During World War I, the U.S. National Research Council was established by President Woodrow Wilson.
- Mass spectrometers that identify the presence of different elements by their signature wavelengths were built in 1918 by Arthur J. Dempster and in 1919 by William Ashton.
- Cast alloy iron parts for farm tractors were being heat treated by Holt Manufacturing Company.
- Directional magnetic properties of single crystals of silicon iron were determined by W.E. Ruder of General Electric Company.
- KLM, the Dutch airline, inaugurated passenger service between Amsterdam and London.
- Silicon-chromium stainless steel for car engine valves was patented by P.A.E. Armstrong.
- The formation of peer groups such as the Steel Treaters Club in Detroit, parent organization to the American Society for Metals and later ASM International, breached age-old barriers to the free exchange of technical knowledge.
- Igor Sikorsky built the first helicopter in Russia, his native land. "It was a very good machine," he recalled, "but it wouldn't fly." Flying came 40 years later. However, Sikorsky is credited with an important first: He built the world's first multiengine plane in 1913.
- A process was developed for coating steel with aluminum, which improved the resistance of the ferrous metal to oxidation at elevated temperatures.
- Steel-studded tires for dirt road service were available on the Corbin Model 30 roadster.
- SAE numbering, a standard numerical indexing system for steels, was established by the Society of Automotive Engineers.

1920–1930

- The hand-held vacuum cleaner for home use was introduced.
- Frank Whittle, an Englishman, conceived and patented the concept for the jet airplane engine. In Germany, development work on a jet engine was roughly concurrent. However, Whittle's progress was held up by a combination of bureaucracy and the need for better high-temperature alloys. By default, the Germans had the first jets to fly. But due to the very short service life of

available high-temperature metals, about 23 hours per engine, their success was limited.

- M.L. Fink and E.D. Campbell demonstrated the martensite body-centered tetragonal (bct) structure—information that is useful in heat treating. Austenite is transformed to martensite; this is notable for the high hardness martensite provides.
- C.L. Clark of the University of Michigan began pioneering studies into the high-temperature properties of materials.
- E.C. Wente developed a way of recording sound directly on film, a technology that led to talking pictures.
- Radar was discovered by A. Hoyt Taylor and Leo C. Young of the U.S. Naval Research Laboratory.
- Radio broadcasting was pioneered in the United States by station KDKA in Pittsburgh.
- X-ray equipment for dental and portable usage was made possible by developments in vacuum tubes and transformers by Coolidge.
- Aluminum forging alloys were developed by R.S. Archer and Zay Jeffries.
- First home television reception took place in the residence of E.F.W. Alexanderson, one of the pioneers in this technology.
- Charles A. Lindberg, Jr., flew nonstop from New York to Paris in a Ryan airplane equipped with a metal tubing fuselage, wood and canvas wings, and a Wright whirlwind engine.
- The Bell System demonstrated long-distance TV transmission in the United States.
- *Principles of Heat Treatment,* one of the first books on heat treating, was published by the metallurgical staff at the U.S. National Bureau of Standards.
- Experimental TV broadcasts were made in the United States and in England.
- United Airlines announced 30-hour service between New York and Los Angeles.
- An artificial larynx was developed by Bell Labs.
- Desk and table-model hearing aids were introduced by the Bell System.
- Magnetic particle testing was invented by William E. Hoke, making it possible to inspect the quality of metal parts without damaging them.
- Transcontinental airmail service was inaugurated in the United States.
- The first continuous mill for rolling sheet steel, the opening wedge to volume production of end products ranging from cars to refrigerators, was developed by John Butler Titus.
- The portable electrocardiograph was developed by H.B. Marvin.
- Chromium plating became a commercial process, due to the work of Colin G. Fink, Marvin J. Udy, Horace E. Haring, and William Blum.
- The first liquid-fueled rocket was launched by Robert H. Goddard.
- The first authoritative book on stainless steel was written by J.H.G. Moneypenny, an English metallurgist.
- In Germany, Fry obtained a patent on nitriding, a heat treating process.
- The Monitor refrigerator was introduced by General Electric.
- Pure columbium metal (now known as niobium) was introduced commercially and was to become a widely used alloying agent in high-performance materials.

1930–1940

- Pressure die casting of copper-base alloys reached commercial status.
- Stainless steel gained in popularity, being used in structures ranging from airplane wings to soda fountain equipment.
- Aluminum was used in collapsible tubes containing products such as face cream.
- Oil-burning furnaces and year-round room air conditioners for the home were introduced.
- Russians published a nine volume encyclopedia of space travel titled *Interplanetary Communications* (1928–1932).
- The first stainless steel railroad cars were built in the United States by the Budd Company.
- All-metal airplanes were flying coast to coast in 18 hours.
- Mobile two-way radios were developed and installed in police cars throughout the United States.
- The Alnico permanent magnet alloy (aluminum-nickel-cobalt-iron) was patented by W.E. Ruder.
- In Germany, Wernher von Braun built two liquid-fueled rockets—forerunners of the V-1 Buzz Bombs and V-2 rockets used by Germany in World War II.
- Night baseball was introduced in 1935 in a game between the Reds and the Phillies in Cincinnati, Ohio.
- General Electric introduced the first electric food disposer.
- The electric blanket was invented by W.K. Kearsley.
- A collection of equilibrium phase diagrams for ferrous metals and alloys was published in Germany by Max Hanson. He later immi-

grated to the United States and headed the department of metallurgy at Armour Research Foundation in Chicago, now the Illinois Institute of Technology Research Institute.

- Howard Hughes set a transcontinental speed record for airplanes: 7 hours, 38 minutes, and 25 seconds.
- The automatic clutch and automated shifting for cars were introduced.
- A thin metallic tape was used in early recording devices.
- In France, Michelin developed steel-belted tires for cars and trucks.
- Steel was replacing wood in golf club shafts. Aluminum was replacing wood in beer barrels.
- Robotic welding was used in the fabrication of large assemblies.
- General Electric broadcast TV programs from the New York World's Fair.

1940–1950

- Magnesium parts were used in trucks, buses, airplanes, conveyors, vacuum cleaners, and typewriters.
- A Ford farm tractor model weighing 2100 lb (950 kg) contained 1155 lb (525 kg) of castings—gray iron, malleable iron, steel, and nonferrous types.
- In Italy, aluminum was replacing copper in bare electrical conductor applications.
- Robert H. Goddard began to develop military-type rockets. He died in 1945.
- Nimonic 73, a nickel-chromium-base alloy for high-temperature service in jet engines, was patented in England.
- The first electron photomicrographs of steel structures were made by R.F. Mehl, E.C. Rambero, and R.F. Baker.
- Window-mounted air conditioners were introduced.
- A process for extracting magnesium from seawater was perfected by Dow Chemical.
- A 1.4 million volt x-ray machine was installed at the U.S. Bureau of Standards.
- Plutonium was made by bombarding uranium atoms in a cyclotron.
- The autogyro, a precursor of the helicopter, was introduced.
- Research was started at Battelle Memorial Institute on a photographic copying process that was to become known as Xerox.
- The first commercial use of radar was demonstrated at General Electric.
- Lead-sheathed coaxial cable between New York and Philadelphia carried 480 telephone messages simultaneously.
- The four-engine Lockheed Constellation, a propjet, carried 60 passengers at a top speed of 300 mph (480 km/h).
- An aluminum-wood rowboat was stamped from a single sheet of 0.064 in. (1.6 mm) gage metal. It weighed 150 lb (68 kg), minus the motor.
- Zippers, called slide fasteners at the time, were being made from nickel silver wire.
- Zirconium reached commercial status.
- International Nickel developed a 9% nickel steel alloy designed for service at low temperatures.
- The first microwave relay system transmitted TV programs from New York City to Schenectady.
- The transistor was invented by John Bardeen, W.H. Brattain, and William Shockley at Bell Labs. It replaced bulky vacuum tubes and opened the door to the high-tech world of computer chips, integrated circuits, and other building blocks for today's magic world of electronics.
- Ductile iron, a casting alloy with steel-like properties, reached commercial status.
- Production in the U.S. auto industry reached 100 million vehicles (since 1900), including passenger cars, trucks, and buses. Between 1940 and 1950 alone, around 40 million vehicles were registered and operating.
- Knolls Atomic Power Laboratory began using digital computers in the design of naval propulsion reactors.
- Blades for jet airplane engines were being made by investment casting, because the alloys in use could not be shaped with traditional technology such as machining and forging.
- Powder metal was used in "talking tape" for sound recorders in devices for radios and TV sets.
- For the first time in the United States, steel was continuously cast, bypassing such traditional operations as rolling.
- The shell molding foundry process was patented by Fiat in Italy, replacing the traditional sand mold technology.
- DuPont announced the first commercial production of titanium using the Kroll process. The material was to become known as the "middleweight champion," because it was both strong and fairly lightweight, weighing less than competing metals.
- Due to its resistance to corrosion, light weight, and appearance, aluminum was being used in a variety of kitchen appliances—ranging from coffeepots to pans to trays for refrigerators.

1950–1960

- General Electric developed a radar system for commercial airports.
- At Calder Hall in England, construction began on the world's first full-scale nuclear power station.
- The solar battery, said to be six times more powerful than the photocell, was invented, as was the atomic battery.
- The first nuclear submarine, USS *Nautilus,* was launched in the United States.
- In France, Stohr invented the electron beam welding machine.
- The Niagara Power System distributed America's first commercial nuclear power from a plant located in New York state.
- Numerical control of machine tools was demonstrated at the Chicago Machine Tool Show.
- The first portable televisions and radios with transistorized circuitry were introduced.
- Two man-made elements—No. 99, Einsteinium (named for Albert Einstein), and No. 100, Fermium (named for Enrico Fermi)—were produced by teams at the University of California and Argonne National Laboratories, respectively.
- General Electric produced synthetic diamonds by simulating temperature and pressure conditions present in nature.
- Copper-clad laminates for printed circuit boards were developed for radio and television applications.
- The first transatlantic telephone cable went into operation. Thirty-six static-free circuits replaced the erratic radiotelephone system.
- The Russians put *Sputnik I* into orbit—launching the Space Age. On the East Coast of the United States, it was reported that garage doors operating on the same frequency as *Sputnik* opened each time the satellite orbited over the area.
- Punched tape input was developed for numerical control of machine tools.
- The first economical steam iron was introduced.
- Computer control of chemical processes gained in acceptance.
- The extra-strong maraging steels were discovered. Early applications included landing gear for airplanes, and rams used in forging and extrusion operations.
- Power steering and power brakes became options on cars.
- Bell Labs developed programming languages for computers.

1960–1970

- *Discoverer XIII* was the first man-made object to be recovered from an orbiting space vehicle.
- The Van Allen radiation belt was explored by NERVA.
- Computers suitable for business purposes and for the processing of scientific data became more affordable.
- Explosive welding was developed in the United States.
- Cemented carbide balls for ballpoint pen tips were developed.
- Long-term research involving fuel cell technology was in its opening phases. In this process, chemical energy is converted directly into electrical energy.
- The X-15, an experimental U.S. airplane, set the air speed record for manned vehicles at 2905 mph (4675 km/h). Titanium was used extensively in the construction of the X-15.
- The Nimbus, an earth-oriented meteorological satellite, was developed.
- The United States successfully launched Tyros, the first weather satellite.
- An electronic system was developed to monitor critical-care patients in hospitals.
- Automatic dial paging systems were developed for both hospital and industrial use.
- Closed-circuit television for viewing medical fluoroscopy was introduced.
- Cable television made its debut.
- Machining centers, which both centralized and improved the efficiency of manufacturing operations, were made possible by combining integrated circuitry and numerical control.
- The U.S. Air Force acquired its first airborne laser ranging system. The operation of this system did not require a vacuum, a requirement for other laser systems.
- Applications for superconductivity technology began to make their appearance. At the required extremely low temperatures, which are close to absolute zero (–273.1 °C, or –459.6 °F), resistance of metals to the passage of electricity is practically eliminated, and conductivity is increased manyfold.
- In 1967 applications soared for the construction of nuclear powerplants in the United States.
- Pontiac introduced the energy-absorbing bumper for cars, a urethane-steel composite.
- Engineers at Chrysler began using computers for selecting and specifying materials.

- The U.S. Bureau of Mines launched studies to determine the feasibility of undersea mining.
- Taking advantage of state-of-the art electronics, the first in a series of high-speed teleprinters capable of 30 characters per second was developed.
- One of the first wide-bodied airplanes, the McDonnell Douglas DC-10, was introduced. Its engines had 40,000 lb (18,150 kg) of thrust.
- Worldwide interest grew in desalination, the conversion of seawater into potable water.
- Four acres of ¼ in. (6.4 mm) stainless steel plates were used to face the St. Louis Memorial Arch.

1970–1980

- An emerging trend in heat treating was greater control and total automation of equipment through use of the computer.
- By 1971 in the United States, the much publicized supersonic transport (SST) had become an "iffy" project.
- Duplexing, a metal-producing process in which a charge is melted in one furnace and refined in another, was being adopted as a way of improving the quality of steels, particularly alloys.
- Japanese foundries with the capability of producing large structures were exporting cast bridge decks and tunnel segments to the United States.
- Photoforming, a photographic-chemical etching process, gained acceptance as a way of making thin-foil metal parts for electronic equipment, aerospace devices, instruments, controls, heads for electric shavers, and so on.
- Induction heat treating systems were being moved into production lines, along with other processes and testing equipment, to improve productivity and quality while lowering manufacturing costs.
- U.S. steelmakers did not have the capacity needed to make 48 in. (122 cm) pipe for the Trans-Alaska Pipeline. The pipe, made from high-strength steel containing copper, chromium, niobium, and vanadium, was supplied by Nippon Steel, a Japanese steelmaker.
- Use of fracture mechanics was becoming common in design and failure analysis. Fracture mechanics allows for the evaluation of structural behavior in terms of applied stress, crack length, and geometry of a part or specimen.
- Automakers began to consider greater use of coated steels, particularly galvanized steel, to provide greater protection against corrosion.
- High-strength low-alloy (HSLA) steels were used to reduce the weight of safety bumpers dictated by law in the United States. Because it has higher strength than the steels previously used, less HSLA steel is needed for a given application.
- Beginning in 1973, titanium honeycomb structures for the B-1 bomber and the Space Shuttle were joined by diffusion bonding. In this process, adjoining surfaces are joined by the application of pressure at elevated temperatures.
- It was found that bubbles in amorphous films of gadolinium-cobalt could serve as magnetic memories for computers.
- During the first half of the 1970s, worldwide shortages were experienced for both materials and energy. Metallurgists were hard pressed to find ways of reducing the use of alloying elements and/or finding substitute materials to get the job done. However, chronic energy shortages spurred work on alternative sources such as coal gasification and geothermal energy.
- With the help of computer-aided design, or CAD, numerical control tapes were being made directly from drawings.
- Tungsten wire of less than 0.0005 in. (0.013 mm) in diameter was being used in windshield defoggers and deicers for cars and in memories for aircraft and space vehicle minicomputers.
- By 1974 in the United States, 49 steel plants, including minimills, had continuous casting facilities. Minimills are small steel mills that specialize in limited product lines and generally are located in areas not served by the major mills.
- Niobium was used in reusable rocket nozzles for the Space Shuttle orbiter.
- A new ferromagnetic material known as metallic glass was produced with rapid solidification technology. The secret is cooling at a rate of 1 million °C (1.8 million °F) per second.
- Nonaerospace markets for titanium were beginning to open up—for example, chemical plant heat exchangers.
- An automaker in Detroit replaced traditional cast iron intake manifolds for its small cars with die-cast aluminum alloy intake manifolds in order to cut down the weight.
- General Electric developed the computer-aided tomography (CAT) scanner to take detailed, cross-sectional x-ray pictures of parts of the human body in less than 5 seconds.

- Titanium aircraft parts normally machined from heavy plate or built up from a number of parts were being superplastically formed from one piece of sheet. This process takes advantage of the unusually high tensile elongation properties of the material.
- Forging of powder metal parts became a commercial reality following years of great promise. Characteristic of this process is that forging improves properties largely through closure of pores. Dies are heated. Properties approximate those of wrought metal parts.
- Laser scanning was used to automate magnetic and fluorescent particle testing.
- Acoustic emission technology was used to monitor nuclear reactor vessels for potential failures. A warning signal is given prior to failure, based on the principle that a limb on a tree makes a cracking sound before it breaks.
- Two-piece drawn and ironed aluminum cans for beer and other beverages were beginning to compete strongly with steel cans. Recycling of aluminum cans as a source of raw material was also catching on.
- A technique called ion implantation was being used to modify the surface properties of metals for electronic applications.
- Window frames and spaces for the inner cabin of the Space Shuttle were made from S-65, a beryllium alloy. Although this metal was of great interest because it is light and extremely hard, its use had been limited due to its brittleness.
- A steelmaking process of German origin was being used to improve toughness and ductility by changing the shape of sulfide inclusions from stringy to nodular form.
- The first U.S. interplanetary mission was launched to investigate the atmosphere and weather of Venus.
- Process simulation, a new function for computers, was beginning to be used by producers of both ferrous and nonferrous metals.
- Shape-memory alloys, copper-aluminum-zinc, were used in thermally activated devices such as constant-temperature controls for bathroom showers and systems for opening and closing greenhouse windows. An alloy of this type can be plastically deformed at a low temperature and upon exposure to a higher temperature will return to its original shape.

1980–1990

- Catalytic converters in auto emission-control systems had become the largest single user of platinum metals.
- Aluminum continued to be the dominant airframe material in commercial airplanes.
- Progress was reported in developmental work on the Tokamak fusion reactor, hoped-for successor to the current atomic reactor.
- A plastic-steel-plastic composite sandwich, 20% steel and 60% polypropylene, was developed as an alternative to standard sheet steel.
- Because of their antifouling properties, copper-nickel alloy panels were being used in cages for fish, oyster, and clam farming.
- Corrosion barrier coatings, said to be comparable in performance to cadmium plating, now banned by pollution-control regulations, were being applied by a process called iron vapor deposition.
- Vikram Sarabhai Space Center in India, developed a method of diffusion bonding titanium to stainless steel. This process was of special interest to those involved in the fabrication of space vehicles because less metal was needed with no sacrifice in strength versus steel alone.
- Wheel rims and spokes for cars were manufactured from HSLA steels to save weight.
- The French steelmaker Creusot-Loire began using clean steel technology, or ladle refining, to improve properties of 9% nickel steel in liquefied natural gas storage applications.
- Hot isostatic pressing was used to reclaim castings that did not pass radiographic inspection. Properties were improved by closing pores in the metal.
- In Japan, as an energy-saving measure, heat retained in forgings was being used to heat treat parts.
- In Sweden, iron ore was directly reduced to iron with the use of a plasma generator system, thereby bypassing the blast furnace step in steelmaking.
- Honda was using fiber-reinforced metal connecting rods in the engine that powered their vehicle designed for use in the city.
- A manufacturer of aircraft engine parts was controlling a group of more than 40 individual heat treating furnaces through a single computer.
- Allegheny Ludlum Steel developed a process for continuous casting of coilable stainless steel strip directly from molten metal, bypassing conventional rolling and associated operations.
- The U.S. Can Company laser welded tinplated steel aerosol cans at the rate of 215 units a minute.
- Technetium, a product of uranium fission in atomic reactors, was found to have a crystal

structure favorable for superconductivity. A mixture of ruthenium and molybdenum has the same crystal structure and becomes superconductive at 10.6 degrees kelvin, the equivalent of –263 °C (–441 °F).

- In France, Peugeot-Citroen was surface hardening shock absorber rods made of AISI 1050 steel with the ion nitriding process at a rate of 1500 units every 3.5 hours.

In closing, the following statement (made circa 1985) keynotes the present decade of 1990 to 2000, as well as the previous one:

"Today, the electronic media using satellites in space to relay microwaves have knitted the world community together as it has never been before. Events in Europe and Asia are instantly known and seen in every city, town, and suburb in America. . . ."

Glossary

A

abrasion. (1) A process in which hard particles or protuberances are forced against and moved along a solid surface. (2) A roughening or scratching of a surface due to *abrasive wear.* (3) The process of grinding or wearing away through the use of abrasives.

abrasive wear. The removal of material from a surface when hard particles slide or roll across the surface under pressure. The particles may be loose or may be part of another surface in contact with the surface being abraded.

absorption. (1) The taking up of a liquid or gas by capillary, osmotic, or solvent action. (2) The capacity of a solid to receive and retain a substance, usually a liquid or gas, with the formation of an apparently homogeneous mixture. (3) Transformation of radiant energy to a different form of energy by interaction with matter. (4) The process by which a liquid is drawn into and tends to fill permeable pores in a porous solid body; also, the increase in mass of a porous solid body resulting from the penetration of a liquid into its permeable pores.

accelerated aging. A process by which the effects of aging are accelerated under extreme and/or cycling temperature and humidity conditions. The process is meant to duplicate long-time environmental conditions in a relatively short space of time.

accuracy. (1) The agreement or correspondence between an experimentally determined value and an accepted reference value for the material undergoing testing. The reference value may be established by an accepted standard (such as those established by ASTM), or in some cases the average value obtained by applying the test method to all the sampling units in a lot or batch of the material may be used. (2) The extent to which the result of a calculation or the reading of an instrument approaches the true value of the calculated or measured quantity.

adhere. To cause two surfaces to be held together by *adhesion.*

adhesion. (1) In frictional contacts, the attractive force between adjacent surfaces. In physical chemistry, adhesion denotes the attraction between a solid surface and a second (liquid or solid) phase. This definition is based on the assumption of a reversible equilibrium. In mechanical technology, adhesion is generally irreversible. In railway engineering, adhesion often means friction. (2) Force of attraction between the molecules (or atoms) of two different phases. (3) The state in which two surfaces are held together by interfacial forces, which may consist of valence forces, interlocking action, or both.

adhesive wear. (1) Wear by transference of material from one surface to another during relative motion due to a process of solid-phase welding. Particles that are removed from one surface are either permanently or temporarily attached to the other surface. (2) Wear due to localized bonding between contacting solid surfaces, leading to material transfer between the two surfaces or loss from either surface.

adsorption. The adhesion of the molecules of gases, dissolved substances, or liquids in more or less concentrated form, to the surfaces of solids or liquids with which they are in contact. The concentration of a substance at a surface or interface of another substance.

age hardening. Hardening by *aging* (heat treatment) usually after rapid cooling or cold working.

aging. (1) The effect on materials of exposure to an environment for a prolonged interval of time. (2) The process of exposing materials to an environment for a prolonged interval of time in order to predict in-service lifetime. (3) Generally, the degradation of properties or function with time. In capacitors, the loss of dielectric constant, K, by dielectric relaxation. Expressed as a percent change per decade of time.

aging (heat treatment). A change in the properties of certain metals and alloys that occurs at ambient or moderately elevated temperatures after hot working or a heat treatment (quench aging in ferrous alloys, natural or artificial aging in ferrous and nonferrous alloys) or after a cold working operation (strain aging). The change in properties is often, but not always, due to a phase change (precipitation), but never involves a change in chemical composition of the metal or alloy.

air-hardening steel. A steel containing sufficient carbon and other alloying elements to harden fully during cooling in air or other gaseous media from a temperature above its transformation range. The term should be restricted to steels that are capable of being hardened by cooling in air in fairly large sections, about 50 mm (2 in.) or more in diameter. Same as self-hardening steel.

alclad. Composite wrought product comprised of an aluminum alloy core having on one or both surfaces a metallurgically bonded aluminum or aluminum alloy coating that is anodic to the core and thus electrochemically protects the core against corrosion.

allotropy. (1) A near synonym for polymorphism. Allotropy is generally restricted to describing polymorphic behavior in elements, terminal phases, and alloys whose behavior closely parallels that of the predominant constituent element. (2) The existence of a substance, especially an element, in two or more physical states (for example, crystals).

alloying element. An element added to and remaining in a metal that changes structure and properties.

alloy steel. Steel containing specified quantities of alloying elements (other than carbon and the commonly accepted amounts of manganese, copper, silicon, sulfur, and phosphorus) within the limits recognized for constructional alloy steels, added to effect changes in mechanical or physical properties.

alpha prime (α') (hexagonal martensite). A supersaturated, nonequilibrium hexagonal α (alpha) phase formed by a diffusionless transformation of the β (beta) phase in titanium alloys. It is often difficult to distinguish from acicular α alpha, although the latter is usually less well defined and frequently has curved, instead of straight, sides.

Alumel. A nickel-base alloy containing about 2.5% Mn, 2% Al, and 1% Si used chiefly as a component of pyrometric thermocouples.

Aluminizing. Forming of an aluminum or aluminum alloy coating on a metal by hot dipping, hot spraying, or diffusion.

amorphous. Not having a crystal structure; noncrystalline.

annealing (metals). A generic term denoting a treatment consisting of heating to and holding at a suitable temperature followed by cooling at a suitable rate, used primarily to soften metallic materials, but also to simultaneously produce desired changes in other properties or in microstructure. The purpose of such changes may be, but is not confined to: improvement of machinability, facilitation of cold work, improvement of mechanical or electrical properties, and/or increase in stability of dimensions. When the term is used unqualifiedly, full annealing is implied. When applied only for the relief of stress, the process is properly called *stress relieving* or stress-relief annealing.

In ferrous alloys, annealing usually is done above the upper critical temperature, but the time-temperature cycles vary widely both in maximum temperature attained and in cooling rate employed, depending on composition, material condition, and results desired. When applicable, the following commercial process names should be used: black annealing, blue annealing, *box annealing, bright annealing,* cycle annealing, *flame annealing, full annealing,* graphiticizing, in-process annealing, isothermal annealing, malleabilizing, orientation annealing, process annealing, quench annealing, *spheroidizing*, subcritical annealing.

In nonferrous alloys, annealing cycles are designed to: (a) remove part or all of the effects of cold working (recrystallization may or may not be involved); (b) cause substantially complete coalescence of precipitates from solid solution in relatively coarse forms; or (c) both, depending on composition and material condition. Specific process names in commercial use are *anneal to temper,* final annealing, *full annealing,* intermediate annealing, *partial annealing, recrystallization annealing,* and stress-relief annealing.

anneal to temper (metals). A final partial anneal that softens a cold-worked nonferrous alloy to a specified level of hardness or tensile strength.

anode. (1) The electrode of an electrolyte cell at which oxidation occurs. Electrons flow away from the anode in the external circuit. It is usually at the electrode that corrosion occurs and metal ions enter solution. (2) The positive (electron-deficient) electrode in an electrochemical circuit.

anodic protection. (1) A technique to reduce the corrosion rate of a metal by polarizing it into its passive region, where dissolution rates are low. (2) Imposing an external electrical potential to protect a metal from corrosive attack. (Applicable only to metals that show active-passive behavior.) Contrast with *cathodic protection.*

arc welding gun. A device used in semiautomatic, machine, and automatic arc welding to transfer current, guide the consumable electrode, and direct shielding gas when used.

artifact. A feature of artificial character, such as a scratch or a piece of dust on a metallographic specimen, that can be erroneously interpreted as a real feature.

artificial aging (heat treatment). Aging above room temperature.

asperity. In tribology, a protuberance in the small-scale topographical irregularities of a solid surface.

as-welded. The condition of weld metal, welded joints, and weldments after welding but prior to any subsequent thermal, mechanical, or chemical treatments.

athermal. Not isothermal. Changing rather than constant temperature conditions.

athermal transformation. A reaction that proceeds without benefit of thermal fluctuations—that is, thermal activation is not required. Such reactions are diffusionless and can take place with great speed when the driving force is sufficiently high. For example, many martensitic transformations occur athermally on cooling, even at relatively low temperatures, because of the progressively increasing drive force. In contrast, a reaction that occurs at constant temperature is an *isothermal transformation;* thermal activation is necessary in this case and the reaction proceeds as a function of time.

atmospheric corrosion. The gradual degradation or alteration of a material by contact with substances present in the atmosphere, such as oxygen, carbon dioxide, water vapor, and sulfur and chlorine compounds.

atom. The smallest particle of an element that retains the characteristic properties and behavior of the element.

austempered ductile iron. A moderately alloyed ductile iron that is austempered for high strength with appreciable ductility.

austempering. A heat treatment for ferrous alloys in which a part is quenched from the austenitizing temperature at a rate fast enough to avoid formation of ferrite or pearlite and then held at a temperature just above M_s until transformation to bainite is complete. Although designated as bainite in both austempered steel and austempered ductile iron (ADI), austempered steel consists of two phase mixtures containing ferrite and carbide, while austempered ductile iron consists of two phase mixtures containing ferrite and austenite.

austenite. A solid solution of one or more elements in face-centered cubic iron (gamma iron). Unless otherwise designated (such as nickel austenite), the solute is generally assumed to be carbon.

austenite steel. An alloy steel whose structure is normally austenitic at room temperature.

austenitizing. Forming austenite by heating a ferrous alloy into the transformation range (partial austenitizing) or above the transformation range (complete austenitizing). When used without qualification, the term implies complete austenitizing.

B

bainite. A metastable aggregate of *ferrite* and *cementite* resulting from the transformation of *austenite* at temperatures below the *pearlite* range but above M_s, the martensite start temperature. Upper bainite is an aggregate that contains parallel lath-shape units of ferrite and produces the so-called "feathery" appearance in optical microscopy.

bainite hardening. Quench-hardening treatment resulting principally in the formation of *bainite*.

bar. (1) A section hot rolled from a *billet* to a form, such as round, hexagonal, octagonal, square, or rectangular, with sharp or rounded corners or edges and a cross-sectional area of less than 105 cm^2 (16 $in.^2$). (2) A solid section that is long in relationship to its cross-sectional dimensions, having a completely symmetrical cross section and a width of greatest distance between parallel faces of 9.5 mm ($3/8$ in.) or more. (3) An obsolete unit of pressure equal to 100 kPa.

base metal. (1) The metal present in the largest proportion in an alloy; brass, for example, is a copper-base alloy. (2) The metal to be brazed, cut, soldered, or welded. (3) After welding, that part of the metal which was not melted. (4) A metal that readily oxidizes, or that dissolves to form ions.

basic oxygen furnace. A large tiltable vessel lined with basic refractory material; the principal type of furnace for modern steelmaking. After the furnace is charged with molten pig iron (which usually comprises 65 to 75% of the charge), scrap steel, and fluxes, a lance is brought down near the surface of the molten metal and a jet of high-velocity oxygen impinges on the metal. The oxygen reacts with carbon and other impurities in the steel to form liquid compounds that dissolve in the slag and gases that escape from the top of the vessel.

batch. A quantity of materials formed during the same process or in one continuous process and having identical characteristics throughout.

beach marks. Macroscopic progression marks on a fatigue fracture or stress-corrosion cracking surface that indicate successive positions of the advancing crack front. The classic appearance is of irregular elliptical or semielliptical rings, radiating outward from one or more origins. Beach marks (also known as clamshell marks or arrest marks) are typically found on service fractures where the part is loaded randomly, intermittently, or with periodic variations in mean stress or alternating stress.

bearing bronzes. Bronzes used for bearing applications. Two common types of bearing bronzes are copper-base alloys containing 5 to 20 wt% Sn and a small amount of phosphorus (phosphor bronzes) and copper-base alloys containing up to 10 wt% Sn and up to 30 wt% Pb (leaded bronzes).

bearing steels. *Alloy steels* used to produce rolling-element bearings. Typically, bearings have been manufactured from both high-carbon (1.00%) and low-carbon (0.20%) steels. The high-carbon steels are used in either a through-hardened or a surface induction-hardened condition. Low-carbon bearing steels are carburized to provide the necessary surface hardness, while maintaining desirable core properties.

bend test. A test for determining relative ductility of metal that is to be formed (usually sheet, strip, plate, or wire) and for determining soundness and toughness of metal (after welding, for example). The specimen is usually bent over a specified diameter through a specified angle for a specified number of cycles. There are four general types of bend test, named according to the manner in which the forces are applied to the specimen to make the bend: free bend, guided bend, semiguided bend, and wraparound bend.

beryllium-copper. Copper-base alloys containing not more than 3% Be. Available in both cast and wrought forms, these alloys rank high among copper alloys in attainable strength, while retaining useful levels of electrical and thermal conductivity. Applications for these alloys include electronic components (connector contacts), electrical equipment (switch and relay blades), antifriction bearings, housings for magnetic sensing devices, and resistance welding contacts.

beta (β). The high-temperature allotrope of titanium with a body-centered cubic crystal structure that occurs above the β transus.

beta annealing. Producing a beta phase by heating certain titanium alloys in the temperature range at which this phase forms, followed by cooling at an appropriate rate to prevent its decomposition.

billet. (1) A semifinished section that is hot rolled from a metal ingot, with a rectangular cross section usually ranging from 105 to 230 cm^2 (16 to 36 in.^{2}), the width being less than twice the thickness. Where the cross section exceeds 230 cm^2 (36 in.^{2}), the term *bloom* is properly but not universally used. Sizes smaller than 105 cm^2 (16. in.^{2}) are usually termed *bars.* (2) A solid semifinished round or square product that has been hot worked by forging, rolling, or extrusion.

blank. (1) In forming, a piece of sheet metal, produced in cutting dies, that is usually subjected to further press operations. (2) A pressed, presintered, or fully sintered powder metallurgy compact, usually in the unfinished condition and requiring cutting, machining, or some other operation to produce the final shape. (3) A piece of stock from which a forging is made, often called a *slug* or multiple. (4) Any article of glass on which subsequent forming or finishing is required.

blast furnace. A shaft furnace in which solid fuel is burned with an air blast to smelt ore in continuous operation. Where the temperature must be high, as in the production of pig iron, the air is preheated. Where the temperature can be lower, as in smelting of copper, lead, and tin ores, a smaller furnace is economical, and preheating of the blast is not required.

bloom. (1) A semifinished hot-rolled product, rectangular in cross section, produced on a blooming mill. See also *billet.* For steel, the width of a bloom is not more than twice the thickness, and the cross-sectional area is usually not less than about 230 cm^2 (36 in.^{2}). Steel blooms are sometimes made by forging. (2) A visible exudation or efflorescence on the surface of an electroplating bath. (3) A bluish fluorescent cast to a painted surface caused by deposition of a thin film of smoke, dust, or oil. (4) A loose, flowerlike corrosion product that forms when certain metals are exposed to a moist environment.

boriding. Thermochemical treatment involving the enrichment of the surface layer of an object with borides. This surface-hardening process is performed below the Ac_1 temperature. Also referred to as boronizing.

box annealing. Annealing a metal or alloy in a sealed container under conditions that minimize oxidation. In box annealing of a ferrous alloy, the charge is usually heated slowly to a temperature below the transformation range, but sometimes above or within it, and is then cooled slowly; this process is also called close annealing or pot annealing.

brale indenter. A conical 120° diamond indenter with a conical tip (a 0.2 mm, or 0.008 in., tip radius is typical) used in certain types of Rockwell and scratch hardness tests.

brass. A copper-zinc alloy containing up to 40% Zn, to which smaller amounts of other elements may be added.

braze. A weld produced by heating an assembly to suitable temperatures and by using a filler metal having a liquidus above 450 °C (840 °F) and below the solidus of the base metal. The filler metal is distributed between the closely fitted faying surfaces of the joint by capillary action.

brazeability. The capacity of a metal to be brazed under the fabrication conditions imposed into a specific suitably designed structure and to perform satisfactorily in the intended service.

braze welding. A method of welding by using a filler metal having a liquidus above 450 °C (840 °F) and below the solidus of the base metals. Unlike *brazing,* in braze welding the filler metal is not distributed in the joint by capillary action.

brazing. A group of welding processes that join solid materials together by heating them to a suitable temperature and using a filler metal having a liquidus above 450 °C (840 °F) and below the solidus of the base materials. The filler metal is distributed between the closely fitted surfaces of the joint by capillary attraction.

bright annealing. Annealing in a protective medium to prevent discoloration of the bright surface.

Brinell hardness number (HB). A number related to the applied load and to the surface area of the permanent impression made by a ball indenter.

Brinell hardness test. A test for determining the hardness of a material by forcing a hard steel or carbide ball of specified diameter (typically, 10 mm, or 0.4 in.) into it under a specified load. The result is expressed as the *Brinell hardness number.*

Brinelling. (1) Indentation of the surface of a solid body by repeated local impact or impacts, or static overfeed. Brinelling may occur especially in a rolling-element bearing. (2) Damage to a solid bearing surface characterized by one or more plastically formed indentations brought about by overload.

brine quenching. A quench in which brine (saltwater chlorides, carbonates, and cyanides) is the quenching medium. The salt addition improves the efficiency of water at the vapor phase or hot stage of the quenching process.

brittle fracture. Separation of a solid accompanied by little or no macroscopic plastic deformation. Typically, brittle fracture occurs by rapid crack propagation with less expenditure of energy than for *ductile fracture.* Brittle tensile fractures have a bright, granular appearance and exhibit little or no necking. A *chevron pattern* may be present on the fracture surface, pointing toward the origin of the crack, especially in brittle fractures in flat platelike components. Examples of brittle fracture include transgranular cracking (*cleavage fracture* and quasi-cleavage fracture) and intergranular cracking (decohesive rupture).

brittleness. The tendency of a material to fracture without first undergoing significant plastic deformation.

bronze. A copper-rich copper-tin alloy with or without small proportions of other elements such as zinc and phosphorus. By extension, certain copper-base alloys containing considerably less tin than other alloying elements, such as manganese bronze (copper-zinc plus manganese, tin, and iron) and leaded tin bronze (copper-lead plus tin and sometimes zinc). Also, certain other essentially binary copper-base alloys containing no tin, such as aluminum bronze (copper-aluminum), silicon bronze (copper-silicon), and beryllium bronze (copper-beryllium). Also trade designations for certain specific copper-base alloys that are actually brasses, such as architectural bronzes (57% Cu, 40% Zn, 3% Pb) and commercial bronze (90% Cu, 10% Zn).

butt joint (welding). A joint between two abutting members lying approximately in the same plane. A welded butt joint may contain a variety of grooves.

C

calibrate. To determine, by measurement or comparison with a standard, the correct value of each scale reading on a measuring (test) instrument.

capillary action. (1) The phenomenon of intrusion of a liquid into interconnected small voids, pores, and channels in the solid, resulting from surface tension. (2) The force by which liquid, in contact with a solid, is distributed between closely fitted faying surfaces of the joint to be brazed or soldered.

carbide. A compound of carbon with one or more metallic elements.

carbide tools. Cutting or forming tools, usually made from tungsten, titanium, tantalum, or niobium carbides, or a combination of them, in a matrix of cobalt, nickel, or other metals. Carbide tools are characterized by high hardnesses and compressive strengths and may be coated to improve wear resistance.

carbon. The element that provides the backbone for all organic polymers. Graphite is a crystalline form of carbon. Diamond is the densest crystalline form of carbon.

carbonitriding. A *case hardening* process in which a suitable ferrous material is heated above the lower transformation temperature in a gaseous atmosphere of such composition as to cause simultaneous absorption of carbon and nitrogen by the surface and, by diffusion, create a concentration gradient. The heat treating process is completed by cooling at a rate that produces the desired properties in the workpiece.

carbon steel. Steel having no specified minimum quantity for any alloying element—other than the commonly accepted amounts of manganese (≤1.65%), silicon (≤0.60%), and copper (≤0.60%)—and containing only an incidental amount of any element other than carbon, silicon, manganese, copper, sulfur, and phosphorus. Low-carbon steels contain up to 0.30% C, medium-carbon steels contain from 0.30 to 0.60% C, and high-carbon steels contain from 0.60 to 1.00% C.

carburizing. Absorption and diffusion of carbon into solid ferrous alloys by heating, to a temperature usually above Ac_3, in contact with a suitable carbonaceous material. A form of *case hardening* that produces a carbon gradient extending inward from the surface, enabling the surface layer to be hardened either by quenching directly from the carburizing temperature or by cooling to room temperature, then reaustenitizing and quenching.

case. In heat treating, the portion of a ferrous alloy, extending inward from the surface, whose composition has been altered during *case hardening*. Typically considered to be the portion of an alloy (a) whose composition has been measurably altered from the original composition, (b) that appears light when etched, or (c) that has a higher hardness value than the core. Contrast with *core*.

case hardening. A generic term covering several processes applicable to steel that change the chemical composition of the surface layer by absorption of carbon, nitrogen, or a mixture of the two and, by diffusion, create a concentration gradient. The processes commonly used are *carburizing* and *quench hardening; cyaniding; nitriding;* and *carbonitriding*. The use of the applicable specific process name is preferred.

casting. (1) Metal object cast to the required shape by pouring of injecting liquid metal into a mold, as distinct from one shaped by a mechanical process. (2) Pouring molten metal into a mold to produce an object of desired shape. (3) Ceramic forming process in which a body slip is introduced into a porous mold, which absorbs sufficient water from the slip to produce a semirigid article.

casting defect. Any imperfection in a casting that does not satisfy one or more of the required design or quality specifications. This term is often used in a limited sense for those flaws formed by improper casting solidification.

catastrophic wear. Sudden surface damage, deterioration, or change of shape caused by wear to such an extent that the life of the part is appreciably shortened or action is impaired.

cathode. The negative *electrode* of an electrolytic cell at which reduction is the principal reaction. (Electrons flow toward the cathode in the external circuit.) Typical cathodic processes are cations taking up electrons and being discharged, oxygen being reduced, and the reduction of an element or group of elements from a higher to a lower valence state.

cathodic corrosion. Corrosion resulting from a cathodic condition of a structure usually caused by the reaction of an amphoteric metal with the alkaline products of electrolysis.

cathodic protection. (1) Reduction of corrosion rate by shifting the corrosion potential of the electrode toward a less oxidizing potential by applying an external electromotive force. (2) Partial or complete protection of a metal from corrosion by making it a *cathode,* using either a galvanic or an impressed current. Contrast with *anodic protection.*

cementite. A hard (~800 HV), brittle compound of iron and carbon, known chemically as iron carbide and having the approximate chemical formula Fe_3C. It is characterized by an orthorhombic crystal structure. When it occurs as a phase in steel, the chemical composition will be altered by the presence of manganese and other carbide-forming elements. The highest cementite contents are observed in white cast irons, which are used in applications where high wear resistance is required.

characteristic. A property of items in a sample or population that when measured, counted, or otherwise observed helps to distinguish between the items.

charge. (1) The materials fed into a furnace. (2) Weights of various liquid and solid materials put into a furnace during one feeding cycle. (3) The weight of plastic material used to load a mold at one time or during one cycle.

Charpy test. An impact test in which a V-notched, keyhole-notched, or U-notched specimen, supported at both ends, is struck behind the notch by a striker mounted at the lower end of a bar that can swing as a pendulum. The energy that is absorbed in fracture is calculated from the height to which the striker would have risen had there been no specimen and the height to which it actually rises after fracture of the specimen.

chevron pattern. A fractographic pattern of radial marks (shear ledges) that look like nested letters "V"; sometimes called a herringbone pattern. Chevron patterns are typically found on brittle fracture surfaces in parts whose widths are considerably greater than their thicknesses. The points of the chevrons can be traced back to the fracture origin.

Chromel. (1) A 90Ni-10Cr alloy used in thermocouples. (2) A series of nickel-chromium alloys, some with iron, used for heat-resistant applications.

chrome plating. (1) Producing a chromate *conversion coating* on magnesium for temporary protection or for a paint base. (2) The solution heat produces the conversion coating.

chromium-molybdenum heat-resistant steels. Alloy steels containing 0.5 to 9% Cr and 0.5 to 1.10% Mo with a carbon content usually below 0.20%. The chromium provides improved oxidation and corrosion resistance, and the molybdenum increases strength at elevated temperatures. Chromium-molybdenum steels are widely used in the oil and gas industries and in fossil fuel and nuclear power plants.

chromizing. A surface treatment at elevated temperature, generally carried out in pack, vapor, or salt baths, in which an alloy is formed by the inward diffusion of chromium into the base metal.

circuit board. In electronics, a sheet insulating material laminated to foil that is etched to produce a circuit pattern on one or both sides. Also called print circuit board or printed wiring board.

cladding. (1) A layer of material, usually metallic, that is mechanically or metallurgically bonded to a substrate. Cladding may be bonded to the substrate by any of several processes, such as roll cladding and explosive forming. (2) A relatively thick layer (>1 mm, or 0.04 in.) of material applied by surfacing for the purpose of improved corrosion resistance or other properties.

clad metal. A composite metal containing two or more layers that have been bonded together. The bonding may have been accomplished by corolling, coextrusion, welding, diffusion bonding, casting, heavy chemical deposition, or heavy electroplating.

cleavage fracture. A fracture, usually of a polycrystalline metal, in which most of the grains have failed by cleavage, resulting in bright reflecting facets. It is one type of crystalline fracture and is associated with low-energy *brittle fracture.*

closed-die forging. The shaping of hot metal completely within the walls or cavities of two dies that come together to enclose the workpiece on all sides. The impression for the forging can be entirely either die or divided between the top and bottom dies. Impression-die forgings, often used interchangeably with the term closed-die forging, refers to a closed-die operation in which the dies contain a provision for controlling the flow of excess material, or flash, that is generated. By contrast, in flashless forging, the material is deformed in a cavity that allows little or no escape of excess material.

close-packed. A geometric arrangement in which a collection of equally sized spheres (atoms) may be packed together in a minimum total volume.

close-tolerance forging. A forging held to unusually close dimensional tolerances so that little or no machining is required after forging.

coalescence. (1) The union of particles of a dispersed phase into larger units, usually effected at temperatures below the fusion point. (2) Growth of grains at the expense of the remainder by absorption or the growth of a phase or particle at the expense of the remainder by absorption or reprecipitation.

coating. A relatively thin layer (<1 mm, or 0.04 in.) of material applied by surfacing for the purpose of corrosion prevention, resistance to high-temperature scaling, wear resistance, lubrication, or other purposes.

coefficient of expansion. A measure of the change in length or volume of an object; specifically, a change measured by the increase in length or volume of an object per unit length or volume.

coefficient of thermal expansion. (1) Change in unit length (or volume) accompanying a unit change of temperature, at a specified temperature. (2) The linear or volume expansion of a given material per degree rise of temperature, expressed as an arbitrary base temperature or as a more complicated equation applicable to a wide range of temperatures.

coining. (1) A closed-die squeezing operation, usually performed cold, in which all surfaces of the work are confined or restrained, resulting in a well-defined imprint of the die upon the work. (2) A restriking operation used to sharpen or change an existing radius or profile. (3) The final pressing of a sintered powder metallurgy compact to obtain a definite surface configuration (not to be confused with re-pressing or sizing).

cold cracking. (1) Cracks in cold or nearly cold cast metal due to excessive internal stress caused by contraction. Often brought about when the mold is too hard or the casting is of unsuitable design. (2) A type of weld cracking that usually occurs below 205 °C (400 °F). Cracking may occur during or after cooling to room temperature, sometimes with a considerable time delay. Three factors combine to produce cold cracks: stress (for example, from thermal expansion and contraction), hydrogen (from hydrogen-containing welding consumables), and a susceptible microstructure (plate martensite is most susceptible to cracking, ferritic and bainitic structures least susceptible).

cold-die quenching. A quench utilizing cold, flat, or shaped dies to extract heat from a part. Cold-die quenching is slow, expensive, and is limited to smaller parts with large surface areas.

cold drawing. Technique for using standard metalworking equipment and systems for forming thermoplastic sheet at room temperature.

cold heading. Working metal at room temperature such that the cross-sectional area of a portion or all of the stock is increased.

cold shortness. Brittleness that exists in some metals at temperatures below the recrystallization temperature.

cold shut. (1) A discontinuity that appears on the surface of cast metal as a result of two streams of liquid meeting and failing to unite. (2) A lap on the surface of a forging or billet that was closed without fusion during deformation. (3) Freezing of the top surface of an ingot before the mold is full.

cold treatment. Exposing steel to suitable subzero temperatures (–85 °C, or –120 °F) for the purpose of obtaining desired conditions or properties such as dimensional or microstructural stability. When the treatment involves the transformation of retained austenite, it is usually followed by tempering.

cold-worked structure. A microstructure resulting from plastic deformation of a metal or alloy below its recrystallization temperature.

cold working. Deforming metal plastically under conditions of temperature and strain rate that induce *strain hardening*. Usually, but not necessarily, conducted at room temperature.

columnar structure. A coarse structure of parallel elongated grains formed by unidirectional growth, most often observed in castings, but sometimes seen in structures resulting from diffusional growth accompanied by a solid-state transformation.

component. (1) One of the elements or compounds used to define a chemical (or alloy) system, including all phases, in terms of the fewest substances possible. (2) One of the individual parts of a vector as referred to a system of coordinates. (3) An individual functional element in a physically independent body that cannot be further reduced or divided without destroying its stated function, for example, a resistor, capacitor, diode, or transistor.

composite material. A combination of two or more materials (reinforcing elements, fillers, and composite matrix binder), differing in form or composition on a macroscale. The constituents retain their identities; that is, they do not dissolve or merge completely into one another although they act in concert. Normally, the components can be physically identified and exhibit an interface between one another.

compound. (1) In chemistry, a substance of relatively fixed composition and properties, whose ultimate structural unit (molecule or repeat unit) is comprised of atoms of two or more elements. The number of atoms of each kind in this ultimate unit is determined by natural laws and is part of the identification of the compound. (2) In reinforced plastics and composites, the intimate admixture of a polymer with other ingredients, such as fillers, softeners, plasticizers, reinforcements, catalysts, pigments, or dyes. A thermoset compound usually contains all the ingredients necessary for the finished product, while a thermoplastic compound may require subsequent addition of pigments, blowing agents, and so forth.

compressive strength. The maximum compressive stress that a material is capable of developing, based on original area of cross section. If a material fails in compression by a shattering fracture, the compressive strength has a very definite value. If a material does not fail in compression by a shattering fracture, the value obtained for compressive strength is an arbitrary value depending upon the degree of distortion that is regarded as indicating complete failure of the material.

compressive stress. A stress that causes an elastic body to deform (shorten) in the direction of the applied load.

computed tomography (CT). The collection of transmission data through an object and the subsequent reconstruction of an image corresponding to a cross section through this object. Also known as computerized axial tomography, computer-assisted tomography, CAT scanning, or industrial computed tomography.

conductivity (electrical). The reciprocal of volume resistivity. The electrical or thermal conductance of a unit cube of any material (conductivity per unit volume).

conductivity (thermal). The time rate of heat flow through unit thickness of an infinite slab of a homogeneous material in a direction perpendicular to the surface, induced by unit temperature difference.

conductor. A wire, cable, or other body capable of carrying an electrical current.

constituent. (1) One of the ingredients that make up a chemical system. (2) A phase or a combination of phases that occurs in a characteristic configuration in an alloy microstructure. (3) In composites, the principal constituents are the fibers and the matrix.

consumable electrode. A general term for any arc welding electrode made chiefly of filler metal. Use of specific names, such as covered electrode, bare electrode, flux-cored electrode, and lightly coated electrode, is preferred.

continuous casting. A casting technique in which a cast shape is continuously withdrawn through the bottom of the mold as it solidifies, so that its length is not determined by mold dimensions. Used chiefly to produce semifinished mill products such as billets, blooms, ingots, slabs, strip, and tubes.

continuous cooling transformation (CCT) diagram. Set of curves drawn using logarithmic time and linear temperature as coordinates, which define, for each cooling curve of an alloy, the beginning and end of the transformation of the initial phase.

conversion coating. A coating consisting of a compound of the surface metal, produced by chemical or electrochemical treatments of the metal. Examples include chromate coatings on zinc, cadmium, magnesium, and aluminum, and oxide and phosphate coatings on steel.

cooling curve. A graph showing the relationship between time and temperature during the cooling of a material. It is used to find the temperatures at which phase changes occur. A property or function other than time may occasionally be used—for example, thermal expansion.

cooling rate. The average slope of the time-temperature curve taken over a specified time and temperature interval.

cooling stresses. Residual stresses in castings resulting from nonuniform distribution of temperature during cooling.

core. (1) A specially formed material inserted in a mold to shape the interior or other part of a casting that cannot be shaped as easily by the pattern. (2) In a ferrous alloy prepared for *case hardening*, that portion of the alloy that is not part of the *case*. Typically considered to be the portion that (a) appears dark (with certain etchants) on an etched cross section, (b) has an essentially unaltered chemical composition, or (c) has a hardness, after hardening, less than a specified value. (3) The central member of a sandwich construction (honeycomb material, foamed plastic, or solid sheet to which the faces of the sandwich are attached).

corrosion. The chemical or electrochemical reaction between a material, usually a metal, and its environment that produces a deterioration of the material and its properties.

corrosion embrittlement. The severe loss of ductility of a metal resulting from corrosive attack, usually intergranular and often not visually apparent.

corrosion fatigue. The process in which a metal fractures prematurely under conditions of simultaneous corrosion and repeated cyclic loading at lower stress levels or fewer cycles than would be required in the absence of the corrosive environment.

corrosion fatigue strength. The maximum repeated stress that can be endured by a metal without failure under definite conditions of corrosion and fatigue and for a specific number of stress cycles and a specified period of time.

corrosion resistance. The ability of a material to withstand contact with ambient natural factors or those of a particular, artificially created atmosphere, without degradation or change in properties. For metals, this could be pitting or rusting; for organic materials, it could be crazing.

corrosive wear. Wear in which chemical or electrochemical reaction with the environment is significant.

crack growth. Rate of propagation of a crack through a material due to a static or dynamic applied load.

cracking. In lubrication technology, that process of converting unwanted long-chain hydrocarbons to shorter molecules by thermal or catalytic action.

crack opening displacement (COD). On a K_{Ic} specimen, the opening displacement of the notch surfaces at the notch and in the direction perpendicular to the plane of the notch and the crack. The displacement at the tip is called the crack tip opening displacement (CTOD); at the mouth, it is called the crack mouth opening displacement (CMOD).

creep. Time-dependent strain occurring under stress. The creep strain occurring at a diminishing rate is called primary creep; that occurring at a minimum and almost constant rate, secondary creep; and that occurring at an accelerating rate, tertiary creep.

creep rate. The slope of the creep-time curve at a given time. Deflection with time under a given static load.

creep-rupture embrittlement. *Embrittlement* under creep conditions of, for example, aluminum alloys and steels that results in abnormally low rupture ductility. In aluminum alloys, iron in amounts above the solubility limit is known to cause such embrittlement; in steels, the phenomenon is related to the amount of impurities (for example, phosphorus, sulfur, copper, arsenic, antimony, and tin) present. In either case, failure occurs by intergranular cracking of the embrittled material.

creep-rupture strength. The stress that causes fracture in a creep test at a given time, in a specified constant environment. This is sometimes referred to as the stress-rupture strength. In glass technology, this is termed the static fatigue strength.

creep-rupture test. A test in which progressive specimen deformation and the time for rupture are both measured. In general, deformation is much greater than that developed during a creep test. Also known as stress-rupture test.

creep stress. The constant load divided by the original cross-sectional area of the specimen.

creep test. A method for determining the extension of metals under a given load at a given temperature. The determination usually involves the plotting of time-elongation curves under constant load; a single test may extend over many months. The results are often expressed as the elongation (in millimeters or inches) per hour on a given gage length (e.g., 25 mm, or 1 in.).

crevice corrosion. *Localized corrosion* of a metal surface at, or immediately adjacent to, an area that is shielded from full exposure to the environment because of close proximity between the metal and the surface of another material.

critical point. (1) The temperature or pressure at which a change in crystal structure, phase, or physical properties occurs. Also termed transformation temperature. (2) In an equilibrium diagram, that combination of composition, temperature, and pressure at which the phases of an inhomogeneous system are in equilibrium.

critical strain. (1) In mechanical testing, the strain at the *yield point*. (2) The strain just sufficient to cause *recrystallization*; because the strain is small, usually only a few percent, recrystallization takes place from only a few nuclei, which produces a recrystallized structure consisting of very large grains.

crystal. (1) A solid composed of atoms, ions, or molecules arranged in a pattern that is repetitive in three dimensions. (2) That form, or particle, or piece of substance in which its atoms are distributed in one specific orderly geometrical array, called a "lattice," essentially throughout. Crystals exhibit characteristic optical and other properties and growth or cleavage surfaces, in characteristic directions.

cyaniding. A case hardening process in which a ferrous material is heated above the lower transformation temperature range in a molten salt containing cyanide to cause simultaneous absorption of carbon and nitrogen at the surface and, by diffusion, create a concentration gradient. *Quench hardening* completes the process.

D

damage tolerance. (1) A design measure of crack growth rate. Cracks in damage-tolerant designed structures are not permitted to grow to critical size during expected service life. (2) The ability of a part component, such as an aerospace engine, to resist failure due to the presence of flaws, cracks, or other damage for a specified period of usage. The damage tolerance approach is used extensively in the aerospace industry.

damping. The loss in energy, as dissipated heat, that results when a material or material system is subjected to an oscillatory load or displacement.

decarburization. Loss of carbon from the surface layer of a carbon-containing alloy due to reaction with one or more chemical substances in a medium that contacts the surface.

deep drawing. Forming deeply recessed parts by forcing sheet metal to undergo plastic flow between dies, usually without substantial thinning of the sheet.

defect. (1) A discontinuity whose size, shape, orientation, or location makes it detrimental to the useful service of the part in which it occurs. (2) A discontinuity or discontinuities that by nature or accumulated effect (for example, total crack length) render a part or product unable to meet minimum applicable acceptance standards or specifications. This term designates rejectability.

deformation. A change in the form of a body due to stress, thermal change, change in moisture, or other causes. Measured in units of length.

degassing (metals). (1) A chemical reaction resulting from a compound added to molten metal to remove gases from the metal. Inert gases are often used in this operation. (2) A fluxing procedure used for aluminum alloys in which nitrogen, chlorine, chlorine and nitrogen, and chlorine and argon are bubbled up through the metal to remove dissolved hydrogen gases and oxides from the alloy.

degradation. A deleterious change in the chemical structure, physical properties, or appearance of a material.

delamination. The separation of layers in a laminate because of failure of the adhesive, either in the adhesive itself or at the interface between the adhesive and the adherend.

delta iron. Solid phase of pure iron that is stable from 1400 to 1540 °C (2550 to 2800 °F) and possesses the body-centered cubic lattice.

dendrite. A crystal that has a treelike branching pattern, being most evident in cast metals slowly cooled through the solidification range.

descaling. (1) Removing the thick layer of oxides formed on some metals at elevated temperatures. (2) A chemical or mechanical process for removing scale or investment material from castings.

desulfurizing. The removal of sulfur from molten metal by reaction with a suitable slag or by the addition of suitable compounds.

dewpoint temperature. The temperature at which condensation of water vapor in a space begins for a given state of humidity and pressure as the vapor temperature is reduced; the temperature corresponding to saturation (100% relative humidity) for a given absolute humidity at constant pressure.

dezincification. Corrosion in which zinc is selectively leached from zinc-containing alloys, leaving a relatively weak layer of copper and copper oxide. Most commonly found in copper-zinc alloys containing less than 85% Cu after extended service in water containing dissolved oxygen.

die. A tool, usually containing a cavity, that imparts shape to solid, molten, or powdered metal primarily because of the shape of the tool itself. Used in many press operations (including blanking, drawing, forging, and forming), in die casting, and in forming green powder metallurgy compacts. Die casting and powder metallurgy dies are sometimes referred to as *molds*.

die casting. (1) A casting made in a die. (2) A casting process in which molten metal is forced under high pressure into the cavity of a metal mold.

die forging. A forging that is formed to the required shape and size through working in machined impressions in specially prepared dies.

die forming. The shaping of solid or powdered metal by forcing it into or through the die cavity.

dielectric. A nonconductor of electricity. The ability of a material to resist the flow of an electrical current.

diffusion. (1) Spreading of a constituent in a gas, liquid, or solid, tending to make the composition of all parts uniform. (2) The spontaneous movement of atoms or molecules to new sites within a material. (3) The movement of a material, such as a gas or liquid, in the body of a plastic. If the gas or liquid is absorbed on one side of a piece of plastic and given off on the other side, the phenomenon is called *permeability*. Diffusion and permeability are not due to holes or pores in the plastic but are caused and controlled by chemical mechanisms.

dimensional stability. (1) A measure of dimensional change caused by such factors as temperature, humidity, chemical treatment, age, or stress (usually expressed as Δ units per unit). (2) Ability of a plastic part to retain the precise shape in which it was molded, cast, or otherwise fabricated.

dip brazing. A brazing process in which the heat required is furnished by a molten chemical or metal bath. When a molten metal bath is used, the bath may act as a flux. When a molten metal bath is used, the bath provides the filler metal.

dip soldering. A soldering process in which the heat required is furnished by a molten metal bath which provides the solder filler metal.

direct quenching. (1) Quenching carburized parts directly from the carburizing operation. (2) Also used for quenching pearlitic malleable parts directly from the malleabilizing operation.

discontinuity. (1) Any interruption in the normal physical structure or configuration of a part, such as cracks, laps, seams, inclusions, or porosity. A discontinuity may or may not affect the utility of the part. (2) An interruption of the typical structure of a weldment, such as a lack of homogeneity in the mechanical, metallurgical, or physical characteristics of the material or weldment. A discontinuity is not necessarily a *defect*.

dislocation. A linear imperfection in a crystalline array of atoms. Two basic types are recognized: (1) An edge dislocation corresponds to the row of mismatched atoms along the edge formed by an extra, partial plane of atoms within the body of a crystal; (2) a screw dislocation corresponds to the axis of a spiral structure in a crystal, characterized by a distortion that joins normally parallel planes together to form a continuous helical ramp (with a pitch or one interplanar distance) winding about the dislocation. Most prevalent is the so-called mixed dislocation, which is any combination of an edge dislocation and a screw dislocation.

distortion. Any deviation from an original size, shape, or contour that occurs because of the application of stress or the release of residual stress.

double-tempering. A treatment in which a quench-hardened ferrous metal is subjected to two complete tempering cycles, usually at substantially the same temperature, for the purpose of ensuring completion of the tempering reaction and promoting stability of the resulting microstructure.

drawability. A measure of the *formability* of a sheet metal subject to a drawing process. The term is usually used to indicate the ability of a metal to be deep drawn.

drawing. (1) A term used for a variety of forming operations, such as *deep drawing* a sheet metal blank; redrawing a tubular part; and drawing rod, wire, and tube. The usual drawing process with regard to sheet metal working in a press is a method for producing a cuplike form from a sheet metal disk by holding it firmly between blankholding surfaces to prevent the formation of wrinkles while the punch travel produces the required shape. (2) The process of stretching a thermoplastic to reduce its cross-sectional area, thus creating a more orderly arrangement of polymer chains with respect to each other. (3) A misnomer for tempering.

drop forging. The forging obtained by hammering metal in a pair of closed dies to produce the form in the finishing impression under a *drop hammer*; forging method requiring special dies for each shape.

drop hammer. A term generally applied to forging hammers in which energy for forging is provided by gravity, steam, or compressed air.

drop hammer forming. A process for producing shapes by the progressive deformation of sheet metal in matched dies under the repetitive blows of a gravity-drop or power-drop hammer. The process is restricted to relatively shallow parts and thin sheet from approximately 0.6 to 1.6 mm (0.024 to 0.064 in.).

dual-phase steels. A new class of *high-strength low-alloy steels* characterized by a tensile strength value of approximately 550 MPa (80 ksi) and by a microstructure consisting of about 20% hard martensite particles dispersed in a soft ductile ferrite matrix. The term dual phase refers to the predominance in the microstructure of two phases, ferrite and martensite. However, small amounts of other phases, such as bainite, pearlite, or retained austenite, may also be present.

ductile fracture. Fracture characterized by tearing of metal accompanied by appreciable gross plastic deformation and expenditure of considerable energy.

ductility. The ability of a material to deform plastically without fracturing.

duplexing. Any two-furnace melting or refining process. Also called duplex melting or duplex processing.

duralumin (obsolete). A term frequently applied to the class of age-hardenable aluminum-copper alloys containing manganese, magnesium, or silicon.

E

earing. The formation of ears or scalloped edges around the top of a drawn shell, resulting from directional differences in the plastic-working properties of rolled metal, with, across, or at angles to the direction of rolling.

885 °F (475 °C) embrittlement. *Embrittlement* of stainless steels upon extended exposure to temperatures between 400 and 510 °C (750 and 950 °F). This type of embrittlement is caused by fine, chromium-rich precipitates that segregate at grain boundaries; time at temperature directly influences the amount of segregation. Grain-boundary segregation of the chromium-rich precipitates increases strength and hardness, decreases ductility and toughness, and changes corrosion resistance. This type of embrittlement can be reversed by heating above the precipitation range.

elastic energy. The amount of energy required to deform a material within the elastic range of behavior, neglecting small heat losses due to internal friction. The energy absorbed by a specimen per unit volume of material contained within the gage length being tested. It is determined by measuring the area under the stress-strain curve up to a specified elastic strain.

elasticity. The property of a material by virtue of which deformation caused by stress disappears upon removal of the stress. A perfectly elastic body completely recovers its original shape and dimensions after release of stress.

electrical resistivity. The electrical resistance offered by a material to the flow of current, times the cross-sectional area of current flow and per unit length of current path; the reciprocal of the conductivity. Also called resistivity or specific resistance.

electric furnace. A metal melting or holding furnace that produces heat from electricity. It may operate on the resistance or induction principle.

electrochemical corrosion. Corrosion that is accompanied by a flow of electrons between cathodic and anodic areas on metallic surfaces.

electrode (electrochemistry). One of a pair of conductors introduced into an electrochemical cell, between which the ions in the intervening medium flow in opposite directions and on whose surfaces reactions occur (when appropriate external connection is made). In direct current operation, one electrode or "pole" is positively charged, the other negatively.

electrode (welding). (1) In arc welding, a current-carrying rod that supports the arc between the rod and work, or between two rods as in twin carbon-arc welding. It may or may not furnish filler metal. (2) In resistance welding, a part of a resistance welding machine through which current and, in most instances, pressure are applied directly to the work. The electrode may be in the form of a rotating wheel, rotating roll, bar, cylinder, plate, clamp, chuck, or modification thereof. (3) In arc and plasma spraying, the current-carrying components that support the arc.

electrodeposition. (1) The deposition of a conductive material from a plating solution by the application of electrical current. (2) The deposition of a substance on an electrode by passing electric current through an electrolyte. Electrochemical (plating), electroforming, electrorefining, and electrotwinning result from electrodeposition.

electrolyte. (1) A chemical substance or mixture, usually liquid containing ions that migrate in an electric field. (2) A chemical compound or mixture of compounds which when molten or in solution will conduct an electric current.

electrolytic copper. Copper that has been refined by the electrolytic deposition, including cathodes that are the direct product of the refining operation, refinery shapes cast from melted cathodes, and, by extension, fabricators' products made therefrom. Usually when this term is used alone, it refers to electrolytic tough pitch copper without elements other than oxygen being present in significant amounts.

electron. An elementary particle that is the negatively charged constituent of ordinary matter. The electron is the lightest known particle possessing an electric charge. Its rest mass is $m_e \cong 9.1 \times 10^{-28}$ g, approximately $1/1836$ of the mass of the proton or neutron, which are, respectively, the positively charged and neutral constituents of ordinary matter.

electron beam heat treating. A selective surface hardening process that rapidly heats a surface by direct bombardment with an accelerated stream of electrons.

electron beam machining. Removing material by melting and vaporizing the workpiece at the point of impingement of a focused high-velocity beam of electrons. The machining is done in high vacuum to eliminate scattering of the electrons due to interaction with gas molecules. The most important use of electron beam machining is for hole drilling.

electron beam welding. A welding process that produces coalescence of metals with the heat obtained from a concentrated beam composed primarily of high-velocity electrons impinging upon the surfaces to be joined. Welding can be carried out at atmospheric pressure (nonvacuum), medium vacuum (approximately 10^{-3} to 25 torr), or high vacuum (approximately 10^{-6} to 10^{-3} torr).

electron micrograph. A reproduction of an image formed by the action of an electron beam on a photographic emulsion.

electron microscope. An electron-optical device that produces a magnified image of an object. Detail may be revealed by selective transmission, reflection, or emission of electrons by the object.

electroslag welding. A fusion welding process in which the welding heat is provided by passing an electric current through a layer of molten conductive slag (flux) contained in a pocket formed by water-cooled dams that bridge the gap between the members being welded. The resistance-heated slag not only melts filler metal electrodes as they are fed into the slag layer, but also provides shielding for the massive weld puddle characteristic of the process.

elongation. (1) A term used in mechanical testing to describe the amount of extension of a testpiece when stressed. (2) In tensile testing, the increase in the gage length, measured after fracture of the specimen within the gage length, usually expressed as a percentage of the original gage length.

embrittlement. The severe loss of *ductility* or *toughness* or both, of a material, usually a metal or alloy. Many forms of embrittlement can lead to *brittle fracture* and can occur during thermal treatment or elevated-temperature service (thermally induced embrittlement). Some of these forms of embrittlement, which affect steels, include blue brittleness, *885 °F (475 °C) embrittlement*, *quench-age embrittlement*, sigma-phase embrittlement, *strain-age embrittlement*, temper embrittlement, tempered martensite embrittlement, and *thermal embrittlement*. In addition, steels and other metals and alloys can be embrittled by environmental conditions (environmentally assisted embrittlement). The forms of environmental embrittlement include acid embrittlement, caustic embrittlement, *corrosion embrittlement*, *creep-rupture embrittlement*, hydrogen embrittlement, liquid metal embrittlement, neutron embrittlement, solder embrittlement, solid metal embrittlement, and *stress-corrosion cracking*.

endothermic reaction. Designating or pertaining to a reaction that involves the absorption of heat.

end-quench hardenability test. A laboratory procedure for determining the hardenability of a steel or other ferrous alloy; widely referred to as the *Jominy test*. Hardenability is determined by heating a standard specimen above the upper critical temperature, placing the hot specimen in a fixture so that a stream of cold water impinges on one end, and, after cooling to room temperature is completed, measuring the hardness near the surface of the specimen at regularly spaced intervals along its length. The data are normally plotted as hardness versus distance from the quenched end.

environment. The aggregate of all conditions (such as contamination, temperature, humidity, radiation, magnetic and electric fields, shock, and vibration) that externally influence the performance of a material or component.

environmental cracking. *Brittle fracture* of a normally ductile material in which the corrosive effect of the environment is a causative factor. Environmental cracking is a general term that includes *corrosion fatigue*, high-temperature hydrogen attack, hydrogen blistering, hydrogen embrittlement, liquid metal embrittlement, solid metal embrittlement, *stress-corrosion cracking*, sulfide corrosion cracking, and sulfide stress cracking. The following terms have been used in the past in connection with environmental cracking, but are becoming obsolete: caustic embrittlement, delayed fracture, season cracking, static fatigue, stepwise cracking, sulfide corrosion cracking, and sulfide stress-corrosion cracking.

equiaxed grain structure. A structure in which the grains have approximately the same dimensions in all directions.

equilibrium. The dynamic condition of physical, chemical, mechanical, or atomic balance that appears to be a condition of rest rather than change.

erosion. (1) Loss of material from a solid surface due to relative motion in contact with a fluid that contains solid particles. Erosion in which the relative motion of particles is nearly parallel to the solid surface is called abrasive erosion. Erosion in which the relative motion of the solid particles is nearly normal to the solid surface is called impingement erosion or impact erosion. (2) Progressive loss of original material from a solid surface due to mechanical interaction between that surface and a fluid, a multicomponent fluid, and impinging liquid, or solid particles. (3) Loss of material from the surface of an electrical contact due to an electrical discharge (arcing).

erosion-corrosion. A conjoint action involving *corrosion* and *erosion* in the presence of a moving corrosive fluid, leading to the accelerated loss of material.

etchant. (1) A chemical solution used to etch a metal to reveal structural details. (2) A solution used to remove, by chemical reaction, the unwanted portion of material from a printed circuit board. (3) Hydrofluoric acid or other agent used to attack the surface of glass for marking or decoration.

etching. (1) Subjecting the surface of a metal to preferential chemical or electrolytic attack in order to reveal structural details for metallographic examination. (2) Chemically or electrochemically removing tenacious films from a metal surface to condition the surface for a subsequent treatment, such as painting or electroplating. (3) A process by which a printed pattern is formed on a printed circuit board by either chemical or chemical and electrolytic removal of the unwanted portion of conductive material bonded to a base.

exfoliation. Corrosion that proceeds laterally from the sites of initiation along planes parallel to the surface, generally at grain boundaries, forming corrosion products that force metal away from the body of the material, giving rise to a layered appearance.

exothermic reaction. A reaction that liberates heat, such as the burning of fuel or when certain plastic resins are cured chemically.

explosive forming. The shaping of metal parts in which the forming pressure is generated by an explosive charge that takes the place of the punch in conventional forming. A single-element die is used with a blank held over it, and the explosive charge is suspended over the blank at a predetermined distance (standoff distance). The complete assembly is often immersed in a tank of water.

extra hard. A *temper* of nonferrous alloys and some ferrous alloys characterized by values of tensile strength and hardness about one-third of the way from those of *full hard* to those of extra spring temper.

extrusion (metals). The conversion of an ingot or billet into lengths of uniform cross section by forcing metal to flow plastically through a die orifice. In forward (direct) extrusion, the die and ram are at opposite ends of the extrusion stock, and the product and ram travel in the same direction. Also, there is relative motion between the extrusion stock and the die. In backward (indirect) extrusion, the die is at the ram end of the stock and the product travels in the direction opposite that of the ram, either around the ram (as in the impact extrusion of cylinders such as cases for dry cell batteries) or up through the center of a hollow ram.

F

face-centered. Having atoms or groups of atoms separated by translations of ½, ½, 0; ½, 0, ½; and 0, ½, ½ from a similar atom or group of atoms. The number of atoms in a face-centered cell must be a multiple of 4.

failure. A general term used to imply that a part in service (a) has become completely inoperable, (b) is still operable but incapable of satisfactorily performing its intended function, or (c) has deteriorated seriously, to the point that it has become unreliable or unsafe for continued use.

fatigue. The phenomenon leading to fracture under repeated or fluctuating stresses having a maximum value less than the ultimate tensile strength of the material. *Fatigue failure* generally occurs at loads that applied statically would produce little perceptible effect. Fatigue fractures are progressive, beginning as minute cracks that grow under the action of the fluctuating stress.

fatigue failure. Failure that occurs when a specimen undergoing *fatigue* completely fractures into two parts or has softened or been otherwise significantly reduced in stiffness by thermal heating or cracking.

fatigue life (*N*). (1) The number of cycles of stress or strain of a specified character that a given specimen sustains before failure of a specified nature occurs. (2) The number of cycles of deformation required to bring about failure of a test specimen under a given set of oscillating conditions (stresses or strains).

fatigue test. A method for determining the range of alternating (fluctuating) stresses a material can withstand without failing.

fatigue wear. (1) Removal of particles detached by fatigue arising from cyclic stress variations. (2) Wear of a solid surface caused by fracture arising from material fatigue.

ferrite. (1) A solid solution of one or more elements in body-centered cubic iron. Unless otherwise designated (for instance, as chromium ferrite), the solute is generally assumed to be carbon. On some equilibrium diagrams, there are two ferrite regions separated by an austenite area. The lower area is alpha ferrite; the upper, delta ferrite. If there is no designation, alpha ferrite is assumed. (2) An essentially carbon-free solid solution in which alpha iron is the solvent, and which is characterized by a body-centered cubic crystal structure. Fully ferritic steels are only obtained when the carbon content is quite low. The most obvious microstructural features in such metals are the ferrite grain boundaries.

ferroalloy. An alloy of iron that contains a sufficient amount of one or more other chemical elements to be useful as an agent for introducing these elements into molten metal, especially into steel or cast iron.

file hardness. Hardness as determined by the use of a steel file of standardized hardness on the assumption that a material that cannot be cut with the file is as hard as, or harder than, the file. Files covering a range of hardnesses may be employed; the most common are files heat treated to approximately 67 to 70 HRC.

filler metal. Metal added in making a brazed, soldered, or welded joint.

fillet weld. A weld, approximately triangular in cross section, joining two surfaces, essentially at right angles to each other in a lap, tee, or corner joint.

finish (metals). (1) Surface condition, quality, or appearance of a metal. (2) Stock on a forging or casting to be removed in finish machining. (3) The forging operation in which the part is forged into its final shape in the finish die. If only one finish operation is scheduled to be performed in the finish die, this operation will be identified simply as finish; first, second, or third finish designations are so termed when one or more finish operations are to be performed in the same finish die.

fisheye (weld defect). A discontinuity found on the fracture surface of a weld in steel, consisting of a small pore or inclusion surrounded by an approximately round, bright area.

fixture. A device designed to hold parts to be joined in proper relation to each other.

flame annealing. Annealing in which the heat is applied directly by a flame.

flame hardening. A process for hardening the surfaces of hardenable ferrous alloys in which an intense flame is used to heat the surface layers above the upper transformation temperature, whereupon the workpiece is immediately quenched.

flame straightening. Correcting distortion in metal structures by localized heating with a gas flame.

flank wear. The loss of relief on the flank of the tool behind the cutting edge due to rubbing contact between the work and the tool during cutting; measured in terms of linear dimension behind the original cutting edge.

flaw. A nonspecific term often used to imply a crack-like discontinuity.

flexural strength. A property of solid material that indicates its ability to withstand a flexural or transverse load.

flow lines. (1) Texture showing the direction of metal flow during hot or cold working. Flow lines can often be revealed by etching the surface or a section of a metal part. (2) In mechanical metallurgy, paths followed by minute volumes of metal during deformation.

fluidized bed. A contained mass of a finely divided solid that behaves like a fluid when brought into suspension in a moving gas or liquid.

fluorescent magnetic particle inspection. Inspection with either dry magnetic particles or those in a liquid suspension, the particles being coated with a fluorescent substance to increase the visibility of the indications.

fluorescent penetrant inspection. Inspection using a fluorescent liquid that will penetrate any surface opening; after the surface has been wiped clean, the location of any surface flaws may be detected by the fluorescence, under ultraviolet light, or back-seepage of the fluid.

forgeability. Term used to describe the relative ability of material to deform without fracture. Also describes the resistance to flow from deformation.

forged structure. The macrostructure through a suitable section of a forging that reveals direction of working.

forging. The process of working metal to a desired shape by impact or pressure in hammers, forging machines (upsetters), presses, rolls, and related forming equipment. Forging hammers, counterblow equipment, and high-energy-rate forging machines apply impact to the workpiece, while most other types of forging equipment apply squeeze pressure in shaping the stock. Some metals can be forged at room temperature, but most are made more plastic for forging by heating.

forging billet. A wrought metal *slug* used as forging stock.

forging dies. Forms for making forgings; they generally consist of a top and bottom die. The simplest will form a completed forging in a single impression; the most complex, consisting of several die inserts, may have a number of impressions for the progressive working of complicated shapes. Forging dies are usually in pairs, with part of the impression in one of the blocks and the rest of the impression in the other block.

formability. The ease with which a metal can be shaped through plastic deformation. Evaluation of the formability of a metal involves measurement of strength, ductility, and the amount of deformation required to cause fracture. The term workability is used interchangeably with formability; however, formability refers to the shaping of sheet metal, while workability refers to shaping materials by bulk forming.

foundry. A commercial establishment or building where metal castings are produced.

fractography. Descriptive treatment of fracture of materials, with specific reference to photographs of the fracture surface. Macrofractography involves photographs at low magnification (<25×); microfractography, photographs as high magnification (>25×).

fracture (metals). The irregular surface produced when a piece of metal is broken.

fracture mechanics. A quantitative analysis for evaluating structural behavior in terms of applied stress, crack length, and specimen or machine component geometry.

fracture strength. The normal stress at the beginning of fracture. Calculated from the load at the beginning of fracture during a tension test and the original cross-sectional area of the specimen.

fracture test. Test in which a specimen is broken and its fracture surface is examined with the unaided eye or with a low-power microscope to determine such factors as composition, grain size, case depth, or discontinuities.

fracture toughness. A generic term for measures of resistance to extension of a crack. The term is sometimes restricted to results of *fracture mechanics* tests, which are directly applicable in fracture control. However, the term commonly includes results from simple tests of notched or precracked specimens not based on fracture mechanics analysis. Results from tests of the latter type are often useful for fracture control, based on either service experience or empirical correlations with fracture mechanics tests.

free machining. Pertains to the machining characteristics of an alloy to which one or more ingredients have been introduced to produce small broken chips, lower power consumption, better surface finish, and longer tool life; among such additions are sulfur or lead to steel, lead to brass, lead and bismuth to aluminum, and sulfur or selenium to stainless steel.

freezing range. That temperature range between *liquidus* and *solidus* temperatures in which molten and solid constituents coexist.

fretting corrosion. (1) The accelerated deterioration at the interface between contacting surfaces as the result of corrosion and slight oscillatory movement between the two surfaces. (2) A form of fretting in which chemical reaction predominates. Fretting corrosion is often characterized by the removal of particles and subsequent formation of oxides, which are often abrasive and so increase the wear. Fretting corrosion can involve other chemical reaction products, which may not be abrasive.

friction welding (metals). A solid-state process in which welds are made by holding a nonrotating workpiece in contact with a rotating workpiece under constant or gradually increasing pressure until the interface reaches the welding temperature and rotation can be stopped.

full annealing. An imprecise term that denotes an annealing cycle to produce minimum strength and hardness. For the term to be meaningful, the composition and starting condition of the material and the time-temperature cycle used must be stated.

full hard. A *temper* of nonferrous alloys and some ferrous alloys corresponding approximately to a cold-worked state beyond which the material can no longer be formed by bending. In specifications, a full hard temper is commonly defined in terms of minimum hardness or minimum tensile strength (or, alternatively, a range of hardness or strength) corresponding to a specific percentage of cold reduction following a full anneal. For aluminum, a full hard temper is equivalent to a reduction of 75% from dead soft; for austenitic stainless steels, a reduction of about 50 to 55%.

furnace brazing. A mass-production *brazing* process in which the filler metal is preplaced on the joint, then the entire assembly is heated to brazing temperature in a furnace. Usually, a protective furnace atmosphere is required, and wetting of the joint surfaces is accomplished without using a brazing flux.

fusion welding. Any welding process in which the filler metal and base metal (substrate), or base metal only, are melted together to complete the weld.

G

gage. (1) The thickness of sheet or the diameter of wire. The various standards are arbitrary and differ with regard to ferrous and nonferrous products as well as sheet and wire. (2) An aid for visual inspection that enables an inspector to determine more reliably whether the size or contour of a formed part meets dimensional requirements. (3) An instrument used to measure thickness or length.

galling. (1) A condition whereby excessive friction between high spots results in localized welding with subsequent *spalling* and a further roughening of the rubbing surfaces of one or both of two mating parts. (2) A severe form of scuffing associated with gross damage to the surfaces, or failure. Galling has been used in many ways in *tribology*; therefore, each time it is encountered its meaning must be ascertained from the specific context of the usage.

galvanic corrosion. Corrosion associated with the current of a galvanic cell consisting of two dissimilar conductors in an electrolyte, or two similar conductors in dissimilar electrolytes. Where the two dissimilar metals are in contact, the resulting reaction is referred to as couple action.

galvanize. To coat a metal surface with zinc using any of various processes.

galvanneal. To produce a zinc-iron alloy coating on iron or steel by keeping the coating molten after hot-dip galvanizing until the zinc alloys completely with the base metal.

galvanometer. An instrument for indicating or measuring a small electric current by means of a mechanical motion derived from electromagnetic or electrodynamic forces produced by the current.

gamma iron. The face-centered cubic form of pure iron, stable from 910 to 1400 °C (1670 to 2550 °F).

gas-shielded arc welding. A general term used to describe gas metal arc welding, gas tungsten arc welding, and flux-cored arc welding when gas shielding is employed. Typical gases employed include argon, helium, argon-hydrogen mixture, or carbon dioxide.

gas tungsten arc cutting. An arc-cutting process in which metals are severed by melting them with an arc between a single tungsten (nonconsumable) electrode and the work. Shielding is obtained from a gas or gas mixture.

gas tungsten arc welding (GTAW). An arc welding process that produces coalescence of metals by heating them with an arc between a tungsten (nonconsumable) electrode and the work. Shielding is obtained from a gas or gas mixture. Pressure may or may not be used and filler metal may or may not be used.

gate (casting). The portion of the runner in a mold through which molten metal enters the mold cavity. The generic term is sometimes applied to the entire network of connecting channels that conduct metal into the mold cavity.

gouging. In welding practice, the forming of a bevel or groove by material removal.

gouging abrasion. A form of high-stress *abrasion* in which easily observable grooves or gouges are created on the surface.

grain. An individual *crystal* in a polycrystalline material; it may or may not contain twinned regions and subgrains.

grain boundary. A narrow zone in a metal or ceramic corresponding to the transition from one crystallographic orientation to another, thus separating one *grain* from another; the atoms in each grain are arranged in an orderly pattern.

grain-boundary etching. In metallography, the development of intersections of grain faces with the polished surface. Because of severe, localized crystal deformation, grain boundaries have higher dissolution potential than grains themselves. Accumulation of impurities in grain boundaries increases this effect.

grain growth. (1) An increase in the average size of the grains in polycrystalline material, usually as a result of heating at elevated temperature. (2) In polycrystalline materials, a phenomenon occurring fairly close below the melting point in which the larger grains grow still larger while the smallest ones gradually diminish and disappear.

grain size. (1) For metals, a measure of the areas or volumes of grains in a polycrystalline material, usually expressed as an average when the individual sizes are fairly uniform. In metals containing two or more phases, grain size refers to that of the matrix unless otherwise specified. Grain size is reported in terms of number of grains per unit area or volume, in terms of average diameter, or as a grain-size number derived from area measurements. (2) For grinding wheels, preferred term is grit size.

graphite. (1) A crystalline allotropic form of carbon. (2) Uncombined carbon in cast irons.

gray iron. A broad class of ferrous casting alloys (cast irons) normally characterized by a microstructure of flake graphite in a ferrous matrix. Gray irons usually contain 2.5 to 4% C, 1 to 3% Si, and additions of manganese, depending on the desired microstructure (as low as 0.1% Mn in ferritic gray irons and as high as 1.2% in pearlitics). Sulfur and phosphorus are also present in small amounts as residual impurities.

H

hardenability. The relative ability of a ferrous alloy to form martensite when quenched from a temperature above the upper critical temperature. Hardenability is commonly measured as the distance below a quenched surface at which the metal exhibits a specific hardness (50 HRC, for example) or a specific percentage of martensite in the microstructure.

hardening. Increasing hardness of metals by suitable treatment, usually involving heating and cooling. When applicable, the following more specific terms should be used: *age hardening, case hardening, flame hardening, induction hardening, precipitation hardening,* and *quench hardening.*

hardfacing. The application of a hard, wear-resistant material to the surface of a component by welding, spraying, or allied welding processes to reduce wear or loss of material by *abrasion,* impact, *erosion, galling,* and cavitation.

hardness. A measure of the resistance of a material to surface indentation or abrasion; may be thought of as a function of the stress required to produce some specified type of surface deformation. There is no absolute scale for hardness; therefore, to express hardness quantitatively, each type of test has its own scale of arbitrarily defined hardness. Indentation hardness can be measured by *Brinell, Rockwell, Vickers, Knoop,* and Scleroscope hardness tests.

heat. A stated tonnage of metal obtained from a period of continuous melting in a cupola or furnace, or the melting period required to handle this tonnage.

heat-affected zone (HAZ). That portion of the base metal that was not melted during brazing, cutting, or welding, but whose microstructure and mechanical properties were altered by the heat.

heat-resistant alloy. An alloy developed for very-high-temperature service where relatively high stresses (tensile, thermal, vibratory, or shock) are encountered and where oxidation resistance is frequently required.

heat sink. A material that absorbs or transfers heat away from a critical element or part.

heat transfer. Flow of heat by conduction, convection, or radiation.

heat-treatable alloy. An alloy that can be hardened by heat treatment.

heat treatment. Heating and cooling a solid metal or alloy in such a way as to obtain desired conditions or properties. Heating for the sole purpose of hot working is excluded from the meaning of this definition.

hexagonal (lattice for crystals). Having two equal coplanar axes, a_1 and a_2, at 120° to each other and a third axis, c, at right angles to the other two; c may or may not equal a_1 and a_2.

hexagonal close-packed. (1) A structure containing two atoms per unit cell located at (0, 0, 0) and ($\frac{1}{3}$, $\frac{2}{3}$, $\frac{1}{2}$) or ($\frac{2}{3}$, $\frac{1}{3}$, $\frac{1}{2}$). (2) One of the two ways in which spherical objects can be most closely packed together so that the close-packed planes are alternately staggered in the order A-B A-B A-B.

high-strength low-alloy (HSLA) steels. Steels designed to provide better mechanical properties and/or greater resistance to *atmospheric corrosion* than conventional carbon steels. They are not considered to be alloy steels in the normal sense because they are designed to meet specific mechanical properties rather than a chemical composition (HSLA steels have yield strengths greater than 275 MPa, or 40 ksi). The chemical composition of a specific HSLA steel may vary for different product thicknesses to meet mechanical property requirements. The HSLA steels have low carbon contents (0.05 to ~0.25% C) in order to produce adequate *formability* and *weldability,* and they have manganese contents up to 2.0%. Small quantities of chromium, nickel, molybdenum, copper, nitrogen, vanadium, niobium, titanium, and zirconium are used in various combinations.

The types of HSLA steels commonly used include: (1) Weathering steels, designed to exhibit superior atmospheric corrosion resistance. (2) Control-rolled steels, hot rolled according to a predetermined rolling schedule designed to develop a highly deformed austenite structure that will transform to a very fine equiaxed ferrite structure on cooling. (3) Pearlite-reduced steels, strengthened by very-fine-grain ferrite and precipitation hardening but with low carbon content and therefore little or no pearlite in the microstructure. (4) Microalloyed steels, with very small additions (generally $< 0.10\%$ each) of such elements as niobium, vanadium, and/or titanium for refinement of grain size and/or precipitation hardening. (5) Acicular ferrite steel, very-low-carbon steels with sufficient hardenability to transform on cooling to a very fine high-strength acicular ferrite (low-carbon bainite) structure rather than the usual polygonal ferrite structure. (6) Dual-phase steels, processed to a microstructure of ferrite containing small, uniformly distributed regions of high-carbon martensite, resulting in a product with low yield strength and a high rate of work hardening, thus providing a high-strength steel of superior formability.

HIP. See *hot isostatic pressing*.

holding. In heat treating of metals, that portion of the thermal cycle during which the temperature of the object is maintained constant.

holding furnace. A furnace into which molten metal can be transferred to be held at the proper temperature until it can be used to make castings.

holding temperature. In heat treating of metals, the constant temperature at which the object is maintained.

homogeneous. A body of material or matter, alike throughout; hence comprised of only one chemical composition and phase, without internal boundaries.

homogenizing. A heat treating practice whereby a metal object is held at high temperature to eliminate or decrease chemical segregation by diffusion.

honeycomb. Manufactured product of resin-impregnated sheet material (paper, fiberglass, and so on) or metal (aluminum, titanium, and corrosion-resistant alloys) foil, formed into hexagonal-shaped cells. Used as a core material in composite sandwich constructions.

hot corrosion. An accelerated corrosion of metal surfaces that results from the combined effect of oxidation and reactions with sulfur compounds and other contaminants, such as chlorides, to form a molten salt on a metal surface that fluxes, destroys, or disrupts the normal protective oxide.

hot cracking. (1) A crack formed in a weldment caused by the segregation at grain boundaries of low-melting constituents in the weld metal. This can result in grain-boundary tearing under thermal contraction stresses. Hot cracking can be minimized by the use of low-impurity welding materials and proper joint design. (2) A crack formed in a cast metal because of internal stress developed upon cooling following solidification. A hot crack is less open than a *hot tear* and usually exhibits less oxidation and decarburization along the fracture surface.

hot-die forging. A hot forging process in which both the dies and the forging stock are heated; typical die temperatures are 110 to 225 °C (200 to 400 °F) lower than the temperature of the stock.

hot extrusion. A process whereby a heated *billet* is forced to flow through a shaped die opening. The temperature at which extrusion is performed depends on the material being extruded. Hot extrusion is used to produce long, straight metal products of constant cross section, such as bars, solid and hollow sections, tubes, wires, and strips, from materials that cannot be formed by cold extrusion.

hot forging. (1) A forging process in which the die and/or forging stock are heated. See also *hot-die forging*. (2) The plastic deformation of a pressed and/or sintered powder compact in at least two directions at temperatures above the *recrystallization temperature*.

hot isostatic pressing. (1) A process for simultaneously heating and forming a compact in which the powder is contained in a sealed flexible sheet metal or glass enclosure and the so-contained powder is subjected to equal pressure from all directions at a temperature high enough to permit *plastic deformation* and *sintering* to take place. (2) A process that subjects a component (casting, powder forging, etc.) to both elevated temperature and isostatic gas pressure in an autoclave. The most widely used pressurizing gas is argon. When castings are hot isostatically pressed, the simultaneous application of heat and pressure virtually eliminates internal voids and microporosity through a combination of plastic deformation, creep, and diffusion.

hot quenching. An imprecise term for various quenching procedures in which a quenching medium is maintained at a prescribed temperature above 70 °C (160 °F).

hot shortness. A tendency for some alloys to separate along grain boundaries when stressed or deformed at temperatures near the melting point. Hot shortness is caused by a low-melting constituent, often present only in minute amounts, that is segregated at grain boundaries.

hot tear. A fracture formed in a metal during solidification because of hindered contraction.

hot-wire test. Method used to test heat extraction rates of various quenchants. Faster heat-extracting quenchants will permit more electric current to pass through a standard wire because it is cooled more quickly.

hot-worked structure. The structure of a material worked at a temperature higher than the *recrystallization temperature*.

hot working. (1) The plastic deformation of metal at such a temperature and strain rate that *recrystallization* takes place simultaneously with the deformation, thus avoiding any *strain hardening*. Also referred to as hot forging and hot forming. (2) Controlled mechanical operations for shaping a product at temperatures above the *recrystallization temperature*.

I

image analysis. Measurement of the size, shape, and distributional parameters of microstructural features by electronic scanning methods, usually automatic or semiautomatic. Image analysis data output can provide individual measurements on each separate feature (feature specific) or field totals for each measured parameter (field specific).

immiscible. (1) Of two phases, the inability to dissolve in one another to form a single solution; mutually insoluble. (2) With respect to two or more fluids, not mutually soluble; incapable of attaining homogeneity.

impact energy. The amount of energy, usually given in joules or foot-pound force, required to fracture a material, usually measured by means of an *Izod test* or *Charpy test*. The type of specimen and test conditions affect the values and therefore should be specified.

impact extrusion. The process (or resultant product) in which a punch strikes a *slug* (usually unheated) in a confining die. The metal flow may be either between punch and die or through another opening. The impact extrusion of unheated slugs is often called cold extrusion.

impact load. An especially severe shock load such as that caused by instantaneous arrest of a falling mass, by shock meeting of two parts (in a mechanical hammer, for example), or by explosive impact in which there can be an exceptionally rapid buildup of stress.

impact strength. A measure of the resiliency or *toughness* of a solid. The maximum force or energy of a blow (given by a fixed procedure) that can be withstood without fracture, as opposed to *fracture strength* under a steady applied force.

impact test. A test for determining the energy absorbed in fracturing a testpiece at high velocity, as distinct from static test. The test may be carried out in tension, bending, or torsion, and the test bar may be notched or unnotched.

impingement attack. *Corrosion* associated with turbulent flow of liquid. May be accelerated by entrained gas bubbles.

impurities. (1) Elements or compounds whose presence in a material is undesirable. (2) In a chemical or material, minor constituent(s) or component(s) not included deliberately; usually to some degree or above some level, undesirable.

inclusion. (1) A physical and mechanical *discontinuity* occurring within a material or part, usually consisting of solid, encapsulated foreign material. Inclusions are often capable of transmitting some structural stresses and energy fields, but to a noticeably different degree than from the parent material. (2) Particles of foreign material in a metallic matrix. The particles are usually compounds, such as oxides, sulfides, or silicates, but may be of any substance that is foreign to (and essentially insoluble in) the matrix.

indentation. In a spot, seam, or projection weld, the depression on the exterior surface of the base metal.

indentation hardness. (1) The resistance of a material to *indentation*. This is the usual type of hardness test, in which a pointed or rounded indenter is pressed into a surface under a substantially static load. (2) Resistance of a solid surface to the penetration of a second, usually harder, body under prescribed conditions. Numerical values used to express indentation hardness are not absolute physical quantities, but depend on the hardness scale used to express hardness.

indication. In inspection, a response to a nondestructive stimulus that implies the presence of an imperfection. The indication must be interpreted to determine if (a) it is a true indication or a false indication and (b) whether or not a true indication represents an unacceptable deviation.

induction bonding. The use of high-frequency (5 to 7 MHz) electromagnetic fields to heat a bonding agent placed between the plastic parts to be joined. The bonding agent consists of microsized ferromagnetic particles dispersed in a thermoplastic matrix, preferably the parent material of the parts to be bonded. When this binder is exposed to the high-frequency source, the ferromagnetic particles respond and melt the surrounding plastic matrix, which in turn melts the interface surfaces of the parts to be joined.

induction brazing. A brazing process in which the surfaces of components to be joined are selectively heated to brazing temperature by electrical energy transmitted to the workpiece by induction, rather than by a direct electrical connection, using an inductor or work coil.

induction hardening. A *surface hardening* process in which only the surface layer of a suitable ferrous workpiece is heated by electromagnetic induction to above the upper critical temperature and immediately quenched.

induction heating. Heating by combined electrical resistance and hysteresis losses induced by subjecting a metal to the varying magnetic field surrounding a coil carrying alternating current.

inert gas. (1) A gas, such as helium, argon, or nitrogen, that is stable, does not support combustion, and does not form reaction products with other materials. (2) In welding, a gas that does not normally combine chemically with the base metal or filler metal. (3) In processing of plastics, a gas (usually nitrogen) that does not absorb or react with ultraviolet light in a curing chamber.

injection molding (metals). A process similar to plastic injection molding using a plastic-coated metal powder of fine particle size (~10 μm).

inoculant. Materials that, when added to molten metal, modify the structure and thus change the physical and mechanical properties to a degree not explained on the basis of the change in composition resulting from their use. Ferrosilicon-base alloys are commonly used to inoculate gray irons and ductile irons.

inorganic. Being or composed of matter other than hydrocarbons and their derivatives, or matter that is not of plant or animal origin.

insulator. A material of such low electrical conductivity that the flow of current through it can usually be neglected. Similarly, a material of low thermal conductivity, such as that used to insulate structures.

interface. The boundary between two phases. Among the three phases (gas, liquid, and solid), there are five types of interfaces: gas-liquid, gas-solid, liquid-liquid, liquid-solid, and solid-solid.

intergranular corrosion. Corrosion occurring preferentially at grain boundaries, usually with slight or negligible attack on the adjacent grains.

interrupted aging. Aging at two or more temperatures, by steps, and cooling to room temperature after each step.

interrupted quenching. A quenching procedure in which the workpiece is removed from the first quench at a temperature substantially higher than that of the quenchant and is then subjected to a second quenching system having a different cooling rate than the first.

investment casting. (1) Casting metal into a mold produced by surrounding, or *investing*, an expendable pattern with a refractory slurry coating that sets at room temperature, after which the wax or plastic pattern is removed through the use of heat prior to filling the mold with liquid metal.

ion. An atom, or group of atoms, which by loss or gain of one or more electrons has acquired an electric charge. If the ion is formed from an atom of hydrogen or an atom of a metal, it is usually positively charged; if the ion is formed from an atom of a nonmetal or from a group of atoms, it is usually negatively charged. The number of electronic charges carried by an ion is termed its electrovalence. The charges are denoted by superscripts that give their sign and number; for example, a sodium ion, which carries one positive charge, is denoted by Na^+; a sulfate ion, which carries two negative charges, by SO_4^{2-}.

ion carburizing. A method of *surface hardening* in which carbon ions are diffused into a workpiece in a vacuum through the use of high-voltage electrical energy. Synonymous with plasma carburizing or glow-discharge carburizing.

ionic bond. (1) A type of chemical bonding in which one or more electrons are transferred completely from one atom to another, thus converting the neutral atoms into electrically charged ions. These ions are approximately spherical and attract each other because of their opposite charges. (2) A primary bond arising from the electrostatic attraction between oppositely charged ions.

ion nitriding. A method of *surface hardening* in which nitrogen ions are diffused into a workpiece in a vacuum through the use of high-voltage electrical energy. Synonymous with plasma nitriding or glow-discharge nitriding.

isostatic pressing. A process for forming a powder metallurgy compact by applying pressure equally from all directions to metal powder contained in a sealed flexible mold.

isothermal transformation. A change in phase that takes place at a constant temperature. The time required for transformation to be completed, and in some instances the time delay before transformation begins, depends on the amount of supercooling below (or superheating above) the equilibrium temperature for the same transformation.

isotropic. Having uniform properties in all directions. The measured properties of an isotropic material are independent of the axis of testing.

Izod test. A type of impact test in which a V-notched specimen, mounted vertically, is subjected to a sudden blow delivered by the weight at the end of a pendulum arm. The energy required to break off the free end is a measure of the *impact strength* or *toughness* of the material.

J

Jominy test. A laboratory test for determining the hardenability of steel that involves heating the test specimen to the proper *austenitizing* temperature and then transferring it to a quenching fixture so designed that the specimen is held vertically 12.7 mm (0.5 in.) above an opening through which a column of water can be directed against the bottom face of the specimen. While the bottom end is being quenched by the column of water, the opposite end is cooling slowly in air, and intermediate positions along the specimen are cooling at intermediate rates. After the specimen has been quenched, parallel flats 180° apart are ground 0.38 mm (0.015 in.) deep on the cylindrical surface. Rockwell C hardness is measured at intervals of $\frac{1}{16}$ in. (1.6 mm) for alloy steels and $\frac{1}{32}$ in. (0.8 mm) for carbon steels, starting from the water-quenched end. A typical plot of these hardness values and their positions on the test bar indicates the relationship between hardness and cooling rate, which in effect is the *hardenability* of the steel.

K

kelvin. A scale of absolute temperatures in which zero is approximately –273.16 °C (–459.69 °F). The color temperature of light is measured in kelvins.

kerf. The width of the cut produced during a cutting process.

killed steel. Steel treated with a strong deoxidizing agent, such as silicon or aluminum, in order to reduce the oxygen content to such a level that no reaction occurs between carbon and oxygen during solidification.

kinetic energy. The energy that a body possesses because of its motion; in classical mechanics, equal to one-half of the body's mass times the square of its speed.

Knoop hardness number (HK). A number related to the applied load and to the projected area of the permanent impression made by a rhombic-based pyramidal diamond indenter having included edge angles of 172° 30′ and 130° 0′. In reporting the hardness numbers, the test load is stated.

Knoop hardness test. An indentation hardness test using calibrated machines to force a rhombic-based pyramidal diamond indenter having specified edge angles, under specified conditions, into the surface of the material under test and to measure the long diagonal after removal of the load.

Kroll process. A process for the production of metallic titanium sponge by the reduction of titanium tetrachloride with a more active metal, such as magnesium or sodium. The sponge is further processed to granules or powder.

L

ladle metallurgy. Degassing processes for steel carried out in a ladle.

lamellar tearing. Occurs in the base metal adjacent to weldments due to high through-thickness strains introduced by weld metal shrinkage in highly restrained joints. Tearing occurs by decohesion and linking along the working direction of the base metal; cracks usually run roughly parallel to the fusion line and are steplike in appearance. Lamellar tearing can be minimized by designing joints to minimize weld shrinkage stresses and joint restraint.

lamination. (1) A type of *discontinuity* with separation or weakness generally aligned parallel to the worked surface of a metal. May be the result of pipe, blisters, seams, inclusions, or segregations elongated and made directional by working. Laminations may also occur in powder metallurgy compacts. (2) In electrical products such as motors, a blanked piece of electrical sheet that is stacked up with several other identical pieces to make a stator or rotor.

laser. A device that emits a concentrated beam of electromagnetic radiation (light). Laser beams are used in metalworking to melt, cut, or weld metals; in less concentrated form they are sometimes used to inspect metal parts.

laser beam cutting. A cutting process that severs materials with the heat obtained from the application of a concentrated coherent light beam impinging upon the workpiece to be cut. The process can be used with (gas-assisted laser beam cutting) or without an externally supplied gas.

laser beam machining. Use of a highly focused monofrequency collimated beam of light to melt or sublime material at the point of impingement on a workpiece.

laser beam welding. A welding process that joins metal parts using the heat obtained by directing a beam from a *laser* onto the weld joint.

laser hardening. A *surface hardening* process that uses a *laser* to quickly heat a surface. Heat conduction into the interior of the part will quickly cool the surface, leaving a shallow martensitic layer.

lath martensite. *Martensite* formed partly in steels containing less than approximately 1.0% C and solely in steels containing less than approximately 0.5% C as parallel arrays of lath-shape units 0.1 to 0.3 μm thick.

lattice. (1) A space lattice is a set of equal and adjoining parallelopipeds formed by dividing space by three sets of parallel planes, the planes in any one set being equally spaced. There are seven ways of so dividing space, corresponding to the seven crystal systems. The unit parallelopiped is usually chosen as the unit cell of the system. (2) A point lattice is a set of points in space located so that each point has identical surroundings. There are 14 ways of so arranging points in space, corresponding to the 14 Bravais lattices.

leaching. Extracting an element or compound from a solid alloy or mixture by preferential dissolution in a suitable liquid.

light metal. One of the low-density metals, such as aluminum (~2.7 g/cm^3), magnesium (~1.7 g/cm^3), titanium (~4.4 g/cm^3), beryllium (~1.8 g/cm^3), or their alloys.

liquid nitriding. A method of *surface hardening* in which molten nitrogen-bearing, fused-salt baths containing both cyanides and cyanates are exposed to parts at subcritical temperatures. A typical commercial bath for liquid nitriding is composed of a mixture of sodium and potassium salts. The sodium salts, which comprise 60 to 70% (by weight) of the total mixture, consist of 96.5% NaCN, 2.5% Na_2CO_3, and 0.5% NaCNO. The potassium salts, 30 to 40% (by weight) of the mixture, consist of 96% KCN, 0.6% K_2CO_3, 0.75% KCNO, and 0.5% KCl. The operating temperature of this salt bath is 565 °C (1050 °F).

liquid nitrocarburizing. A *nitrocarburizing* process (where both carbon and nitrogen are absorbed into the surface) utilizing molten liquid salt baths below the lower critical temperature. Liquid nitrocarburizing processes are used to improve wear resistance and fatigue properties of steels and cast irons.

liquid penetrant inspection. A type of *nondestructive inspection* that locates discontinuities that are open to the surface of a metal by first allowing a penetrating dye or fluorescent liquid to infiltrate the discontinuity, removing the excess penetrant, and then applying a developing agent that causes the penetrant to seep back out of the discontinuity and register as an indication. Liquid penetrant inspection is suitable for both ferrous and nonferrous materials, but is limited to the detection of open surface discontinuities in nonporous solids.

liquidus (metals). (1) The lowest temperature at which a metal or an alloy is completely liquid. (2) In a *phase diagram*, the locus of points representing the temperatures at which the various compositions in the system begin to freeze on cooling or finish melting on heating. See also *solidus*.

load. (1) In the case of testing machines, a force applied to a testpiece that is measured in units such as pound-force, newton, or kilogram-force. (2) In *tribology*, the force applied normal to the surface of one body by another contacting body or bodies. The term normal force is more precise and therefore preferred; however, the term normal load is also in use. If applied vertically, the load can be expressed in mass units, but it is preferable to use force units such as newtons (N).

localized corrosion. Corrosion at discrete sites, for example, *crevice corrosion*, pitting, and *stress-corrosion cracking*.

lot. (1) A specific amount of material produced at one time using one process and constant conditions of manufacture, and offered for sale as a unit quantity. (2) A quantity of material that is thought to be uniform in one or more stated properties, such as isotopic, chemical, or physical characteristics. (3) A quantity of bulk material of similar composition whose properties are under study. (4) A definite quantity of a product or material accumulated under conditions that are considered uniform for sampling purposes.

low-alloy steels. A category of ferrous materials that exhibit mechanical properties superior to plain carbon steels as the result of additions of such alloying elements as nickel, chromium, and molybdenum. Total alloy content can range from 2.07% up to levels just below that of stainless steels, which contain a minimum of 10% Cr. For many low-alloy steels, the primary function of the alloying elements is to increase hardenability in order to optimize mechanical properties and toughness after heat treatment. In some cases, however, alloy additions are used to reduce environmental degradation under certain specified service conditions.

low-cycle fatigue. *Fatigue* that occurs at relatively small numbers of cycles (<10^4 cycles). Low-cycle fatigue may be accompanied by some plastic, or permanent, deformation.

Lüders lines. Elongated surface markings or depressions, in sheet metal, often visible with the unaided eye, caused by discontinuous (inhomogeneous) yielding. Also known as Lders bands, Hartmann lines, Piobert lines, or stretcher strains.

M

machinability index. A relative measure of the machinability of an engineering material under specified standard conditions. Also known as machinability rating.

machining stress. *Residual stress* caused by machining.

macrograph. A graphic representation of the surface of a prepared specimen at a magnification not exceeding 25×. When photographed, the reproduction is known as a *photomacrograph*.

macrohardness test. A term applied to such hardness testing procedures as the *Rockwell* or *Brinell* hardness tests to distinguish them from microindentation hardness tests such as the *Knoop* or *Vickers* tests.

magnetic-particle inspection. A *nondestructive inspection* method for determining the existence and extent of surface cracks and similar imperfections in ferromagnetic materials. Finely divided magnet particles, applied to the magnetized part, are attracted to and outline the pattern of any magnetic-leakage fields created by discontinuities.

malleability. The characteristic of metals that permit *plastic deformation* in compression without fracture.

maraging. A *precipitation hardening* treatment applied to a special group of iron-base alloys to precipitate one or more intermetallic compounds in a matrix of essentially carbon-free *martensite*.

maraging steels. A special class of high-strength steels that differ from conventional steels in that they are hardened by a metallurgical reaction that does not involve carbon. Instead, these steels are strengthened by the *precipitation* of intermetallic compounds at temperatures of about 480 °C (900 °F). The term maraging is derived from martensite age hardening of a low-carbon, iron-nickel *lath martensite* matrix. Commercial maraging steels are designed to provide specific levels of yield strength from 1030 to 2420 MPa (150 to 350 ksi), with some having yield strengths as high as 3450 MPa (500 ksi). These steels typically have very high nickel, cobalt, and molybdenum contents and very low carbon contents.

martempering. (1) A *hardening* procedure in which an austenitized ferrous material is quenched into an appropriate medium at a temperature just above the martensite start temperature of the material, held in the medium until the temperature is uniform throughout, although not long enough for *bainite* to form, then cooled in air. The treatment is frequently followed by tempering. (2) When the process is applied to carburized material, the controlling martensite start temperature is that of the *case*. This variation of the process is frequently called marquenching.

martensite. A generic term for microstructures formed by a diffusionless phase transformation in which the parent and product phases have a specific crystallographic relationship. Martensite is characterized by an acicular pattern in the microstructure in both ferrous and nonferrous alloys. In alloys where the solute atoms occupy interstitial positions in the martensitic lattice (such as carbon in iron), the structure is hard and highly strained; but where the solute atoms occupy substitutional positions (such as nickel in iron), the martensite is soft and ductile. The amount of high-temperature phase that transforms to martensite on cooling depends to large extent on the lowest temperature attained, there being a rather distinct beginning temperature (M_s) and a temperature at which the transformation is essentially complete (M_f).

mass spectrometry. An analytical technique for identification of chemical structures, analysis of mixtures, and quantitative elemental analysis, based on application of the mass spectrometer.

material characterization. The use of various analytical methods (spectroscopy, microscopy, chromatography, etc.) to describe those features of composition (both bulk and surface) and structure (including defects) of a material that are significant for a particular preparation, study of properties, or use. Test methods that yield information primarily related to materials properties, such as thermal, electrical, and mechanical properties, are excluded from this definition.

matrix (metals). The continuous or principal phase in which another constituent is dispersed.

mechanical metallurgy. The science and technology dealing with the behavior of metals when subjected to applied forces; often considered to be restricted to plastic working or shaping of metals.

mechanical properties. The properties of a material that reveal its elastic and inelastic behavior when force is applied, thereby indicating its suitability for mechanical applications; for example, modulus of elasticity, *tensile strength, elongation, hardness,* and fatigue limit.

mechanical testing. The methods by which the *mechanical properties* of a metal are determined.

mechanical wear. Removal of material due to mechanical processes under conditions of sliding, rolling, or repeated impact. The term mechanical wear includes *adhesive wear, abrasive wear*, and *fatigue wear*.

mechanical working. The subjecting of metals to pressure exerted by rolls, hammers, or presses, in order to change the shape or physical properties of the metal.

melt. (1) To change a solid to a liquid by the application of heat. (2) A charge of molten metal or plastic.

melting point (metals). The temperature at which a pure metal, compound, or eutectic changes from solid to liquid; the temperature at which the liquid and the solid are at equilibrium.

melting range (metals). The range of temperatures over which an alloy other than a compound or eutectic changes from solid to liquid; the range of temperatures from *solidus* to *liquidus* at any given composition on a *phase diagram*.

metal. (1) An opaque lustrous elemental chemical substance that is a good conductor of heat and electricity and, when polished, a good reflector of light. Most elemental metals are malleable and ductile and are, in general, denser than the other elemental substances. (2) As to structure, metals may be distinguished from nonmetals by their atomic binding and electron availability. Metallic atoms tend to lose electrons from the outer shells, the positive ions thus formed being held together by the electron gas produced by the separation. The ability of these "free electrons" to carry an electric current, and the fact that this ability decreases as temperature increases, establish the prime distinctions of a metallic solid. (3) From a chemical viewpoint, an elemental substance whose hydroxide is alkaline. (4) An alloy.

metal-arc welding. Any of a group of arc welding processes in which metals are fused together using the heat of an arc between the metal electrode and the work. Use of the specific process name is preferred.

metallic glass. A noncrystalline metal or alloy, commonly produced by drastic supercooling of a molten alloy, by molecular deposition, which involves growth from the vapor phase (e.g., thermal evaporation and sputtering) or from a liquid phase (e.g., electroless deposition and electrodeposition), or by external action techniques (e.g., ion implantation and ion beam mixing). Glassy alloys can be grouped into two major categories. The first group includes the transition metal/metal binary alloy systems, such as Cu-Zr, Ni-Ti, W-Si, and Ni-Nb. The second class consists of transition metal/metalloid alloys. These alloys are usually iron-, nickel-, or cobalt-base systems, may contain film formers (such as chromium and titanium), and normally contain approximately 20 at.% P, B, Si, and/or C as the metalloid component. Also called amorphous alloy or metal.

metallizing. (1) Forming a metallic coating by atomized spraying with molten metal or by vacuum deposition. Also called spray metallizing. (2) Applying an electrically conductive metallic layer to the surface of a nonconductor.

metallograph. An optical instrument designed for visual observation and photomicrography of prepared surfaces of opaque materials at magnifications of 25 to approximately 2000×. The instrument consists of a high-intensity illuminating source, a microscope, and a camera bellows. On some instruments, provisions are made for examination of specimen surfaces using polarized light, phase contrast, oblique illumination, dark-field illumination, and bright-field illumination.

metallography. The study of the structure of metals and alloys by various methods, especially by optical and electron microscopy.

metallurgy. The science and technology of metals and alloys. Process metallurgy is concerned with the extraction of metals from their ores and with refining of metals; physical metallurgy, with the physical and mechanical properties of metals as affected by composition, processing, and environmental conditions; and mechanical metallurgy, with the response of metals to applied forces.

metastable. (1) Of a material not truly stable with respect to some transition, conversion, or reaction but stabilized kinetically either by rapid cooling or by some molecular characteristics as, for example, by the extremely high viscosity of polymers. (2) Possessing a state of pseudoequilibrium that has a free energy higher than that of the true equilibrium state.

microcracking. Cracks formed in composites when thermal stresses locally exceed the strength of the matrix. Since most microcracks do not penetrate the reinforcing fibers, microcracks in a cross-plied laminate or in a laminate made from cloth prepreg are usually limited to the thickness of a single ply.

micrograph. A graphic reproduction of the surface of a specimen at a magnification greater than 25×. If produced by photographic means, it is called a photomicrograph (not a microphotograph).

microhardness. The *hardness* of a material as determined by forcing an indenter such as a Vickers or Knoop indenter into the surface of a material under very light load; usually, the indentations are so small that they must be measured with a microscope. Capable of determining hardnesses of different microconstituents within a structure, or of measuring steep hardness gradients such as those encountered in *case hardening*.

microhardness test. A microindentation hardness test using a calibrated machine to force a diamond indenter of specific geometry, under a test load of 1 to 1000 gram-force, into the surface of the test material and to measure the diagonal or diagonals optically.

microradiography. The technique of passing x-rays through a thin section of a material in contact with a fine-grained photographic film and then viewing the radiograph at 50 to 100× to observe the distribution of constituents and/or defects.

migration. Movement of entities (such as electrons, ions, atoms, molecules, vacancies, and grain boundaries) from one place to another under the influence of a driving force (such as an electrical potential or a concentration gradient).

mild steel. *Carbon steel* with a maximum of about 0.25% C and containing 0.4 to 0.7% Mn, 0.1 to 0.5% Si, and some residuals of sulfur, phosphorus, and/or other elements.

mischmetal. A natural mixture of rare-earth elements (atomic numbers 57 through 71) in metallic form. It contains about 50% Ce, the remainder being principally lanthanum and neodymium. Mischmetal is used as an alloying additive in ferrous alloys to scavenge sulfur, oxygen, and other impurities, and in magnesium alloys to improve high-temperature strength.

miscible. Of two phases, the ability of each to dissolve in the other. May occur in a limited range of ratios of the two, or in any ratio.

mold. (1) The form, made of sand, metal, or refractory material, that contains the cavity into which molten metal is poured to produce a casting of desired shape. (2) A *die*.

molecular fluorescence spectroscopy. An analytical technique that measures the fluorescence emission characteristic of a molecular, as opposed to an atomic, species. The emission results from electronic transitions between molecular states and can be used to detect and/or measure trace amounts of molecular species.

molecular structure. The manner in which electrons and nuclei interact to form a molecule, as elucidated by quantum mechanics and the study of molecular spectra.

molecule. A molecule may be thought of either as a structure built of *atoms* bound together by chemical forces, or as a structure in which two or more positively charged nuclei are maintained in some definite geometrical configuration by attractive forces from the surrounding cloud of *electrons*. Besides chemically stable molecules, short-lived molecular fragments termed free radicals can be observed under special circumstances.

morphology. The characteristic shape, form, or surface texture or contours of the crystals, grains, or particles of (or in) a material, generally on a microscopic scale.

mounting. A means by which a specimen for metallographic examination may be held during preparation of a section surface. The specimen can be embedded in plastic or secured mechanically in clamps.

N

natural aging. Spontaneous *aging* of a supersaturated solid solution at room temperature.

necking. (1) The reduction of the cross-sectional area of a material in a localized area by uniaxial tension or by *stretching*. (2) The reduction of the diameter of a portion of the length of a cylindrical shell or tube.

net shape. The shape of a powder metallurgy part, casting, or forging that conforms closely to specified dimensions. Such a part requires no secondary machining or finishing. A near-net-shape part can be either one in which some but not all of the surfaces are net or one in which the surfaces require only minimal machining or finishing.

nitriding. Introducing nitrogen into the surface layer of a solid ferrous alloy by holding at a suitable temperature (below Ac_1 for ferritic steels) in contact with a nitrogenous material, usually ammonia or molten cyanide of appropriate composition. *Quenching* is not required to produce a hard case.

nitrocarburizing. Any of several processes in which both nitrogen and carbon are absorbed into the surface layers of a ferrous material at temperatures below the lower critical temperature and, by diffusion, create a concentration gradient. Nitrocarburizing is performed primarily to provide an antiscuffing surface layer and to improve fatigue resistance.

noble metal. (1) A metal whose potential is highly positive relative to the hydrogen electrode. (2) A metal with marked resistance to chemical reaction, particularly to oxidation and to solution by inorganic acids. The term as often used is synonymous with precious metal.

nodular graphite. *Graphite* in the nodular form as opposed to flake form. Nodular graphite is characteristic of malleable iron. The graphite of nodular or ductile iron is spherulitic in form, but called nodular.

nondestructive evaluation (NDE). Broadly considered synonymous with *nondestructive inspection* (NDI). More specifically, the quantitative analysis of NDI findings to determine whether the material will be acceptable for its function, despite the presence of discontinuities. With NDE, a *discontinuity* can be classified by its size, shape, type, and location, allowing the investigator to determine whether or not the flaw(s) is acceptable. Damage-tolerant design approaches are based on the philosophy of ensuring safe operation in the presence of flaws.

nondestructive inspection (NDI). A process or procedure, such as ultrasonic or radiographic inspection, for determining the quality or characteristics of a material, part, or assembly, without permanently altering the subject or its properties. Used to find internal anomalies in a structure without degrading its properties or impairing its serviceability.

normal. An imaginary line forming right angles with a surface or other lines; sometimes called the perpendicular. It is used as a basis for determining angles of incidence reflection and refraction.

normalizing. Heating a ferrous alloy to a suitable temperature above the transformation range and then cooling in air to a temperature substantially below the transformation range.

notch brittleness. Susceptibility of a material to *brittle fracture* at points of stress concentration. For example, in a notch tensile test, the material is said to be notch brittle if the notch strength is less than the tensile strength of an unnotched specimen. Otherwise, it is said to be notch ductile.

notch depth. The distance from the surface of a test specimen to the bottom of the notch. In a cylindrical test specimen, the percentage of the original cross-sectional area removed by machining an annular groove.

notched specimen. A test specimen that has been deliberately cut or notched, usually in a V-shape, to induce and locate point of failure.

notch sensitivity. The extent to which the sensitivity of a material to fracture is increased by the presence of stress concentration, such as a notch, a sudden change in cross section, a crack, or a scratch. Low notch sensitivity is usually associated with ductile materials, and high notch sensitivity is usually associated with brittle materials.

nuclear magnetic resonance (NMR). A phenomenon exhibited by a large number of atomic nuclei that is based on the existence of nuclear magnetic moments associated with quantized nuclear spins. These nuclear moments, when placed in a magnetic field, give rise to distinct nuclear Zeeman energy levels between which spectroscopic transitions can be induced by radio-frequency radiation. Plots of these transition frequencies, termed spectra, furnish important information about molecular structure and sample composition.

O

offal. The material trimmed from blanks or formed panels.

oil hardening. *Quench hardening* treatment of steels involving cooling in oil.

oil quenching. *Hardening* of carbon steel in an oil bath. Oils are categorized as conventional, fast, and martempering.

Olsen ductility test. A cupping test in which a piece of sheet metal, restrained except at the center, is deformed by a standard steel ball until fracture occurs. The height of the cup at the time of fracture is a measure of the ductility.

open-die forging. The hot mechanical forming of metals between flat or shaped *dies* in which metal flow is not completely restricted. Also known as hand or smith forging.

open dies. *Dies* with flat surfaces that are used for preforming stock or producing hand forgings.

open hearth furnace. A reverberatory melting furnace with a shallow hearth and a low roof. The flame passes over the charge on the hearth, causing the charge to be heated both by direct flame and by radiation from the roof and sidewalls of the furnace.

optical emission spectroscopy. Pertaining to emission spectroscopy in the near-ultraviolet, visible, or near-infrared wavelength regions of the electromagnetic spectrum.

optical microscope. An instrument used to obtain an enlarged image of a small object, utilizing visible light. In general it consists of a light source, a condenser, an objective lens, an ocular or eyepiece, and a mechanical stage for focusing and moving the specimen. Magnification capability of the optical microscope ranges from 1 to 1500×.

optical pyrometer. An instrument for measuring the temperature of heated material by comparing the intensity of light emitted with a known intensity of an incandescent lamp filament.

orange peel (metals). A surface roughening in the form of a pebble-grained pattern that occurs when a metal of unusually coarse grain size is stressed beyond its elastic limit. Also called pebbles and alligator skin.

oxidation. (1) A reaction in which there is an increase in valence resulting from a loss of electrons. (2) A corrosion reaction in which the corroded metal forms an oxide; usually applied to reaction with a gas containing elemental oxygen, such as air. (3) A chemical reaction in which one substance is changed to another by oxygen combining with the substance. Much of the dross from holding and melting furnaces is the result of oxidation of the alloy held in the furnace.

oxidative wear. (1) A *corrosive wear* process in which chemical reaction with oxygen or an oxidizing environment predominates. (2) A type of wear resulting from the sliding action between two metallic components that generates oxide films on the metal surfaces. These oxide films prevent the formation of a metallic bond between the sliding surfaces, resulting in fine wear debris and low wear rates.

oxyfuel gas welding (OFW). Any of a group of processes used to fuse metals together by heating them with gas flames resulting from combustion of a specific fuel gas, such as acetylene, hydrogen, natural gas, or propane. The process may be used with or without the application of pressure to the joint, and with or without addition of filler metal.

oxygen probe. An atmosphere-monitoring device that electronically measures the difference between the partial pressure of oxygen in a furnace or furnace supply atmosphere and the external air.

P

pack carburizing. A method of *surface hardening* of steel in which parts are packed in a steel box with a carburizing compound and heated to elevated temperatures. Common carburizing compounds contain 1 to 10% alkali or alkaline earth metal carbonates (for example, barium carbonate, $BaCO_3$) bound to a hardwood charcoal or to coke by oil, tar, or molasses. This process has been largely supplanted by gas and liquid carburizing processes.

partial annealing. An imprecise term used to denote a treatment given cold-worked metallic material to reduce its strength to a controlled level or to effect stress relief. To be meaningful, the type of material, the degree of cold work, and the time-temperature schedule must be stated.

particle-induced x-ray emission (PIXE). A method of trace elemental analysis in which a beam of ions (usually protons) is directed at a thin foil on which the sample to be analyzed has been deposited; the energy spectrum of the resulting x-rays is measured.

parts per billion. A measure of proportion by weight, equivalent to one unit weight of a material per billion (10^9) unit weights of compound.

parts per million. A measure of proportion by weight, equivalent to one unit weight of a material per million (10^6) unit weights of compound.

passivation. (1) A reduction of the anodic reaction rate of an electrode involved in corrosion. (2) The process in metal corrosion by which metals become *passive*. (3) The changing of a chemically active surface of a metal to a much less reactive state. (4) The formation of an insulating layer directly over the semiconductor surface to protect the surface from contaminants, moisture, and so forth.

passivator. A type of corrosion inhibitor that appreciably changes the potential of a metal to a more noble (positive) value.

passive. (1) A metal corroding under the control of a surface reaction product. (2) The state of the metal surface characterized by low corrosion rates in a potential region that is strongly oxidizing for the metal.

patenting. In wiremaking, a heat treatment applied to medium-carbon or high-carbon steel before drawing of wire or between drafts. This process consists of heating to a temperature above the transformation range and then cooling to a temperature below Ae_1 in air or in a bath of molten lead or salt.

patina. The coating, usually green, that forms on the surface of metals such as copper and copper alloys exposed to the atmosphere. Also used to describe the appearance of a weathered surface of any metal.

pearlite. A metastable lamellar aggregate of *ferrite* and *cementite* resulting from the transformation of *austenite* at temperatures above the *bainite* range.

pellet. In powder metallurgy, a small rounded or spherical solid body that is similar to a shotted particle. See also *shotting*.

penetrant. A liquid with low surface tension used in *liquid penetrant inspection* to flow into surface openings of parts being inspected.

permanent set. The deformation remaining after a specimen has been stressed a prescribed amount in tension, compression, or shear for a specified time period and released for a specified time period. For creep tests, the residual unrecoverable deformation after the load causing the creep has been removed for a substantial and specified period of time. Also, the increase in length, expressed as a percentage of the original length, by which an elastic material fails to return to its original length after being stressed for a standard period of time.

permeability. (1) The passage or diffusion (or rate of passage) of a gas, vapor, liquid, or solid through a material (often porous) without physically or chemically affecting it; the measure of fluid flow (gas or liquid) through a material. (2) A general term used to express various relationships between magnetic induction and magnetizing force. These relationships are either "absolute permeability," which is a change in magnetic induction divided by the corresponding change in magnetizing force, or "specific (relative) permeability," the ratio of the absolute permeability to the permeability of free space. (3) In metal casting, the characteristics of molding materials that permit gases to pass through them. "Permeability number" is determined by a standard test.

pewter. A tin-base white metal containing antimony and copper. Originally, pewter was defined as an alloy of tin and lead, but to avoid toxicity and dullness of finish, lead is excluded from modern pewter. These modern compositions contain 1 to 8% Sb and 0.25 to 3% Cu. Typical pewter products include coffee and tea services, trays, steins, mugs, candy dishes, jewelry, bowls, plates, vases, candlesticks, compotes, decanters, and cordial cups.

pH. The negative logarithm of the hydrogen-ion activity; it denotes the degree of acidity or basicity of a solution. At 25 °C (77 °F), 7.0 is the neutral value. Decreasing values below 7.0 indicates increasing acidity; increasing values above 7.0, increasing basicity. The pH values range from 0 to 14.

phase. A physically homogeneous and distinct portion of a material system.

phase change. The transition from one physical state to another, such as gas to liquid, liquid to solid, gas to solid, or vice versa.

phase diagram. A graphical representation of the temperature and composition limits of phase fields in an alloy or ceramic system as they actually exist under the specific conditions of heating or cooling. A phase diagram may be an equilibrium diagram, an approximation to an equilibrium diagram, or a representation of metastable conditions or phases. Synonymous with constitution diagram.

photomacrograph. A *macrograph* produced by photographic means.

photomicrograph. A *micrograph* produced by photographic means.

physical metallurgy. The science and technology dealing with the properties of metals and alloys, and of the effects of composition, processing, and environment on those properties.

physical properties. Properties of a material that are relatively insensitive to structure and can be measured without the application of force; for example, density, electrical conductivity, coefficient of thermal expansion, magnetic permeability, and lattice parameter. Does not include chemical reactivity.

physical testing. Methods used to determine the entire range of the *physical properties* of a material. In addition to density and thermal, electrical, and magnetic properties, physical testing methods may be used to assess simple fundamental physical properties such as color, crystalline form, and melting point.

pickup. (1) Transfer of metal from tools to part or from part to tools during a forming operation. (2) Small particles of oxidized metal adhering to the surface of a mill product.

piercing. The general term for cutting (shearing or punching) openings, such as holes and slots, in sheet material, plate, or parts. This operation is similar to blanking; the difference is that the *slug* or pierce produced by piercing is scrap, while the *blank* produced by blanking is the useful part.

pig. A metal casting used in remelting.

pig iron. (1) High-carbon iron made by reduction of iron ore in the blast furnace. (2) Cast iron in the form of *pigs*.

pinhole porosity. Porosity consisting of numerous small gas holes distributed throughout a metal; found in weld metal, castings, and electrodeposited metal.

pinholes. (1) Very small holes that are sometimes found as a type of porosity in a casting because of the microshrinkage or gas evolution during solidification. In wrought products, due to removal of inclusions or microconstituents during macroetching of transverse sections. (2) Small cavities that penetrate the surface of a cured composite or plastic part. (3) In photography, a very small circular aperture.

pipe. (1) The central cavity formed by contraction in metal, especially ingots, during solidification. (2) An imperfection in wrought or cast products resulting from such a cavity. (3) A tubular metal product cast or wrought.

plane (crystal). An idiomorphic face of a *crystal*. Any atom-containing plane in a crystal.

plane strain. The stress condition in linear elastic fracture mechanics in which there is zero strain in a direction normal to both the axis of applied tensile stress and the direction of crack growth (that is, parallel to the crack front); most nearly achieved in loading thick plates along a direction parallel to the plate surface. Under plane-strain conditions, the plane of fracture instability is normal to the axis of the principal tensile stress.

plane-strain fracture toughness (K_{Ic}). The crack extension resistance under conditions of crack-tip plane strain.

plane stress. The stress condition in linear elastic fracture mechanics in which the stress in the thickness direction is zero; most nearly achieved in loading very thin sheet along a direction parallel to the surface of the sheet. Under plane-stress conditions, the plane of fracture instability is inclined 45° to the axis of the principal tensile stress.

plasma. A gas of sufficient energy so that a large fraction of the species present is ionized and thus conducts electricity. Plasmas may be generated by the passage of a current between electrodes, by induction, or by a combination of these methods.

plasma arc welding (PAW). An arc welding process that produces coalescence of metals by heating them with a constricted arc between an electrode and the workpiece (transferred arc) or the electrode and the constricting nozzle (nontransferred arc). Shielding is obtained from hot, ionized gas issuing from an orifice surrounding the electrode and may be supplemented by an auxiliary source of shielding gas, which may be an inert gas or a mixture of gases. Pressure may or may not be used, and filler metal may or may not be supplied.

plastic deformation. The permanent (inelastic) distortion of materials under applied stresses that strain the material beyond its elastic limit.

plasticity. The property of a material that allows it to be repeatedly deformed without rupture when acted upon by a force sufficient to cause deformation and that allows it to retain its shape after the applied force has been removed.

plastic memory. The tendency of a thermoplastic material that has been stretched while hot to return to its unstretched shape upon being reheated.

plate. A flat-rolled metal product of some minimum thickness and width arbitrarily dependent on the type of metal. Plate thicknesses commonly range from 6 to 300 mm (0.25 to 12 in.); widths from 200 to 2000 mm (8 to 80 in.).

plate martensite. *Martensite* formed partly in steel containing more than approximately 0.5% C and solely in steel containing more than approximately 1.0% C that appears as lenticular-shape plates (crystals).

plating. Forming an adherent layer of metal on an object; often used as a shop term for electroplating.

plating rack. A fixture used to hold work and conduct current to it during electroplating.

plowing. In *tribology*, the formation of grooves by *plastic deformation* of the softer of two surfaces in relative motion.

polished surface. A surface prepared for metallographic inspection that reflects a large proportion of the incident light in a specular manner.

polishing. (1) A surface finishing process for ceramics and metal utilizing successive grades of abrasive. (2) Smoothing metal surfaces, often to a high luster, by rubbing the surface with a fine abrasive, usually contained in a cloth or other soft lap. Results in microscopic flow of some surface metal together with actual removal of a small amount of surface metal. (3) Removal of material by the action of abrasive grains carried to the work by a flexible support, generally either a wheel or a coated abrasive belt. (4) A mechanical, chemical, or electrolytic process or combination thereof used to prepare a smooth, reflective surface suitable for microstructural examination that is free of artifacts or damage introduced during prior sectioning or grinding.

polishing artifact. A false structure introduced during a polishing stage of a surface preparation sequence.

polycrystalline. Pertaining to a solid comprised of many *crystals* or crystallites, intimately bonded together. May be homogeneous (one substance) or heterogeneous (two or more crystal types or compositions).

population. In statistics, a generic term denoting any finite or infinite collection of individual samples or data points in the broadest concept; an aggregate determined by some property that distinguishes samples that do and do not belong.

pore. (1) A small opening, void, interstice, or channel within a consolidated solid mass or agglomerate, usually larger than atomic or molecular dimensions. (2) A minute cavity in a powder metallurgy compact, sometimes added intentionally.

porous P/M parts. Powder metallurgy components that are characterized by interconnected porosity. Primary application areas for porous P/M parts are filters, damping devices, storage reservoirs for liquids (including self-lubricating bearings), and battery elements. Bronzes, stainless steels, nickel-base alloys, titanium, and aluminum are used in porous P/M applications.

postheating. Heating weldments immediately after welding, for tempering, for stress relieving, or for providing a controlled rate of cooling to prevent formation of a hard or brittle structure.

potentiometer. An instrument that measures electromotive force by balancing against it an equal and opposite electromotive force across a calibrated resistance carrying a definite current.

powder metallurgy part. A shaped object that has been formed from metal powders and sintered by heating below the melting point of the major constituent. A structural or mechanical component made by the powder metallurgy process.

powder production. The process by which a metal powder is produced, such as machining, milling, atomization, condensation, reduction, oxide decomposition, carbonyl decomposition, electrolytic deposition, or precipitation from a solution.

precipitation. In metals, the separation of a new phase from solid or liquid solution, usually with changing conditions of temperature, pressure, or both.

precipitation hardening. *Hardening* in metals caused by the precipitation of a constituent from a supersaturated solid solution.

precipitation heat treatment. *Artificial aging* of metals in which a constituent precipitates from a supersaturated solid solution.

precision casting. A metal casting of reproducible, accurate dimensions, regardless of how it is made. Often used interchangeably with *investment casting*.

preheating metals. (1) Heating before some further thermal or mechanical treatment. For tool steel, heating to an intermediate temperature immediately before final austenitizing. For some nonferrous alloys, heating to a high temperature for a long time, in order to homogenize the structure before working. (2) In welding and related processes, heating to an intermediate temperature for a short time immediately before welding, brazing, soldering, cutting, or thermal spraying. (3) In powder metallurgy, an early stage in the sintering procedure when, in a continuous furnace, lubricant or binder burnoff occurs without atmosphere protection prior to actual sintering in the protective atmosphere of the high heat chamber.

pressure casting. (1) Making castings with pressure on the molten or plastic metal, as in *injection molding*, *die casting*, centrifugal casting, cold chamber pressure casting, and squeeze casting. (2) A casting made with pressure applied to the molten or plastic metal.

printed circuit. An electronic circuit produced by printing an electrically conductive pattern, wiring, or components on a supporting dielectric substrate, which may be either rigid or flexible.

process metallurgy. The science and technology of winning metals from their ores and purifying metals; sometimes referred to as chemical metallurgy. Its two chief branches are extractive metallurgy and refining.

protective atmosphere. (1) A gas envelope surrounding the part to be brazed, welded, or thermal sprayed, with the gas composition controlled with respect to chemical composition, dew point, pressure, flow rate, and so forth. Examples are inert gases, combusted fuel gases, hydrogen, and vacuum. (2) The atmosphere in a heat treating or sintering furnace designed to protect the parts or compacts from oxidation, nitridation, or other contamination from the environment.

Q

qualification test. A series of tests conducted by the procuring activity, or an agent thereof, to determine the conformance of materials, or materials systems, to the requirements of a specification, normally resulting in a qualified products list under the specification. Generally, qualification under a specification requires a conformance to all tests in the specification, or it may be limited to conformance to a specific type or class, or both, under the specification.

qualified products list (QPL). A list of commercial plastic products that have been pretested and found to meet the requirements of a specification, especially a government specification.

qualitative analysis. An analysis in which some or all of the components of a sample are identified.

quality. (1) The totality of features and characteristics of a product or service that bear on its ability to satisfy a given need (fitness-for-use concept of quality). (2) Degree of excellence of a product or service (comparative concept). Often determined subjectively by comparison against an ideal standard or against similar products or services available from other sources. (3) A quantitative evaluation of the features and characteristics of a product or service (quantitative concept).

quality characteristics. Any dimension, mechanical property, physical property, functional characteristic, or appearance characteristic that can be used as a basis for measuring the quality of a unit of product or service.

quantitative analysis. A measurement in which the amount of one or more components of a sample is determined.

quarter hard. A *temper* of nonferrous alloys and some ferrous alloys characterized by tensile strength about midway between that of dead soft and half hard tempers.

quench-age embrittlement. *Embrittlement* of low-carbon steels resulting from precipitation of solute carbon at existing dislocations and from precipitation hardening of the steel caused by differences in the solid solubility of carbon in ferrite at different temperatures. Quench-age embrittlement usually is caused by rapid cooling of the steel from temperatures slightly below Ac_1 (the temperature at which *austenite* begins to form), and can be minimized by quenching from lower temperatures.

quench cracking. Fracture of a metal during *quenching* from elevated temperature. Most frequently observed in hardened carbon steel, alloy steel, or tool steel parts of high hardness and low toughness. Cracks often emanate from fillets, holes, corners, or other stress raisers and result from high stresses due to the volume changes accompanying transformation to *martensite*.

quench hardening. (1) Hardening suitable alpha-beta alloys (most often certain copper to titanium alloys) by solution treating and quenching to develop a martensitic like structure. (2) In ferrous alloys, hardening by austenitizing and then cooling at a rate such that a substantial amount of *austenite* transforms to *martensite*.

quenching. Rapid cooling of metals (often steels) from a suitable elevated temperature. This generally is accomplished by immersion in water, oil, polymer solution, or salt, although forced air is sometimes used.

quenching crack. A crack formed in a metal as a result of thermal stresses produced by rapid cooling from a high temperature.

R

racking. A term used to describe the placing of metal parts to be heat treated on a rack or tray. This is done to keep parts in a proper position to avoid heat-related distortions and to keep the parts separated.

radial crack. Damage produced in brittle materials by a hard, sharp object pressed onto the surface. The resulting crack shape is semielliptical and generally perpendicular to the surface.

radiograph. A photographic shadow image resulting from uneven absorption of penetrating radiation in a test object.

radiography. A method of *nondestructive inspection* in which a test object is exposed to a beam of x-rays or gamma rays and the resulting shadow image of the object is recorded on photographic film placed behind the object, or displayed on a viewing screen or television monitor (real-time radiography). Internal *discontinuities* are detected by observing and interpreting variations in the image caused by differences in thickness, density, or absorption within the test object. Variations of radiography include *computed tomography*, fluoroscopy, and neutron radiography.

radiology. The general term given to material inspection methods that are based on the differential absorption of penetrating radiation—either electromagnetic radiation of very short wavelength or particulate radiation—by the part or testpiece (object) being inspected. Because of differences in density and variations in thickness of the part or differences in absorption characteristics caused by variations in composition, different portions of a testpiece absorb different amounts of penetrating radiation. These variations in the absorption of the penetrating radiation can be monitored by detecting the unabsorbed radiation that passes through the testpiece.

ratchet marks. Lines or markings on a fatigue fracture surface that result from the intersection and connection of fatigue fractures propagating from multiple origins. Ratchet marks are parallel to the overall direction of crack propagation and are visible to the unaided eye or at low magnification.

reactive metal. A metal that readily combines with oxygen at elevated temperatures to form very stable oxides—for example, titanium, zirconium, and beryllium. Reactive metals may also become embrittled by the interstitial absorption of oxygen, hydrogen, and nitrogen.

reagent. A substance, chemical, or solution used in the laboratory to detect, measure, or react with other substances, chemicals, or solutions.

recrystallization. (1) The formation of a new, strain-free grain structure from that existing in cold-worked metal, usually accomplished by heating. (2) The change from one crystal structure to another, as occurs on heating or cooling through a critical temperature. (3) A process, usually physical, by which one crystal species is grown at the expense of another, or at the expense of others of the same substance but smaller in size.

recrystallization annealing. Annealing cold-worked metal to produce a new grain structure without phase change.

recrystallization temperature. (1) The lowest temperature at which the distorted grain structure of a cold-worked metal is replaced by a new, strain-free grain structure during prolonged heating. Time, purity of the metal, and prior deformation are important factors. (2) The approximate minimum temperature at which complete *recrystallization* of a cold-worked metal occurs within a specified time.

reducing atmosphere. (1) A furnace atmosphere which tends to remove oxygen from substances or materials placed in the furnace. (2) A chemically active protective atmosphere that at elevated temperature will reduce metal oxides to their metallic state. Reducing atmosphere is a relative term and such an atmosphere may be reducing to one oxide, but not to another.

refractory alloy. (1) A heat-resistant alloy. (2) An alloy having an extremely high melting point.

reliability. A quantitative measure of the ability of a product or service to fulfill its intended function for a specified period of time.

repeatability. A term used to refer to the test result variability associated with a limited set of specifically defined sources of variability within a single laboratory.

reproducibility. A term used to describe test result variability associated with specifically defined components of variance obtained both from within a single laboratory and between laboratories.

residual stress. (1) The stress existing in a body at rest, in equilibrium, at uniform temperature, and not subjected to external forces. Often caused by the forming or thermal processing curing process. (2) An internal stress not depending on external forces resulting from such factors as cold working, phase changes, or temperature gradients. (3) Stress present in a body that is free of external forces or thermal gradients. (4) Stress remaining in a structure or member as a result of thermal or mechanical treatment or both. Stress arises in *fusion welding* primarily because the weld metal contracts on cooling from the *solidus* to room temperature.

resistance brazing. A resistance joining process in which the workpieces are heated locally and filler metal that is preplaced between the workpieces is melted by the heat obtained from resistance to the flow of electric current through the electrodes and the work. In the usual application of resistance brazing, the heating current is passed through the joint itself.

resistance seam welding. A resistance welding process that produces coalescence at the faying surfaces by the heat obtained from resistance to electric current through workpieces that are held together under pressure by electrode wheels. The resulting weld is a series of overlapping resistance spot welds made progressively along a joint by rotating the electrodes.

resistance soldering. Soldering in which the joint is heated by electrical resistance. Filler metal is either face-fed into the joint or preplaced in the joint.

river pattern (metals). A term used in fractography to describe a characteristic pattern of cleavage steps running parallel to the local direction of crack propagation on the fracture surfaces of grains that have separated by *cleavage*.

riveting. Joining of two or more members of a structure by means of metal rivets, the unheaded end being *upset* after the rivet is in place.

Rockwell hardness number. A number derived from the net increase in the depth of impression as the load on an indenter is increased from a fixed minor load to a major load and then returned to the minor load. Various scales of Rockwell hardness numbers have been developed based on the hardness of the materials to be evaluated. The scales are designated by alphabetic suffixes to the hardness designation. For example, 64 HRC represents the Rockwell hardness number of 64 on the Rockwell C scale.

Rockwell hardness test. An indentation hardness test using a calibrated machine that utilizes the depth of indentation, under constant load, as a measure of hardness. Either a 120° diamond cone with a slightly rounded point, or a 1.6 or 3.2 mm ($1/16$ or $1/8$ in.) diam steel ball is used as the indenter.

Rockwell superficial hardness number. Like the *Rockwell hardness number,* the superficial Rockwell number is expressed by the symbol HR followed by a scale designation. For example, 81 HR 30N represents the Rockwell superficial hardness number of 81 on the Rockwell 30N scale.

Rockwell superficial hardness test. The same test as used to determine the *Rockwell hardness number* except that smaller minor and major loads are used. In Rockwell testing, the minor load is 10 kgf, and the major load is 60, 100, or 150 kgf. In superficial Rockwell testing, the minor load is 3 kgf, and the major loads are 15, 30, or 45 kgf. In both tests, the indenter may be either a diamond cone or a steel ball, depending principally on the characteristics of the material being tested.

rolling. The reduction of the cross-sectional area of metal stock, or the general shaping of metal products, through the use of rotating rolls.

room temperature. A temperature in the range of ~20 to 30 °C (~70 to 85 °F). The term room temperature is usually applied to an atmosphere of unspecified relative humidity.

roughness. (1) Relatively finely spaced surface irregularities, the heights, widths, and direction of which establish the predominant surface pattern. (2) The microscopic peak-to-valley distances of surface protuberances and depressions.

S

sacrificial protection. Reduction of corrosion of a metal in an *electrolyte* by galvanically coupling it to a more anodic metal; a form of *cathodic protection.*

salt bath heat treatment. *Heat treatment* for metals carried out in a bath of molten salt.

sample. (1) One or more units of a product (or a relatively small quantity of a bulk material) withdrawn from a *lot* or process stream and then tested or inspected to provide information about the properties, dimensions, or other quality characteristics of the lot or process stream. (2) A portion of a material intended to be representative of the whole.

sample average. The sum of all the observed values in a sample divided by the sample size. It is a point estimate of the population mean. Also known as arithmetic mean.

sample median. The middle value when all observed values in a sample are arranged in order of magnitude. If an even number of samples are tested, the average of the two middlemost values is used. It is a point estimate of the population median, or 50% point.

sample percentage. The percentage of observed values between two stated values of the variable under consideration. It is a point estimate of the percentage of the population between the same two stated values.

scale (metals). Surface oxidation consisting of partially adherent layers of corrosion products, left on metals by heating or casting in air or in other oxidizing atmospheres.

scanning acoustic microscopy (SAM). The use of a reflection-type acoustic microscope to generate very-high-resolution images of surface and near-surface features or defects in a material. The images are created by mechanically scanning a transducer with an acoustic lens in a raster pattern over the sample. Compared with conventional ultrasound imaging techniques, which operate in the 1 to 10 MHz range, SAM is carried out at 100 to 2000 MHz.

scanning Auger microscopy (SAM). An analytical technique that measures the lateral distribution of elements on the surface of a material by recording the intensity of their Auger electrons versus the position of the electron beam.

scanning electron microscope. A high-power magnifying and imaging instrument using an accelerated electron beam as an optical device and containing circuitry that causes the beam to traverse or scan an area of sample in the same manner as does an oscilloscope or TV tube. May utilize reflected (*scanning electron microscopy*) or transmitted (*scanning transmission electron microscopy*) electron optics. The scanning electron microscope provides two outstanding improvements over the *optical microscope*: It extends the resolution limits so that picture magnifications can be increased from 1000 to 2000× up to 30,000 to 60,000×, and it improves the depth-of-field resolution more dramatically, by a factor of approximately 300, thus facilitating its use in fracture studies.

scanning electron microscopy (SEM). An analytical technique in which an image is formed on a cathode-ray tube whose raster is synchronized with the raster of a point beam of electrons scanned over an area of the sample surface. The brightness of the image at any point is proportional to the scattering by or secondary emissions from the point on the sample being struck by the electron beam.

scanning transmission electron microscopy (STEM). An analytical technique in which an image is formed on a cathode-ray tube whose raster is synchronized with the raster of a point beam of electrons scanned over an area of the sample. The brightness of the image at any point is proportional to the number of electrons that are transmitted through the sample at the point where it is struck by the beam.

scarfing. Cutting surface areas of metal objects, ordinarily by using an oxyfuel gas torch. The operation permits surface imperfections to be cut from ingots, billets, or the edges of plate that are to be beveled for butt welding.

scrap. (1) Products that are discarded because they are defective or otherwise unsuitable for sale. (2) Discarded metallic material, from whatever source, that may be reclaimed through melting and refining.

scratch. A groove produced in a solid surface by the cutting and/or plowing action of a sharp particle or protuberance moving along the surface.

seam. (1) On a metal surface, an unwelded fold or lap that appears as a crack, usually resulting from a *discontinuity*. (2) A surface *defect* on a casting related to but of lesser degree than a *cold shut*. (3) A ridge on the surface of a casting caused by a crack in the mold face.

seam weld. A continuous weld made between or upon overlapping members, in which coalescence may start and occur on the faying surfaces, or may have proceeded from the surface of one member. The continuous weld may consist of a single weld bead or a series of overlapping spot welds. Common seam weld types include (a) lap seam welds joining flat sheets, (b) flange-joint lap seam welds with at least one flange overlapping the mating piece, and (c) mash seam welds with work metal compressed at the joint to reduce joint thickness.

semiconductor. A solid crystalline material whose electrical conductivity is intermediate between that of a metal and an insulator, ranging from about 10^5 siemens to 10^{-7} siemens per meter, and is usually strongly temperature dependent.

shear strength. The maximum shear stress that a material is capable of sustaining. Shear strength is calculated from the maximum load during a shear or torsion test and is based on the original cross-sectional area of the specimen.

shelf life. The length of time a material, substance, product, or reagent can be stored under specified environmental conditions and continue to meet all applicable specification requirements and/or remain suitable for its intended function.

shielded metal arc cutting. A metal arc cutting process in which metals are severed by melting them with the heat of an arc between a covered metal electrode and the base metal.

shielding gas. (1) Protective gas used to prevent atmospheric contamination during welding. (2) A stream of inert gas directed at the substrate during thermal spraying so as to envelop the plasma flame and substrate, intended to provide a barrier to the atmosphere in order to minimize oxidation.

shotblasting. Blasting with metal shot; usually used to remove deposits or mill scale more rapidly or more effectively than can be done by sandblasting.

shot peening. A method of cold working metals in which compressive stresses are induced in the exposed surface layers of parts by the impingement of a stream of shot, directed at the metal surface at high velocity under controlled conditions. It differs from blast cleaning in primary purpose and in the extent to which it is controlled to yield accurate and reproducible results. Although shot peening cleans the surface being peened, this function is incidental. The major purpose of shot peening is to increase fatigue strength. Shot for peening is made of iron, steel, or glass.

shotting. The production of shot by pouring molten metal in finely divided streams. Solidified spherical particles are formed during descent in a tank of water.

significant. Statistically significant. An effect of difference between populations is said to be present if the value of a test statistic is significant, that is, lies outside the predetermined limits.

sintering. The bonding of adjacent surfaces of particles in a mass of powder or a compact by heating. Sintering strengthens a powder mass and normally produces densification and, in powdered metals, recrystallization.

slack quenching. The incomplete hardening of steel due to quenching from the austenitizing temperature at a rate slower than the critical cooling rate for the particular steel, resulting in the formation of one or more transformation products in addition to martensite.

slag. A nonmetallic product resulting from the mutual dissolution of flux and nonmetallic impurities in smelting, refining, and certain welding operations (see, for example, *electroslag welding*). In steelmaking operations, the slag serves to protect the molten metal from the air and to extract certain impurities.

slug. (1) A short piece of metal to be placed in a die for forging or extrusion. (2) A small piece of material produced by piercing a hole in sheet material.

slurry. (1) A thick mixture of liquid and solids, the solids being in suspension in the liquid. (2) Any pourable or pumpable suspension of a high content of insoluble particulate solids in a liquid medium, most often water.

snap temper. A precautionary interim stress-relieving treatment applied to high-hardenability steels immediately after quenching to prevent cracking because of delay in tempering them at the prescribed higher temperature.

***S-N* curve for 50% survival.** A curve fitted to the median value of *fatigue life* at each of several stress levels. It is an estimate of the relationship between applied stress and the number of cycles-to-failure that 50% of the *population* would survive.

soaking. In heat treating of metals, prolonged holding at a selected temperature to effect homogenization of structure or composition. See also *homogenizing*.

soak time. The length of time a ceramic material is held at the peak temperature of the firing cycle.

solder. A filler metal used in soldering, which has a *liquidus* not exceeding 450 °C (840 °F). The most commonly used solders are tin-lead alloys. Other solder alloys include tin-antimony, tin-silver, tin-zinc, cadmium-silver, cadmium-zinc, zinc-aluminum, indium-base alloys, bismuth-base alloys (fusible alloys), and gold-base alloys.

soldering. A group of processes that join metals by heating them to a suitable temperature below the *solidus* of the base metals and applying a filler metal having a *liquidus* not exceeding 450 °C (840 °F). Molten filler metal is distributed between the closely fitted surfaces of the joint by *capillary action*.

solidification. The change in state from liquid to solid upon cooling through the melting temperature or melting range.

solidus. (1) The highest temperature at which a metal or alloy is completely solid. (2) In a *phase diagram*, the locus of points representing the temperatures at which various compositions stop freezing upon cooling or begin to melt upon heating.

solute. The component of either a liquid or solid solution that is present to a lesser or minor extent; the component that is dissolved in the *solvent*.

solution. In chemistry, a homogeneous dispersion of two or more types of molecular or ionic species. Solutions may be composed of any combination of liquids, solids, or gases, but they always consist of a single phase.

solution heat treatment. Heating an alloy to a suitable temperature, holding at that temperature long enough to cause one or more constituents to enter into solid solution, and then cooling rapidly enough to hold these constituents in solution.

solvent. The component of either a liquid or solid solution that is present to a greater or major extent; the component that dissolves the *solute*.

solvus. In a phase or equilibrium diagram, the locus of points representing the temperature at which solid phases with various compositions coexist with other solid phases, that is, the limits of solid solubility.

sonic testing. Any inspection method that uses sound waves (in the audible frequency range, about 20 to 20,000 Hz) to induce a response from a part or test specimen. Sometimes, but inadvisably, used as a synonym for ultrasonic testing.

space lattice. A regular, periodic array of points (lattice points) in space that represents the location of atoms of the same kind in a perfect *crystal*. The concept may be extended, where appropriate, to crystalline compounds and other substances, in which case the lattice points often represent locations of groups of *atoms* of identical composition, arrangement, and orientation.

spalling (metals). (1) Separation of particles from a surface in the form of flakes. The term spalling is commonly associated with rolling-element bearings and with gear teeth. Spalling is usually a result of the subsurface fatigue and is more extensive than pitting. (2) In *tribology*, the separation of macroscopic particles from a surface in the form of flakes or chips, usually associated with rolling-element bearings and gear teeth, but also resulting from impact events. (3) The spontaneous chipping, fragmentation, or separation of a surface or surface coating. (4) A chipping or flaking of a surface due to any kind of improper heat treatment or material dissociation.

spark testing. A method used for the classification of ferrous alloys according to their chemical compositions, by visual examination of the spark pattern or stream that is thrown off when the alloys are held against a grinding wheel rotating at high speed.

specimen. A test object, often of standard dimensions and/or configuration, that is used for destructive or nondestructive testing. One or more specimens may be cut from each unit of a *sample*.

spectrophotometry. A method for identification of substances and determination of their concentration by measuring light transmittance in different parts of the spectrum.

spectroscopy. The branch of physical science treating the theory, measurement, and interpretation of spectra.

spheroidal graphite. *Graphite* of spheroidal shape with a polycrystalline radial structure. This structure can be obtained, for example, by adding cerium or magnesium to the melt.

spheroidizing. Heating and cooling to produce a spheroidal or globular form of carbide in steel. Spheroidizing methods frequently used are:

1. Prolonged holding at a temperature just below Ae_1
2. Heating and cooling alternately between temperatures that are just above and just below Ae_1
3. Heating to a temperature above Ae_1 or Ae_3 and then cooling very slowly in the furnace or holding at a temperature just below Ae_1
4. Cooling at a suitable rate from the minimum temperature at which all carbide is dissolved to prevent the re-formation of a carbide network, and then reheating in accordance with method 1 or 2 above. (Applicable to hypereutectoid steel containing a carbide network.)

spot weld. A weld made between or upon overlapping members in which coalescence may start and occur on the faying surfaces or may proceed from the surface of one member. The weld cross section is approximately circular.

spray quenching. A quenching process using spray nozzles to spray water or other liquids on a part. The quench rate is controlled by the velocity and volume of liquid per unit area per unit of time of impingement.

springback. (1) The elastic recovery of metal after stressing. (2) The extent to which metal tends to return to its original shape or contour after undergoing a forming operation. This is compensated for by overbending or by a secondary operation of restriking. (3) In flash, upset, or pressure welding, the deflection in the welding machine caused by the upset pressure.

spring temper. A *temper* of nonferrous alloys and some ferrous alloys characterized by tensile strength and hardness about two-thirds of the way from *full hard* to extra spring temper.

stabilizing treatment. (1) Before finishing to final dimensions, repeatedly heating a ferrous or nonferrous part to or slightly above its normal operating temperature and then cooling to room temperature to ensure dimensional stability in service. (2) Transforming retained austenite in quenching hardenable steels, usually by *cold treatment.* (3) Heating a solution-treated stabilized grade of austenitic stainless steel to 870 to 900 °C (1600 to 1650 °F) to precipitate all carbon as TiC, NbC, or TaC so that sensitization is avoided on subsequent exposure to elevated temperature.

stamping. The general term used to denote all sheet metal pressworking. It includes blanking, shearing, hot or cold forming, drawing, bending, and coining.

standardization. (1) The process of establishing, by common agreement, engineering criteria, terms, principles, practices, materials, items, processes, and equipment parts and components. (2) The adoption of generally accepted uniform procedures, dimensions, materials, or parts that directly affect the design of a product or a facility. (3) In analytical chemistry, the assignment of a compositional value to one standard on the basis of another standard.

standard reference material. A reference material, the composition or properties of which are certified by a recognized standardizing agency or group.

statistical quality control. The application of statistical techniques for measuring and improving the quality of processes and products (includes statistical process control, diagnostic tools, sampling plans, and other statistical techniques).

steam treatment. The treatment of a sintered ferrous part in steam at temperatures between 510 and 595 °C (950 and 1100 °F) in order to produce a layer of black iron oxide (magnetite, or ferrous-ferric oxide, $FeO \cdot Fe_2O_3$) on the exposed surface for the purpose of increasing hardness and wear resistance.

stereophotogrammetry. A method of generating topographic maps of fracture surfaces by the use of a stereoscopic microscope interfaced to a microcomputer that calculates the three-dimensional coordinates of the fracture surface and produces the corresponding profile map, contour plot, or carpet plot.

sterling silver. A silver alloy containing at least 92.5% Ag, the remainder being unspecified but usually copper. Sterling silver is used for flat and hollow tableware and for various items of jewelry.

stiffness. (1) The rate of stress with respect to strain; the greater the stress required to produce a given strain, the stiffer the material is said to be. (2) The ability of a material or shape to resist elastic deflection. For identical shapes, the stiffness is proportional to the modulus of elasticity. For a given material, the stiffness increases with increasing moment of inertia, which is computed from cross-sectional dimensions.

stock. A general term used to refer to a supply of metal in any form or shape and also to an individual piece of metal that is formed, forged, or machined to make parts.

straightening. (1) Any bending, twisting, or stretching operation to correct any deviation from straightness in bars, tubes, or similar long parts or shapes. This deviation can be expressed as either camber (deviation from a straight line) or as total indicator reading (TIR) per unit of length. (2) A finishing operation for correcting misalignment in a forging or between various sections of a forging.

strain. The unit of change in the size or shape of a body due to force. Also known as nominal strain.

The term is also used in a broader sense to denote a dimensionless number that characterizes the change in dimensions of an object during a deformation or flow process.

strain-age embrittlement. A loss in *ductility* accompanied by an increase in hardness and strength that occurs when low-carbon steel (especially rimmed or capped steel) is aged following *plastic deformation*. The degree of *embrittlement* is a function of aging time and temperature, occurring in a matter of minutes at about 200 °C (400 °F), but requiring a few hours to a year at room temperature.

strain aging. (1) *Aging* following *plastic deformation*. (2) The changes in ductility, hardness, yield point, and tensile strength that occur when a metal or alloy that has been cold worked is stored for some time. In steel, strain aging is characterized by loss of ductility and a corresponding increase in hardness, yield point, and tensile strength.

strain gage. A device for measuring small amounts of strain produced during tensile and similar tests on metal. A coil of fine wire is mounted on a piece of paper, plastic, or similar carrier matrix (backing material), which is rectangular in shape and usually about 25 mm (1 in.) long. This is glued to a portion of the metal under test. As the coil extends with the specimen, its electrical resistance increases in direct proportion. This is known as a bonded resistance-strain gage. Other types of gages measure the actual deformation. Mechanical, optical, or electronic devices are sometimes used to magnify the strain for easier reading.

strain hardening. An increase in hardness and strength of metals caused by *plastic deformation* at temperatures below the recrystallization range. Also known as work hardening.

strand casting. A generic term describing *continuous casting* of one or more elongated shapes such as billets, blooms, or slabs; if two or more shapes are cast simultaneously, they are often of identical cross section.

stress. The intensity of the internally distributed forces or components of forces that resist a change in the volume or shape of a material that is or has been subjected to external forces. Stress is expressed in force per unit area. Stress can be normal (tension or compression) or shear.

stress corrosion. Preferential attack of areas under stress in a corrosive environment, where such an environment alone would not have caused corrosion.

stress-corrosion cracking (SCC). A cracking process that requires the simultaneous action of a corrodent and sustained tensile stress. This excludes corrosion-reduced sections that fail by fast fracture. It also excludes intercrystalline or transcrystalline corrosion, which can disintegrate an alloy without applied or residual stress. Stress-corrosion cracking may occur in combination with hydrogen embrittlement.

stress crack. External or internal cracks in a plastic caused by tensile stresses less than that of its short-time mechanical strength, frequently accelerated by the environment to which the plastic is exposed. The stresses that cause cracking may be present internally or externally or may be combinations of these stresses.

stress-relief cracking. Cracking in the *heat-affected zone* or weld metal that occurs during the exposure of weldments to elevated temperatures during postweld heat treatment, in order to reduce *residual stresses* and improve toughness, or high temperature service. Stress-relief cracking occurs only in metals that can precipitation-harden during such elevated-temperature exposure; it usually occurs as stress raisers, is intergranular in nature, and is generally observed in the coarse-grained region of the weld heat-affected zone. Also called postweld heat treatment cracking or stress-relief embrittlement.

stress-relief heat treatment. Uniform heating of a structure or a portion thereof to a sufficient temperature to relieve the major portion of the *residual stresses*, followed by uniform cooling.

Stress relieving. Heating to a suitable temperature, holding long enough to reduce *residual stresses*, and then cooling slowly enough to minimize the development of new residual stresses.

stretching. The extension of the surface of a metal sheet in all directions. In stretching, the flange of the flat blank is securely clamped. Deformation is restricted to the area initially within the die. The stretching limit is the onset of metal failure.

stringer. In wrought materials, an elongated configuration of microconstituents or foreign material aligned in the direction of working. The term is commonly associated with elongated oxide or sulfide inclusions in steel.

strip. (1) A flat-rolled metal product of some maximum thickness and width arbitrarily dependent on the type of metal; narrower than sheet. (2) A roll-compacted metal powder product.

structural shape. A piece of metal of any of several designs accepted as standard by the structural branch of the iron and steel industries.

structure. As applied to a *crystal,* the shape and size of the unit cell and the location of all atoms within the unit cell. As applied to microstructure, the size, shape, and arrangement of phases.

stud welding. An arc welding process in which the contact surfaces of a stud, or similar fastener, and a workpiece are heated and melted by an arc drawn between them. The stud is then plunged rapidly onto the workpiece to form a weld. Partial shielding may be obtained by the use of a ceramic ferrule surrounding the stud. Shielding gas or flux may or may not be used. The two basic methods of stud welding are known as stud arc welding, which produces a large amount of weld metal around the stud base and a relatively deep penetration into the base metal; and capacitor discharge stud welding, which produces a very small amount of weld metal around the stud base and shallow penetration into the base metal.

submerged arc welding. Arc welding in which the arc between a bare metal electrode and the work is shielded by a blanket of granular, fusible material overlying the joint. Pressure is not applied to the joint, and filler metal is obtained from the consumable electrode (and sometimes from a supplementary welding rod).

substrate. (1) The material, workpiece, or substance on which a coating is deposited. (2) A material upon the surface of which an adhesive-containing substance is spread for any purpose, such as bonding or coating. A broader term than adherend. (3) In electronic devices, a body, board, or layer of material on which some other active or useful material(s) or component(s) may be deposited or laid—for example, electronic circuitry laid on an alumina ceramic board. (4) In catalysts, the formed, porous, high-surface area carrier on which the catalytic agent is widely and thinly distributed for reasons of performance and economy.

subsurface corrosion. Formation of isolated particles of corrosion products beneath a metal surface. This results from the preferential reactions of certain alloy constituents to inward diffusion of oxygen, nitrogen, or sulfur.

sulfidation. The reaction of a metal or alloy with a sulfur-containing species to produce a sulfur compound that forms on or beneath the surface on the metal or alloy.

sulfide stress cracking (SSC). Brittle fracture by cracking under the combined action of *tensile stress* and *corrosion* in the presence of water and hydrogen sulfide.

sulfide-type inclusions. In steels, nonmetallic inclusions composed essentially of manganese iron sulfide solid solutions, (Fe,Mn)S. They are characterized by plasticity at hot-rolling and forging temperatures and, in the hot-worked product, appear as dove-gray elongated inclusions varying from a threadlike to oval outline.

superconductivity. A property of many metals, alloys, compounds, oxides, and organic materials at temperatures near absolute zero by virtue of which their electrical resistivity vanishes and they become strongly diamagnetic.

superplasticity. The ability of certain metals (most notably aluminum- and titanium-base alloys) to develop extremely high tensile elongations at elevated temperatures and under controlled rates of deformation.

supersaturated. A metastable solution in which the dissolved material exceeds the amount the solvent can hold in normal equilibrium at the temperature and other conditions that prevail.

surface hardening. A generic term covering several processes applicable to a suitable ferrous alloy that produces, by *quench hardening* only, a surface layer that is harder or more wear resistant than the core. There is no significant alteration of the chemical composition of the surface layer. The processes commonly used are *carbonitriding, carburizing, induction hardening, flame hardening, nitriding,* and *nitrocarburizing*. Use of the applicable specific process name is preferred.

surface tension. (1) The force acting on the surface of a liquid, tending to minimize the area of the surface. (2) The force existing in a liquid/vapor phase interface that tends to diminish the area of the interface. This force acts at each point on the interface in the plane tangent to that point.

swage. (1) The operation of reducing or changing the cross-sectional area of stock by the fast impact of revolving dies. (2) The tapering of bar, rod, wire, or tubing by forging, hammering, or squeezing; reducing a section by progressively tapering lengthwise until the entire section attains the smaller dimension of the taper.

T

tack welds. (1) Small, scattered welds made to hold parts of a weldment in proper alignment while the final welds are being made. (2) Intermittent welds to secure weld backing bars.

temper (metals). (1) In *heat treatment,* reheating hardened steel or hardened cast iron to some temperature below the eutectoid temperature for the purpose of decreasing *hardness* and increasing *toughness*. The process also is sometimes applied to normalized steel. (2) In tool steels, temper is sometimes used, but inadvisedly, to denote the carbon content. (3) In nonferrous alloys and in some ferrous alloys (steels that cannot be hardened by heat treatment), the hardness and strength produced by mechanical or thermal treatment, or both, and characterized by a certain structure, mechanical properties, or reduction in area during cold working. (4) To moisten green sand for casting molds with water.

temper color. A thin, tightly adhering oxide skin (only a few molecules thick) that forms when steel is tempered at a low temperature, or for a short time, in air or a mildly oxidizing atmosphere. The color, which ranges from straw to blue depending on the thickness of the oxide skin, varies with both tempering time and temperature.

tempered martensite. The decomposition products that result from heating *martensite* below the ferrite-austenite transformation temperature. Under the optical microscope, darkening of the martensite needles is observed in the initial stages of tempering. Prolonged tempering at high temperatures produces spheroidized carbides in a matrix of ferrite. At the higher resolution of the electron microscope, the initial stage of tempering is observed to result in a structure containing a precipitate of fine iron carbide particles. At approximately 260 °C (500 °F), a transition occurs to a structure of larger and elongated cementite particles in a ferrite matrix. With further tempering at high temperatures, the cementite particles become spheroidal, decreased in number, and increased in size.

tensile strength. In tensile testing, the ratio of maximum load to original cross-sectional area. Also called ultimate strength.

tensile stress. A stress that causes two parts of an elastic body, on either side of a typical stress plane, to pull apart.

thermal conductivity. (1) Ability of a material to conduct heat. (2) The rate of heat flow under steady conditions, through unit area, per unit temperature gradient in the direction perpendicular to the area. Usually expressed in English units as Btu per square feet per degrees Fahrenheit (Btu/ft^2 · °F). It is given in SI units as watts per meter kelvin (W/m · K).

thermal embrittlement. Intergranular fracture of maraging steels with decreased toughness resulting from improper processing after hot working. Thermal embrittlement occurs upon heating above 1095 °C (2000 °F) and then slow cooling through the temperature range of 980 to 815 °C (1800 to 1500 °F), and has been attributed to precipitation of titanium carbides and titanium carbonitrides at austenite grain boundaries during cooling through the critical temperature range.

thermal expansion. The change in length of a material with change in temperature.

thermocouple. A device for measuring temperatures, consisting of lengths of two dissimilar metals or alloys that are electrically joined at one end and connected to a voltage-measuring instrument at the other end. When one junction is hotter than the other, a thermal electromotive force is produced that is roughly proportional to the difference in temperature between the hot and cold junctions. Nonstandard materials include nickel-molybdenum, nickel-cobalt, iridium-rhodium, platinum-molybdenum, gold-palladium, palladium-platinum, and tungsten-rhenium alloys.

tolerance. The specified permissible deviation from a specified nominal dimension, or the permissible variation in size or other quality characteristic of a part.

torsion. (1) A twisting deformation of a solid or tubular body about an axis in which lines that were initially parallel to the axis become helices. (2) A twisting action resulting in shear stresses and strains.

toughness. Ability of a material to absorb energy and deform plastically before fracturing. Toughness is proportional to the area under the stress-strain curve from the origin to the breaking point. In metals, toughness is usually measured by the energy absorbed in a notch impact test.

tramp alloys. Residual alloying elements that are introduced into steel when unidentified alloy steel is present in the scrap charge to a steelmaking furnace.

transistor. An active semiconductor device capable of providing power amplification and having three or more terminals.

tribology. (1) The science and technology of interacting surfaces in relative motion and of the practices related thereto. (2) The science concerned with the design, friction, lubrication, and wear of contacting surfaces that move relative to each other (as in bearings, cams, or gears, for example).

U–V

ultrasonic inspection. A *nondestructive inspection* method in which beams of high-frequency sound waves are introduced into materials for the detection of surface and subsurface flaws in the material. The sound waves travel through the material with some attendant loss of energy (attenuation) and are reflected at interfaces. The reflected beam is displayed and then analyzed to define the presence and location of flaws or discontinuities. Most ultrasonic inspection is done at frequencies between 0.1 and 25 MHz—well above the range of human hearing, which is about 20 Hz to 20 kHz.

upset. (1) The localized increase in cross-sectional area of a workpiece or weldment resulting from the application of pressure during mechanical fabrication or welding. (2) That portion of a welding cycle during which the cross-sectional area is increased by the application of pressure. (3) Bulk deformation resulting from the application of pressure in welding. The upset may be measured as a percent increase in interfacial area, a reduction in length, or a percent reduction in thickness (for lap joints).

vacancy. A structural imperfection in which an individual atom site is temporarily unoccupied.

vacuum carburizing. A high-temperature gas carburizing process using furnace pressures between 13 and 67 kPa (0.1 and 0.5 torr) during the carburizing portion of the cycle. Steels undergoing this treatment are austenitized in a rough vacuum, carburized in a partial pressure of hydrocarbon gas, diffused in a rough vacuum, and then quenched in either oil or gas. Both batch and continuous furnaces are used.

vacuum furnace. A furnace using low atmospheric pressures instead of a protective gas atmosphere like most heat treating furnaces. Vacuum furnaces are categorized as hot wall or cold wall, depending on the location of the heating and insulating components.

Vickers hardness test. A microindentation hardness test employing a 136° diamond pyramid indenter (Vickers) and variable loads, enabling the use of one hardness scale for all ranges of hardness—from very soft lead to tungsten carbide. Also known as diamond pyramid hardness test.

W–Z

warpage (metals). (1) Deformation other than contraction that develops in a casting between solidification and room temperature. (2) The distortion that occurs during annealing, stress relieving, and high-temperature service.

water quenching. A quench in which water is the quenching medium. The major disadvantage of water quenching is its poor efficiency at the beginning or hot stage of the quenching process.

weld. A localized coalescence of metals or nonmetals produced either by heating the materials to suitable temperatures, with or without the application of pressure, or by the application of pressure alone and with or without the use of filler material.

weldability. A specific or relative measure of the ability of a material to be welded under a given set of conditions. Implicit in this definition is the ability of the completed weldment to fulfill all functions for which the part was designed.

welding. (1) Joining two or more pieces of material by applying heat or pressure, or both, with or without filler material, to produce a localized union through fusion or recrystallization across the interface. The thickness of the filler material is much greater than the capillary dimensions encountered in *brazing*. (2) May also be extended to include brazing and *soldering*. (3) In *tribology*, adhesion between solid surfaces in direct contact at any temperature.

wetting agent. (1) A substance that reduces the surface tension of a liquid, thereby causing it to spread most readily on a solid surface. (2) A surface-active agent that produces wetting by decreasing the cohesion within the liquid.

x-ray. A penetrating electromagnetic radiation, usually generated by accelerating electrons to high velocity and suddenly stopping them by collision with a solid body. Wavelengths of x-rays range from about 10^{-1} to 10^{2} Å, the average wavelength used in research being about 1 Å. Also known as roentgen ray or x-radiation.

yield. (1) Evidence of *plastic deformation* in structural materials. Also known as plastic flow or creep. (2) The ratio of the number of acceptable items produced in a production run to the total number that were attempted to be produced. (3) Comparison of casting weight to the total weight of metal poured into the mold.

yield point. The first stress in a material usually less than the maximum attainable stress, at which an increase in strain occurs without an increase in stress. Only certain materials—those which exhibit a localized, heterogeneous type of transition from elastic to plastic deformation—produce a yield point. If there is a decrease in stress after yielding, a distinction may be made between upper and lower yield points. The load at which a sudden drop in the flow curve occurs is called the upper yield point. The constant load shown on the flow curve is the lower yield point.

Bibliography

- *ASM Handbook,* Vol 1, *Properties and Selection: Irons, Steels, and High-Performance Alloys*, ASM International, 1990
- *ASM Handbook,* Vol 2, *Properties and Selection: Nonferrous Alloys and Special-Purpose Materials*, ASM International, 1991
- *ASM Handbook,* Vol 4, *Heat Treating*, ASM International, 1991
- *ASM Metals Reference Book*, 3rd ed., ASM International, 1993
- H.R. Clauser, Ed., *Encyclopedia/Handbook of Materials, Parts, and Finishes,* Technomic Publishing, 1976
- J.R. Davis, Ed., *ASM Materials Engineering Dictionary*, ASM International, 1992
- *Heat Treater's Guide: Practices and Procedures for Irons and Steels*, 2nd ed., ASM International, 1995
- *Heat Treater's Guide: Practices and Procedures for Nonferrous Alloys*, ASM International , 1996
- J.A. Jacobs and T.F. Kilduff, *Engineering Materials Technology*, 2nd ed., Prentice Hall, 1994
- MEI Course, "Metallurgy for the Nonmetallurgist," ASM International
- MEI Course, "Elements of Metallurgy," ASM International
- D.J. Wulpi, *Understanding How Components Fail*, American Society for Metals, 1980

Index

F

G

H

I

J

K

L

M

N

O

P

Q

R

S

U

V